KT-551-065

Progress in Probabilty
Volume 25

Stable Processes and Related Topics:

A Selection of Papers from the Mathematical Sciences Institute Workshop, January 9–13, 1990

Stamatis Cambanis
Gennady Samorodnitsky
Murad S. Taqqu
Editors

1991

Birkhäuser
Boston · Basel · Berlin

Stamatis Cambanis
Department of Statistics
University of North Carolina
Chapel Hill, NC 27599
U.S.A.

Gennady Samorodnitsky
Engineering Theory Center
Cornell University
Ithaca, NY 14853
U.S.A.

Murad S. Taqqu
Department of Mathematics
Boston University
111 Cummington Street
Boston, MA 02215

Library of Congress Cataloging-in-Publication Data

Stable processes and related topics: a selection of papers from the
Mathematical Sciences Institute Workshop, January 9–13, 1990 /
Stamatis Cambanis, Gennady Samorodnitsky, Murad S. Taqqu, editors.
p. cm. – (Progress in probability ; 25)
Selection of papers from the Workshop on Stable Processes and
Related Topics held at Cornell University.
Includes bibliographical references.
ISBN (invalid) 0-8176-3584-1
1. Stochastic processes–Congresses. 2. Distribution (Probabilty
theory)–Congresses. I. Cambanis, S. (Stamatis), 1943–
II. Samorodnitsky, Gennady. III. Taqqu, Murad S. IV. Cornell
University, Mathematical Sciences Institute. V. Cornell
University. Mathematical Sciences Institute. Workshop (1990)
VI. Workshop on Stable Processes and Related Topics (1990: Cornell
University) VII. Series.
QA274.A1S73 1991
519.2–dc20 90-29883
CIP

ISBN 0-8176-3485-1
ISBN 3-7643-3485-1

Camera-ready copy prepared by the authors.
Printed and bound in Germany

9 8 7 6 5 4 3 2 1

CONTENTS

PREFACE

The Workshop on Stable Processes and Related Topics took place at Cornell University in January 9-13, 1990, under the sponsorship of the Mathematical Sciences Institute. It attracted an international roster of probabilists from Brazil, Japan, Korea, Poland, Germany, Holland and France as well as the U.S. This volume contains a sample of the papers presented at the Workshop. All the papers have been refereed.

Gaussian processes have been studied extensively over the last fifty years and form the bedrock of stochastic modeling. Their importance stems from the Central Limit Theorem. They share a number of special properties which facilitates their analysis and makes them particularly suitable to statistical inference. The many properties they share, however, is also the seed of their limitations. What happens in the real world away from the ideal Gaussian model?

The non-Gaussian world may contain random processes that are close to the Gaussian. What are appropriate classes of nearly Gaussian models and how typical or robust is the Gaussian model amongst them? Moving further away from normality, what are appropriate non-Gaussian models that are sufficiently different to encompass distinct behavior, yet sufficiently simple to be amenable to efficient statistical inference?

The very Central Limit Theorem which provides the fundamental justification for approximate normality, points to stable and other infinitely divisible models. Some of these may be close to and others very different from Gaussian models. In the last twenty years there have been increasing efforts to better understand stable and other infinitely divisible processes. These efforts have been directed towards delineating the extent to which they share the features of the Gaussian models, and even more significantly towards discovering their own distinguishing and frequently surprising features. While it is expected that the heavy tailed stable models may be substantially different from their Gaussian counterparts, both their share of similarities and contrasting features have offered surprises and unexpected insights to investigators.

The main goal of the workshop has been to review the state of the art on the structure of stable processes as models for random phenomena. This was successfully accomplished during a week filled with a wide variety of lively presentations and informal exchanges.

Looking ahead, while a great deal still remains to be done to improve our understanding of stable models, two directions emerge as next stages in this reseach program. The first is the pursuit of the structural analysis of other infinitely divisible models, of which some will be close and some sharply different from stable. The second is the development of statistical inferential methodology for the heavy tailed stable models.

DESCRIPTION OF CONTENTS

One may divide the papers in this volume into three groups. The papers in the first group deal with the theory of stable processes, those in the second group deal with the theory of more general infinitely divisible processes, while the third group deals with applications of stable and infinitely divisible processes. Of course, as with any classification,some papers could be classified in different ways.

The first group is the largest. A major theme of discussion is the dependence structure of stable processes. The paper "Conditional variance of symmetric stable variables" by S. Cambanis and W. Wu gives necessary and sufficient conditions for the existence of conditional variance in a symmetric stable distribution in $\mathbf{R}^2$ and gives the explicit form of this conditional variance. V. Mandrekar's and B. Thelen's paper "On multiple Markov $S\alpha S$ processes" discusses symmetric α stable processes which are nth order Markov and gives conditions under which the solutions of nth order stochastic differential equations driven by symmetric stable noise are themselves nth order Markov. The best linear predictors for harmonizable stable processes are described in the paper "An extremal problem in H^p of the upper half plane with application to prediction of stochastic processes" by B. Rajput, K. Rama-Murthy and C. Sundberg.

The paper by J. Levy and M. Taqqu "A characterization of the asymptotic behavior of stationary stable processes" provides a measure of asymtotic dependence for processes, when the covariance does not exist, and applies it to various stable processes. The asymptotic behavior of stable processes, namely their ergodic properties, is also studied in K. Podgorsky's and A. Weron's paper "Characterization of ergodic stationary stable processes via the dynamical functional".

J. Nolan shows in his contribution "Bounded stationary stable processes and entropy" that the classical necessary condition for continuity of stationary Gaussian processes extends to their stable counterparts. M. Konō and M. Maejima provide a survey and new results on self-similar stable processes. L. Giraitis and D. Surgailis show in "On shot noise processes attracted to fractional Lévy motion" that certain shot noise processes converge to stable processes with long range dependence.

The papers "Construction of multiple stable measures and integrals using LePage representation" by G. Samorodnitsky and M. Taqqu and "Multiple stable integrals as weak limits" by J. Szulga discuss representations and properties of the stable counterparts to multiple Wiener integrals.

In the second group of papers we find the paper "Gaussian measures of large balls in $\mathbf{R}^n$" by W. Linde which studies the asymtotics of the distribution of a norm of Gaussian random vectors in a finite-dimensional Euclidian space. J. Rosinski describes in his contibution "On a class of infinitely divisible processes represented as mixtures of Gaussian processes" a series representation of such processes which include symmetric stable processes. G. OBrien and W. Vervaat study in their paper, "Capacities, large deviations and loglog laws", large deviations and laws of iterated logarithm for a Poisson random measure with a view to applications to stable and infinitely divisble processes.

In the third group of papers, we find the paper by S. Mittnik and S. Rachev "Alternative multivariate stable distributions and their applications to financial modeling" whose title is self-descriptive. J. Kinateder describes in his contribution "A stochastic integral representation for the bootstrap of the sample mean" what happens when the statistical procedure known as bootsrapping is applied to random variables in the domain of attraction of a stable law. Finally, C. Hardin, G. Samorodnitsky and M. Taqqu provide in "Numerical computation of non-linear stable regression functions" an algorithm and a computer program to evaluate regressions of skewed stable distributions.

Gaussian measures of large balls in $\mathbb{R}^n$

WERNER LINDE*

Let μ be a symmetric Gaussian measure on $\mathbb{R}^n$. Then we investigate the asymptotic behaviour of the function $u \longrightarrow \mu\{x \in \mathbb{R}^n;\ \|x-x_0\| > u\}$ as $u \longrightarrow \infty$ for some norms $\|\cdot\|$ and $x_0 \in \mathbb{R}^n$. The basic tool for those investigations is a generalization of Laplace's method to a larger class of functions. The general results are applied to ℓ_p-norms where we obtain new results for $0<p<2$.

1. Introduction

Let $(X_t)_{t\in T}$ be a separable centered Gaussian process with

$$\mathbb{P}\{\sup_t X_t < \infty\} = 1.$$

Then the asymptotic behaviour of

$$\mathbb{P}\{\sup_t X_t > u\} \quad \text{as } u \longrightarrow \infty \tag{1}$$

is of special interest. Let Φ be defined by

$$\Phi(u) := (2\pi)^{-1/2} \int_{-\infty}^{u} e^{-s^2/2} ds, \quad u \in \mathbb{R},$$

and set

$$\sigma := \sup_{t\in T} (\mathcal{E}|X_t|^2)^{1/2} .$$

Then (1) behaves "almost" like $1-\Phi(u/\sigma)$ as $u \longrightarrow \infty$. Here "almost" means that on one hand

---

*Supported in part by Ministerio de Educatión y Ciencia Español (Dirección General de Investigación Cientifica y Técnica)

$$1-\Phi(u/\sigma) \leq \mathbb{P}\{\sup_t X_t > u\}$$

and on the other hand we have

$$\mathbb{P}\{\sup_t X_t > u\}/(1-\Phi(u/\sigma)) = O(e^{\beta u^2})$$

for any $\beta>0$ (cf. [10], [13], [15], [2]). Recall that $1-\Phi(u)$ behaves like $(2\pi)^{-1/2}u^{-1}e^{-u^2/2}$ as $u \longrightarrow \infty$.

Let E be a Banach space and let μ be a Radon Gaussian centered measure on E, i.e. all the one-dimensional distributions $a(\mu)$ are Gaussian centered on $\mathbb{R}$, $a \in E^*$, (topological dual space). Then it follows (use Hahn-Banach theorem)

$$\mu\{x \in E; \|x\|>u\} = \mathbb{P}\{\sup_{a\in A} X_a > u\}$$

where $(X_a)_{a\in A}$ is the Gaussian process on (E,μ) defined by $X_a(x) := \langle x,a\rangle$ and $A := \{a \in E^*;\ \|a\|\leq 1\}$. Consequently, all results about the behaviour of the supremum of a Gaussian process apply. For instance,

$$2(1-\Phi(u/\sigma)) \leq \mu\{x \in E;\ \|x\|>u\},\ u>0,$$

by symmetry of $A \subseteq E^*$, and

$$\exp(u^2/2\sigma^2)\mu\{\|x\| > u\} = O(e^{\beta u^2}),\ \beta>0,$$

which recently has been improved to $O(e^{\beta u})$, $\beta>0$ (cf. [19]). Observe that for general μ (not necessarily centered) this is only valid for some $\beta>0$.

Moreover, this result is the best possible, in general. But for special Banach spaces or special Gaussian measures this upper bound is too large. For instance, as shown in [21] in Hilbert spaces we always have

$$\exp(u^2/2\sigma^2)\mu\{\|x\| > u\} = O(u^N)$$

for some natural number N depending on μ. Or if μ on ℓ_p, $2<p<\infty$, has diagonal form (the unit vectors e_i, $i=1,\dots$, are independent random variables on (ℓ_p,μ)), then we even have

$$\exp(u^2/2\sigma^2)\mu\{\|x\|>u\} = O(u^{-1})$$

(cf. [4] or [14]). These results heightened our interest in the investigation of the asymptotic behaviour of $\mu\{\|x\| > u\}$ in some concrete Banach space. During our investigation we mentioned that this question is even non-trivial in $\mathbb{R}^n$ endowed with a concrete norm. For example, the exact behaviour of $\mu\{x \in \mathbb{R}^n; \|x\|_p > u\}$,

$\|x\|_p := (\sum_1^n |\xi_i|^p)^{1/p}$, $x=(\xi_1,\ldots,\xi_n)$, seems to be unknown for $2<p<\infty$ provided that μ is not of diagonal form (cf. final remarks of this paper for the case $p<2$). So the aim of this paper is to investigate the function

$$u \longrightarrow \mu\{x \in \mathbb{R}^n; \|x-x_o\| > u\} \tag{2}$$

as $u \longrightarrow \infty$ where $x_o \in \mathbb{R}^n$, μ is Gaussian and $\|\cdot\|$ is a norm (quasi-norm) possessing some special properties. From a different point of view similar problems have been treated in [1], [5], [6], [3], [8], [16], [17] and [18].

2. Rough estimates

Let $\|\cdot\|$ be an arbitrary norm on $\mathbb{R}^n$ and let $|\cdot|$ be the Euclidean norm. By μ we always denote a Gaussian centered measure on $\mathbb{R}^n$. We assume that μ has full support, i.e. we have $\mu = S(\gamma_n)$ for a one-to-one linear map S from $\mathbb{R}^n$ into $\mathbb{R}^n$. Here and in the sequel γ_n denotes the canonical Gaussian measure on $\mathbb{R}^n$. Then (2) may be written as

$$\gamma_n\{\|Sx-x_o\|>u\} = \gamma_n\{\|x-y_o\|_S>u\}$$

where $\|x\|_S := \|Sx\|$ and $y_o := S^{-1}x_o$.

Consequently, by a suitable change of the norm it suffices to investigate (2) for $\mu = \gamma_n$. Furthermore, if $\sigma>0$ is defined by

$$\sigma := \sup\{\|x\|;\ |x| \le 1\},$$

then replacing $\|\cdot\|$ by $\|\cdot\|/\sigma$ we obtain a new norm with $\sigma=1$, i.e. we have $\|x\| \le |x|$, $x \in \mathbb{R}^n$, and the set $C := \{x \in \mathbb{R}^n;\ \|x\| = |x| = 1\}$ of contact points between the boundaries of the corresponding unit balls is non-empty. Under these assumptions we easily obtain

$$\sqrt{2/\pi}\int_u^\infty e^{-s^2/2}ds \le \gamma_n\{\|x\| > u\} \le \frac{2^{1-n/2}}{\Gamma(n/2)}\int_u^\infty s^{n-1}e^{-s^2/s}ds \tag{3}$$

for $u>0$. The left hand side of (3) behaves like

$$\sqrt{2/\pi}\, u^{-1}e^{-u^2/2}$$

and the right hand side like

$$\frac{2^{1-n/2}}{\Gamma(n/2)}\, u^{n-2}e^{-u^2/s} \quad \text{as} \quad u \longrightarrow \infty.$$

Thus the behaviour of $e^{u^2/2}\gamma_n\{\|x\|>u\}$ is up to a constant between u^{-1} and u^{n-2}. For $x_o \neq 0$ we have

$$\gamma_n\{x\in\mathbb{R}^n;\ \|x-x_o\|>u\} \leq \gamma_n\{x\in\mathbb{R}^n;\ |x-x_o|>u\}$$

and it is known (cf. [12] or Corollary 3 below) that the right hand side behaves like

$$(2\pi)^{-1/2}|x_o|^{-(n-1)/2}u^{(n-3)/2}\exp(-(u-|x_o|)^2/2)$$

as $u \longrightarrow \infty$. To obtain a lower estimate define $\alpha \geq 0$ by

$$\alpha := \sup\{\langle x_o,z\rangle;\ z \in C\}.$$

If $\alpha>0$ and $z_o \in C$ satisfies $\langle x_o,z_o\rangle = \alpha$, then

$$\gamma_n\{\|x-x_o\|>u\} \geq \gamma_n\{|\langle x,z_o\rangle - \alpha| > u\}$$

and the behaviour of the right hand side is

$$(2\pi)^{-1/2}u^{-1}\exp(-(u-\alpha)^2/2) \quad \text{as } u \longrightarrow \infty.$$

For $\alpha=0$ we get no better lower estimate than for $x_o=0$.

3. Laplace's method

Laplace's method is the basic tool for determining the asymptotic behaviour of a function defined by an integral. For instance, if h maps a compact set $K \subseteq \mathbb{R}^n$ with non-void interior $\mathring{K}$ continuously into $\mathbb{R}$ and h possesses exactly one minimum at $x_o \in \mathring{K}$, then

$$\int_K e^{-u^2h(x)}dx \sim (2\pi)^{n/2}|h''(x_o)|^{-1/2}\, u^{-n}\, e^{-u^2h(x_o)}$$

as $u \longrightarrow \infty$. Clearly we have to assume that $h''(x_o)$ exists and that it is positive definite (cf. [7] or [9]). We write $f \sim g$ whenever $\lim_{u\to\infty} f(u)/g(u) = 1$.

Later on we have to apply Laplace's method in a more general situation, namely, to the integral

$$\int_K e^{-u^2h(x)+ug(x)}dx \qquad (4)$$

for some functions f and g on K. We could not find this generalization of Laplace's method in the literature.

Proposition 1. Let g and h be functions mapping $K \subseteq \mathbb{R}^n$ continuously into $\mathbb{R}$. Suppose that

(i) $h(x) > h(x_o)$ for some $x_o \in \mathring{K}$ and $x \neq x_o$,

(ii) $A := h''(x_o)$ exists and $A > 0$,

(iii) g is differentiable at x_o with derivative d_{x_o}.

Then (4) behaves like

$$(2\pi)^{n/2}|A|^{-1/2}\exp(\langle A^{-1}d_{x_o}, d_{x_o}\rangle/2)u^{-n}e^{-u^2h(x_o)+ug(x_o)}$$

as $u \longrightarrow \infty$.

Proof. Without losing generality we may assume $h(x_o) = g(x_o) = 0$, i.e. we have $h(x) > 0$ for all $x \in K$, $x \neq x_o$. Hence for any $\delta > 0$

$$u^n \int_{\substack{|x-x_o|\geq\delta \\ x\in K}} e^{-u^2h(x)+ug(x)}\,dx$$

tends to zero as $u \longrightarrow \infty$. For given $\epsilon, \rho > 0$ by (ii) and (iii) there exist $\delta_1, \delta_2 > 0$ such that

$$h(x+x_o) \geq \langle Ax,x\rangle/2 - \epsilon\langle x,x\rangle/2, \quad |x|<\delta_1$$

and

$$g(x+x_o) \leq \langle x,d_{x_o}\rangle + \rho|x| \qquad |x|<\delta_2.$$

If $\epsilon > 0$ is small enough, then

$$A_\epsilon := A - \epsilon I$$

is positive definitive as well and the Gaussian measure μ_ϵ with covariance matrix A_ϵ^{-1} is well-defined. So we obtain

$$u^n \int_{|x-x_o|<\delta} e^{-u^2h(x)+ug(x)}\,dx \leq u^n \int_{|x|<\delta} e^{-u^2\langle A_\epsilon x,x\rangle/2+u\langle x,d_{x_o}\rangle + u\rho|x|}\,dx$$

$$= \int_{|x|<\delta u} e^{-\langle A_\epsilon x,x\rangle/2+\langle x,d_{x_o}\rangle + \rho|x|}\,dx$$

provided that $\delta<\delta_1, \delta_2$. If $u \longrightarrow \infty$, the last-written integral tends to

$$(2\pi)^{n/2}|A_\epsilon|^{-1/2} \int_{\mathbb{R}^n} e^{\langle x,d_{x_o}\rangle + \rho|x|}\,d\mu_\epsilon(x). \tag{5}$$

This lets us conclude that

$$\lim_{\delta\downarrow 0} \overline{\lim_{u\to\infty}} u^n \int_{|x-x_o|<\delta} e^{-u^2h(x) + ug(x)} dx \tag{6}$$

is less than or equal to (5). Next we take first the limit $\rho \longrightarrow 0$ using Lebesgue's theorem with respect to μ_ϵ and then $\epsilon \longrightarrow 0$ which shows that (6) is less than

$$(2\pi)^{n/2}|A|^{-1/2} \int_{\mathbb{R}^n} e^{\langle x, d_{x_o}\rangle} d\mu(x) = (2\pi)^{n/2}|A|^{-1/2} \exp(\langle A^{-1}d_{x_o}, d_{x_o}\rangle/2)$$

where μ possesses the covariance matrix A^{-1}. Similar arguments show that

$$\lim_{\delta\downarrow 0} \underline{\lim}_{u\to\infty} \int_{|x-x_o|<\delta} e^{-u^2h(x) + ug(x)} dx$$

is larger than the last-written expression which completes the proof.

Remark. It is easy to see that Proposition 1 remains true in the more general case

$$\int_K \varphi(x)\, e^{-u^2h(x) + ug(x)} dx$$

provided that

$$\lim_{x\to x_o} \varphi(x) = 1.$$

4. Gaussian measures of balls.

We want to apply Proposition 1 to the function

$$u \longrightarrow \gamma_n\{x \in \mathbb{R}^n;\ \|x-x_o\| > u\}.$$

As above we assume $\|x\| \le |x|$, $x \in \mathbb{R}^n$, and

$$C := \{x \in \mathbb{R}^n;\ |x| = \|x\| = 1\} \neq \emptyset.$$

By $S^{n-1} \subseteq \mathbb{R}^n$ we denote the unit sphere in $\mathbb{R}^n$, i.e.

$$S^{n-1} := \{x \in \mathbb{R}^n:\ |x| = 1\}$$

and endow it with the (non-normalized) Lebesgue measure dz,

$$\int_{S^{n-1}} dz = 2\pi^{n/2}[\Gamma(n/2)]^{-1}.$$

Lemma 2. Under these assumptions we have

$$\gamma_n\{\|x-x_o\| > u\} \sim (2\pi)^{-n/2} e^{-|x_o|^2/2} \int_{S^{n-1}} (u/\|z\|)^{n-2} \exp(-u^2/2\|z\|^2 + u\langle x_o, z\rangle/\|z\|)dz$$

as $u \longrightarrow \infty$.

Proof. We start with the identity

$$\gamma_n\{\|x-x_o\|>u\} = (2\pi)^{-n/2} e^{-|x_o|^2/2} \int_{S^{n-1}} \int_{u/\|z\|}^{\infty} r^{n-1} e^{-r^2/2 + r\langle z,x_o\rangle} dr\, dz. \tag{7}$$

The function

$$s \longrightarrow \int_s^{\infty} r^{n-1} e^{-r^2/2 + \alpha r} dr$$

is asymptotically equivalent to

$$s^{n-2} e^{-s^2/2 + \alpha s}$$

where this equivalence is uniform for all α in a compact interval of $\mathbb{R}$. Applying this to (7) with $s := u/\|z\| \geq u$ and $\alpha := \langle z, x_o\rangle$ ($|\alpha| \leq |x_o|$) completes the proof.

Corollary 3. ([12]). If $x_o \neq 0$, then we have

$$\gamma_n\{|x-x_o|>u\} \sim 1/(2\pi)^{-1/2}\, u^{(n-3)/2} |x_o|^{-(n-1)/2}\, e^{-(u-|x_o|)^2/2}$$

Proof. By Lemma 2 we obtain

$$\gamma_n\{|x-x_o|>u\} \sim (2\pi)^{-n/2} e^{-|x_o|^2/2} u^{n-2} e^{-u^2/2} \int_{S^{n-1}} e^{u\langle x_o, z\rangle} dz. \tag{8}$$

The function $z \longrightarrow \langle x_o, z\rangle$ attains its maximal value at $z_o := x_o/|x_o|$ and the integral in (8) behaves like

$$\int_{\substack{|y|<\delta \\ y\perp z_o}} (1-|y|^2)^{-1/2} e^{u|x_o|(1-|y|^2)^{1/2}} dy$$

for any $\delta > 0$. Using Laplace's method this is asymptotically equivalent to

$$(2\pi)^{(n-1)/2} u^{-(n-1)/2} |x_o|^{-(n-1)/2} e^{u|x_o|}$$

and the corollary is proved.

Combining Proposition 1 with Lemma 2 we obtain the main result of this paper (cr. [1] or [3] for $x_o = 0$).

Theorem 4. Let $\|\cdot\|$ be a norm on $\mathbb{R}^n$ with $\|x\| \le |x|$ such that there exists one and only one $z_o \in S^{n-1}$ with $\|\pm z_o\| = 1$. Suppose that the norm $\|\cdot\|$ is twice differentiable at z_o with second derivative A satisfying $I-A > 0$. Then

$$\gamma_n\{\|x-x_o\|>u\} \sim (2\pi)^{-1/2}|I-A|^{-1/2}e^{\langle Ax_o,(I-A)^{-1}x_o\rangle/2}u^{-1}e^{(u-|\langle x_o,z_o\rangle|)^2/2}$$

if $\langle x_o,z_o\rangle \ne 0$ and it behaves like

$$(2/\pi)^{1/2}|I-A|^{-1/2}e^{\langle Ax_o,(I-A)^{-1}x_o\rangle/2}\; u^{-1}e^{-u^2/2}$$

for $\langle x_o,z_o\rangle = 0$.

Proof. In view of Lemma 2 and the remark after Proposition 1 it suffices to investigate the behaviour of

$$\int_{S^{n-1}} \exp(-u^2/2\|z\|^2 + u\langle x_o,z\rangle/\|z\|)dz \tag{9}$$

as $u \longrightarrow \infty$.

Since the function $z \longrightarrow \|z\|^{-2}$ possesses exactly two minimal values on S^{n-1}, namely z_o and $-z_o$, (9) is asymptotically equivalent to

$$\int_{\substack{y\perp z_o\\ |y|<\delta}} \exp\Big[-\frac{u^2}{2\|y+(1-|y|^2)^{1/2}z_o\|^2}$$

$$+\frac{u(\langle y,x_o\rangle + (1-|y|^2)^{1/2}\langle x_o,z_o\rangle)}{\|y+(1-|y|^2)^{1/2}z_o\|}\Big](1-|y|^2)^{-1/2}dy \tag{10}$$

plus a second term with $-z_o$ instead of z_o. Here we have to choose $\delta > 0$ small enough. The second derivative of

$$y \longrightarrow \|y + (1-|y|^2)^{1/2}z_o\|^{-2}/2$$

at zero is $I-A|_{\{z_o\}^\perp}$ while the first derivative of the coefficient for u in the exponent of (10) equals $\langle y,v_o\rangle$ where

$$v_o = x_o - \langle x_o, z_o\rangle\, z_o .$$

Hence Proposition 1 applies and (9) behaves like

$$(2\pi)^{(n-1)/2} \big| I-A|_{\{z_o\}^\perp}\big|^{-\frac{1}{2}}\, e^{\langle (I-A|_{\{z_o\}^\perp})^{-1} v_o, v_o\rangle /2}$$

$$u^{-(n-1)} e^{-u^2/2} \{ e^{u\langle x_o, z_o\rangle} + e^{-u\langle x_o, z_o\rangle} \} . \tag{11}$$

If $\langle x_o, z_o\rangle = 0$, then the term in the brackets is equal to 2, while for $\langle x_o, z_o\rangle \neq 0$ the behaviour of this term is

$$e^{u|\langle x_o, z_o\rangle|} .$$

Since $Az_o = 0$ we have

$$\big|(I-A)|_{\{z_o\}^\perp}\big| = |I-A|$$

and

$$\langle (I-A|_{\{z_o\}^\perp})^{-1} v_o, v_o\rangle - \langle x_o, x_o\rangle = \langle (I-A)^{-1} x_o, x_o\rangle - |\langle x_o, z_o\rangle|^2 -$$

$$\langle x_o, x_o\rangle$$

$$= \langle (I-A)^{-1} x_o, Ax_o\rangle - |\langle x_o, z_o\rangle|^2 .$$

Inserting this in (11) Lemma 2 proves the theorem.

Remark 1. Observe that the second derivative A always satisfies $0 \leq A \leq I$. So the theorem doesn't apply iff A possesses an eigenvalue equal to 1.

Remark 2. If $C = \{x \in \mathbb{R}^n;\ |x| = \|x\| = 1\}$ consists of 2N points $\pm z_1, \ldots, \pm z_n$, then we may apply Theorem 4 to each of these points provided that the norm is twice differentiable at z_i, $1 \leq i \leq N$, with second derivative A_i satisfying $I-A_i > 0$. In this case we have

$$\gamma_n\{\|x-x_o\|>u\} \sim (2\pi)^{-\frac{1}{2}} u^{-1} e^{-(u-\alpha)^2/2} \sum_{i\in J} |I-A_i|^{-\frac{1}{2}} e^{\langle (I-A_i)^{-1} x_o, A_i x_o\rangle /2}$$

if $\alpha := \sup_{1\leq i\leq N} |\langle x_o, z_i\rangle| > 0$ and where

$$J := \{i \in \{1,\ldots,N\};\ |\langle x_o, z_i\rangle| = \alpha\}.$$

For $\alpha = 0$ we get two times this expression. Observe that $J=\{1,\ldots,N\}$ in this case.

Remark 3. If $\mu = S(\gamma_n)$ is Gaussian on $\mathbb{R}^n$ with full support, then

$$\mu\{\|x-x_o\| > u\} = \gamma_n\{\|x-y_o\|_S > u\}$$

where $\|x\|_S := \|Sx\|$, and $y_o := S^{-1}x_o$. Moreover

$$\sigma = \sup_{|x|\leq 1} \|x\|_S = \sup_{|x|\leq 1} \|Sx\| = \|S\|$$

and we again assume $\sigma = 1$, i.e. $\|S\| = 1$. Otherwise we have to replace u by u/σ and x_o by x_o/σ. Then

$$C = \{x \in \mathbb{R}^n;\ |x| = \|Sx\| = 1\},$$

that is, C contains exactly those points of S^{n-1} where S attains its norm.

Let $z \in C$ be given and assume that the original norm is twice differentiable at Sz with second derivative A. Then $\|\cdot\|_S$ is twice differentiable at z and the second derivative equals S^*AS. Thus, if $C = \{\pm z_o\}$ and A is the second derivative of $\|\cdot\|$ at Sz_o with $I-S^*AS > 0$, then Theorem 4 applies and

$$\mu\{\|x-x_o\|>u\} \sim(2\pi)^{-\frac{1}{2}}u^{-1}e^{-(u-\alpha)^2/2}|I-S^*AS|^{-\frac{1}{2}}e^{\langle(I-S^*AS)^{-1}y_o,S^*ASy_o\rangle/2} \tag{12}$$

provided that $\alpha := |\langle y_o,z_o\rangle| > 0$, $y_o = S^{-1}x_o$.

Observe that the exponent in the second e-term of (12) coincides with

$$\langle(I-RA)^{-1}x_o,Ax_o\rangle/2$$

where $R=SS^*$ denotes the covariance matrix of μ.

For $\alpha = 0$ we obtain two times (12) and one may easily extend these results to the case that C consists of a finite number of points.

5. Applications to special norms.

Let $a_1,\dots,a_N$ be elements of $\mathbb{R}^n$ with

$$\sup_{1\leq i\leq N} |a_i| = 1$$

and with $a_i \neq \pm a_j$, $i\neq j$. Without losing generality we assume that

$$\|x\| := \sup_{1\leq i\leq N} |\langle x,a_i\rangle|$$

defines a norm on $\mathbb{R}^n$. Indeed, we are interested in the behaviour of $\gamma_n\{\sup_{1\leq i\leq N} |\langle x-x_o,a_i\rangle| > u\}$ as $u \longrightarrow \infty$. Hence we may restrict γ_n to the subspace

$$\{x \in \mathbb{R}^n;\ \langle x,a_i\rangle = 0,\quad 1\le i\le N\}^{\perp}$$

if necessary.

By assumption the norm defined above satisfies $\|x\| \le |x|$ and the set C of contact points coincides with

$$C = \{\pm a_i;\ |a_i| = 1\}.$$

The norm is twice differentiable at those points and the second derivative is equal to zero. Hence Theorem 4 (Remark 2) applies and we obtain:

Proposition 5. Let $a_1,\dots,a_N$ be as above. Then it holds

$$\gamma_n\{\sup_{1\le i\le N} |\langle x-x_o,a_i\rangle| > u\} \sim (2\pi)^{-1/2}\, k\, u^{-1} e^{-(u-\alpha)^2/2}$$

iff

$$\sup\{|\langle x_o,a_i\rangle|\ ;\ |a_i| = 1\} = \alpha > 0$$

and

$$\mathrm{card}\{i;\ |\langle x_o,a_i\rangle| = \alpha,\ |a_i| = 1\} = k.$$

The measure of the supremum behaves like

$$(2/\pi)^{1/2} m\, u^{-1} e^{-u^2/2}$$

iff $\alpha = 0$, i.e. $\langle x_o,a_i\rangle = 0$ for $i \in C$ and

$$\mathrm{card}(C)/2 = \mathrm{card}\{i;\ |a_i| = 1\} = m.$$

We formulate the next two applications of Proposition 5 in the language of Gaussian random variables.

Proposition 6. Let $(X_1,\dots,X_N)$ be a Gaussian symmetric vector with

$$\sup_{1\le i\le N} \mathcal{E}|X_i|^2 = 1 \quad\text{and}\quad |\mathcal{E}X_iX_j| < 1,\quad i\ne j.$$

If $x_o = (\xi_1,\dots,\xi_N)$ belongs to the support of the distribution of $(X_1,\dots,X_N)$, i.e.

$$\sum_{i=1}^N \alpha_i X_i = 0 \quad \text{a.e. implies} \quad \sum_{i=1}^N \alpha_i \xi_i = 0, \tag{13}$$

then

$$\mathbb{P}\{\sup_{1\le i\le N} |X_i-\xi_i| > u\} \sim c u^{-1} e^{-(u-\alpha)^2/2}.$$

Here

$$\alpha := \sup_{i\in I_o} |\xi_i|, \qquad I_o := \{i;\ \mathcal{E}X_i^2 = 1\},$$

$$c := (2/\pi)^{1/2}\mathrm{card}(I_o) \qquad \text{for } \alpha = 0 \text{ and}$$

$$c := (2\pi)^{-1/2}\mathrm{card}\{i \in I_o;\ |\xi_i| = \alpha\} \qquad \text{for } \alpha \neq 0.$$

Proof. Let $R = (r_{ij})_{i,j=1}^{N}$ be the covariance matrix of $(X_1,\dots,X_N)$, i.e.

$$r_{ij} = \mathcal{E}X_iX_j, \qquad 1\leq i,j\leq N.$$

Writing R as $R = SS^*$ where $S:\mathbb{R}^n \longrightarrow \mathbb{R}^N$ is injective ($n=\mathrm{rank}(R)$) we obtain

$$\mu := \mathrm{dist}(X_1,\dots,X_N) = S(\gamma_n).$$

Then (13) is equivalent to $x_o \in \mathrm{supp}(\mu) = S(\mathbb{R}^n)$, that is, $x_o = Sy_o$ for some $y_o \in \mathbb{R}^n$ and it follows

$$\mathbb{P}(\sup_{1\leq i\leq N} |X_i-\xi_i|>u\} = \mu\{x\in\mathbb{R}^N;\ \sup_{1\leq i\leq N} |\langle x-x_o,e_i\rangle|>u\}$$

$$= \gamma_n\{y\in\mathbb{R}^n;\ \sup_{1\leq i\leq N} |\langle y-y_o,S^*e_i\rangle|>u\}.$$

Here e_i denotes the i-th unit vector in $\mathbb{R}^N$. Consequently we are in the situation of Proposition 5 with

$$C = \{\pm S^*e_i;\ i \in I_o\}.$$

Observe that we have

$$\pm S^*e_j \neq S^*e_i,\ i,j \in I_o,\ i\neq j,$$

by virtue of

$$|\mathcal{E}X_iX_j| < 1, \qquad i\neq j.$$

An application of Proposition 5 completes the proof because of

$$\langle y_o,S^*e_i\rangle = \langle Sy_o,e_i\rangle = \langle x_o,e_i\rangle = \xi_i.$$

Remark 1. For $x_o = 0$ and $I_o = \{1,\dots,N\}$ this has been proved in [4] by different methods.

Remark 2. Observe that we have $\mathcal{E}|X_i|^2 = \mathcal{E}|X_j|^2 = |\mathcal{E}X_iX_j| = 1$ for some pair (i,j) of different numbers iff $\mathcal{E}|X_i|^2 = 1$ and $X_j = \pm X_i$. Choosing a maximal subset $J \subseteq \{i,\dots,N\}$ such that for $i \in J$ either $\mathcal{E}|X_i|^2 < 1$

or, if $\mathcal{E}|X_i|^2 = 1$, then $X_j \neq \pm X_i$, $j \in J\setminus\{i\}$, Proposition 6 applies to the Gaussian sequence $(X_i)_{i\in J}$.

In order to state a second application of Proposition 5 let $E_N \subseteq \mathbb{R}^N$ be defined by $E_N := \{-1,1\}^N$. If $(X_1,\dots,X_N)$ is a symmetric Gaussian vector and $e = (\epsilon_1,\dots,\epsilon_N) \in E_N$, then the (Gaussian) random variable X_e is defined by

$$X_e := \sum_{i=1}^{N} \epsilon_i X_i .$$

Furthermore, we assume for simplicity

$$\sup_{e\in E_N} \mathcal{E}|X_e|^2 = 1 .$$

Lemma 7. Let $e = (\epsilon_1,\dots,\epsilon_N)$ and $d=(\delta_1,\dots,\delta_N)$ be elements of E_N and define J by $J := \{i;\ \mathcal{E}_i \neq \delta_i\}$. If $\mathcal{E}|X_e|^2 = 1$ and $X_e = X_d$ a.e., then we have necessarily $X_j = 0$ for all $j \in J$.

Proof. Since $-e=d$ implies $X_e = 0$ contradicting $\mathcal{E}|X_e|^2 = 1$ it follows $e \neq -d$ and $J \neq \{1,\dots,N\}$. Hence

$$0 = X_e - X_d = \sum_{j\in J}(\epsilon_j-\delta_j)X_j = 2 \sum_{j\in J} \epsilon_j X_j$$

yields

$$\mathcal{E}\Big|\sum_{j\notin J} \epsilon_j X_j\Big|^2 = 1.$$

Next we choose $\tilde{\epsilon}_j = \pm 1$, $j \in J$, arbitrarily and define the random variable Y by

$$Y := \sum_{j\in J} \tilde{\epsilon}_j X_j .$$

Then we have

$$\mathcal{E}\Big|Y + \sum_{j\notin J} \epsilon_j X_j\Big|^2 \leq 1$$

which implies

$$\mathcal{E}|Y|^2 + 2 \sum_{j\notin J} \epsilon_j \mathcal{E}YX_j \leq 0.$$

Since this is true for $-Y$ as well we arrive at $Y=0$ a.e., i.e. we have $\sum_{j\in J} \tilde{\epsilon}_j X_j = 0$ a.e. for all choices of $\tilde{\epsilon}_j = \pm 1$, $j \in J$.

It readily follows that $X_j = 0$, $j \in J$.

Proposition 8. Let $(X_1,\dots,X_N)$ be a symmetric Gaussian vector with $X_i \neq 0$, $1 \leq i \leq N$. If

$$\sup\{\mathcal{E}|\sum_{i=1}^{N} \epsilon_i X_i|^2 = 1;\ \epsilon_i = \pm 1\} = 1$$

and $x_o = (\xi_1,\dots,\xi_N)$ satisfies (13), then

$$\mathbb{P}\{\sum_{i=1}^{N} |X_i - \xi_i| > u\} \sim (2/\pi)^{1/2} m\, u^{-1} e^{-u^2/2}$$

iff card $\{e \in E_N;\ \mathcal{E}|X_e|^2 = 1\} = 2m$ and $\langle x_o, e\rangle = 0$ for all $e \in E_N$ with $\mathcal{E}|X_e|^2 = 1$. Its behaviour is

$$(2\pi)^{-1/2}\, k\, u^{-1} e^{-(u-\alpha)^2/2}$$

iff

$$\sup\{\langle x_o, e\rangle;\ \mathcal{E}|X_e|^2 = 1\} = \alpha > 0$$

and

$$\text{card}\ \{e \in E_N;\ \mathcal{E}|X_e|^2 = 1 \quad \text{and}\ \langle x_o, e\rangle = \alpha\} = k.$$

Proof. Let S, μ and y_o be defined as in Proposition 6. Then we have

$$\mathbb{P}\{\sum_{i=1}^{N} |X_j - \xi_i| > u\} = \mu\{x \in \mathbb{R}^N;\ \sup_{e \in E_N} |\langle x - x_o, e\rangle| > u\} = \gamma_n\{y \in \mathbb{R}^n;$$

$$\sup_{e \in E_N} |\langle y - y_o, S^* e\rangle| > u\}$$

and combined with Proposition 5 this finishes the proof. Observe that in view of Lemma 7 $\mathcal{E}|X_e|^2 = \mathcal{E}|X_d|^2 = 1$, $e \neq d$, necessarily implies $S^* e \neq S^* d$ by $X_i \neq 0$, $i \leq i \leq N$.

Remark 1. Let $R = (r_{ij})_{i,j=1}^{N}$ be the covariance matrix of $(X_1,\dots,X_N)$. By assumption

$$\sup_{e \in E_N} \langle Re, e\rangle = \sup\{\sum_{i,j=1}^{N} \epsilon_i \epsilon_j r_{ij};\ \epsilon_i = \pm 1\} = 1$$

and

$$2m = \text{card}\{e \in E_N;\ \langle Re, e\rangle = 1\},$$

i.e. 2m is the number of contact points between the ellipsoid

$$\mathcal{E} = \{x \in \mathbb{R}^N;\ \langle Rx, x\rangle = 1\}$$

($\mathcal{E}$ may be degenerated) and the symmetric cube of volume 2^N. We have $m = 2^{N-1}$ iff the main axis of $\mathcal{E}$ are the unit vectors in $\mathbb{R}^N$, that is, iff

R is a diagonal matrix with trace(R) = 1 and this happens iff the X_i's are independent. Especially we have $m \leq 2^{N-2}$ for dependent sequences $(X_1,\dots,X_N)$.

Remark 2. Using exactly the same method as in the proof of Lemma 7 one can easily verify the following: The number k of Proposition 8 is always less or equal to 2^{ℓ} where

$$\ell = \text{card}\{i;\ \xi_i = 0\}.$$

If the X_i's are independent, then we have

$$\alpha = \|x_o\|_1 = \sum_{i=1}^{N} |\xi_i|$$

and $k=2^{\ell}$ in this case.

6. Applications to ℓ_p-norms.

Given $x=(\xi_1,\dots,\xi_n)\in\mathbb{R}^n$ the ℓ_p-norm is as usual defined by

$$\|x\|_p := \{ \sum_{i=1}^{n} |\xi_i|^p\}^{1/p}.$$

Here we assume $0<p<\infty$. Observe that for $0<p<1$ we only obtain a quasi-norm, i.e. we only have a triangle inequality with some constant $c>1$. But it is easy to see that Theorem 4 applies in this case as well. The only difference is that the second derivative of the norm may be negative. Let μ be a Gaussian symmetric measure on $\mathbb{R}^n$. The preceding results (Proposition 8) describe the behaviour of

$$\mu\{x \in \mathbb{R}^n:\ \|x-x_o\|_p>u\} \tag{14}$$

for p=1. Furthermore, for p=2 the behaviour of (14) is known as well (cf. Corollary 3 or [12]). Unfortunately we were not able to determine the behaviour of (14) for arbitrary μ and $p>0$. Here only some partial results can be proved. Before stating them we need some well-known properties of the norm (quasi-norm) $\|\cdot\|_p$.

Lemma 9. Let p be a positive real number and let $y = (\eta_1,\dots,\eta_n)$ be an element of $\mathbb{R}^n$ with $\|y\|_p = 1$. Suppose, furthermore, $\eta_i \neq 0$ if $0<p<2$. Then $\|\cdot\|_p$ is twice differentiable at y and the second derivative A_y coincides with

$$A_y = (p-1)[D_y-(d_y\otimes d_y)].$$

Here D_y denotes a diagonal matrix with $(|\eta_i|^{p-2})_{i=1}^n$ on the diagonal and d_y is the first derivative at y, i.e.

$$d_y = (|\eta_i|^{p-1}\mathrm{sgn}(\eta_i))_{i=1}^n .$$

Let S be an n×n matrix of rank n such that

$$1 = \|S\| = \sup_{|x|\leq 1} \|Sx\|_p = \sup\{\|Ra\|_p ; \langle Ra,a\rangle = 1\}$$

where $R := SS^*$.

Recall tht $\|x\|_S = \|Sx\|$ and $C=\{x\in\mathbb{R}^n;\ \|x\|_S = |x| = 1\}$ consists of those points where S attains its norm on S^{n-1}.

Proposition 10. Let S be as above and let z belong to C. Then $\|\cdot\|_S$ is twice differentiable at z with derivative

$$B_z = (p-1)[S^*D_{Sz}S - z \otimes z].$$

Especially, we have $B_z(z) = 0$ and the restriction of B_z to $\{z\}^\perp$ is equal to $(p-1)S^*D_{Sz}S$.

Proof. We start with the proof of the existence of B_z. To do so suppose $0<p<2$ an let us assume that

$$\langle Sz,e_i\rangle = 0 \qquad \text{for some } i\in\{1,\dots,n\}.$$

Setting $f_i := S^{-1}e_i$ we obtain

$$\|S(tf_i\pm z)\|_p = (1+|t|^p)^{1/p}, \qquad t \in \mathbb{R},$$

yielding

$$2(1+|t|^p)^{2/p} \leq |tf_i+z|^2 + |tf_i-z|^2 = 2t^2|f_i|^2 + 2,$$

or, equivalently,

$$(1+|t|^p)^{1/p} \leq (1+t^2|f_i|^2)^{1/2}$$

for all $t \in \mathbb{R}$. This is surely false for $0<p<2$ which proves $\langle Sz,e_i\rangle \neq 0$ for all $i\in\{1,\dots,n\}$. Consequently, the ℓ_p-norm is twice differentiable at Sz with derivative A_{Sz}. Since

$$B_z = S^*A_{Sz}S = (p-1)[S^*D_{Sz}S - (S^*d_{Sz}) \otimes (S^*d_{Sz})]$$

that completes the proof because of

$$S^*d_{Sz} = z.$$

The last-written equality follows from the fact that $|\cdot|$ and $\|\cdot\|_S$

possess the same first derivatives at z.

Corollary 11. If $z \in C$, then we have necessarily

$$I-B_z \geq 0$$

which is equivalent to

$$(p-1)\langle S^* D_{Sz} Sx, x\rangle \leq |x|^2 \tag{15}$$

for all $x \in \{z\}^\perp$.

Remark 1. Especially, for $0<p\leq 1$ we always have $B_z \leq 0$ which implies

$$I-B_z > 0 \tag{16}$$

in this case. We do not know whether or not this remains true for $p>2$ (cf. Corollary 17 for $p<2$).

Remark 2. Observe that 1 is always an eigenvalue of $S^* D_{Sz} S$. Indeed, it holds $S^* D_{Sz} S(z) = z$. Let $\lambda_1, \ldots, \lambda_{n-1}$ be the remaining eigenvalues of $S^* D_{Sz} S$. Then (15) is equivalent to

$$0 \leq \lambda_i \leq (p-1)^{-1}, \; 1 \leq i \leq n-1,$$

$1<p<\infty$, while (16) happens iff $\lambda_i < (p-1)^{-1}$.

The next lemma is an easy consequence of the fact that the function $t \longrightarrow t^p$, $t>0$, is concave for $0<p<1$

Lemma 12. Let S be an $n \times n$ matrix with

$$\|S\| = \sup_{|x|\leq 1} \|Sx\|_p = 1.$$

If $0<p<1$, then for each $e \in E_n$ there exists at most one $z \in S^{n-1}$ satisfying

$$\|Sz\|_p = 1 \quad \text{and} \quad (\operatorname{sgn}\langle Sz, e_i\rangle)_{i=1}^n = e.$$

In particular, the number of points of S^{n-1} where S attains its norm is not larger than 2^n.

Combining this lemma with Remark 1 after Proposition 10 we can determine the asymptotic behaviour of (14) for those p via Theorem 4. But we prefer to formulate our results in the language of random variables. Let $(X_1, \ldots, X_n)$ be a symmetric Gaussian r.v. with full support. For $a=(\alpha_1, \ldots, \alpha_n) \in \mathbb{R}^n$ the (Gaussian) random variable X_a is defined by

$$X_a := \sum_{i=1}^{n} \alpha_i X_i .$$

Let S be an n×n matrix with $SS^* = R$ where R denotes the covariance matrix of $(X_1,\ldots,X_n)$. Then we have

$$\sigma = \|S\| = \sup_{|x|\leq 1} \|Sx\|_p = \sup\{(\sum_{j=1}^{n} |\mathcal{E}X_j X_a|^p)^{1/p};\ \mathcal{E}|X_a|^2=1\}. \qquad (17)$$

Let $\mathcal{A} \subseteq \mathbb{R}^n$ be the set of those $a \in \mathbb{R}^n$ with $\mathcal{E}|X_a|^2 = 1$ for which the supremum in (17) is attained. Assuming $\sigma = 1$ and S of rank n(X_a=0 implies a=0) it follows that

$$C = S^*(\mathcal{A})$$

where C is defined as before. Especially, in view of Lemma 12 we get

$$\text{card}(\mathcal{A}) = \text{card}(C) \leq 2^n$$

provided that $0<p<1$.

Now we are in a position to formulate the announced result about the behaviour of (14) in the case $0<p<1$.

Proposition 13. Let $(X_1,\ldots,X_n)$ be a symmetric Gaussian vector with full support. Suppose that $\sigma=1$ where σ is defined by (17). If $0<p<1$, then it follows

$$\mathbb{P}\{(\sum_{i=1}^{n} |X_i-\xi_i|^p)^{1/p}>u\} \sim c\, u^{-1} e^{-(u-\alpha)^2/2}$$

for each $x_o = (\xi_1,\ldots,\xi_n)\in\mathbb{R}^n$. Here

$$\alpha = \sup\{\langle x_o,a\rangle;\ a \in \mathcal{A}\}$$

and

$$c=(2/\pi)^{1/2}|R|^{-1/2} \sum_{a\in\mathcal{A}} |R^{-1}-A_{Ra}|^{-1/2}\exp(\langle A_{Ra}x_o,(I-RA_{Ra})^{-1}x_o\rangle/2) \qquad (18)$$

provided that $\alpha=0$. Here R denotes the covariance matrix of the Gaussian vector and the second derivative A_{Ra} coincides with $(p-1)[D_{Ra}-a \otimes a]$ in this case. For $\alpha\neq 0$ one has to divide (18) by two and to replace $\mathcal{A}$ by $\mathcal{A}_o$ where

$$\mathcal{A}_o := \{a\in\mathcal{A};\ \langle x_o,a\rangle = \alpha\}.$$

Proof. This easily follows from the preceding results using the following equalities:

$$B_{S^*a} S^{-1}x_o = S^* A_{SS^*a} x_o = S^* A_{Ra} x_o$$

which implies

$$\langle B_{S^*a} S^{-1}x_o, (I-B_{S^*a})^{-1} S^{-1} x_o\rangle = \langle A_{Ra}x_o, (I-RA_{Ra})^{-1} x_o\rangle$$

and

$$|I-B_{S^*a}| = |I-S^* A_{Ra} S| = |R|\ |R^{-1}-A_{Ra}|.$$

Observe that

$$A_{Ra} = (p-1)[D_{Ra} - d_{Ra} \otimes d_{Ra}]$$
$$= (p-1)[D_{Ra} - a \otimes a]$$

because of $d_{Ra} = (S^*)^{-1}(S^*a) = a$.

Remark. The assumption $0<p<1$ is only used to guarantee that $\mathrm{card}(\mathcal{A})<\infty$ and that $I-B_{S^*a} = I-S^*A_{Ra}S > 0$, i.e. $R^{-1}-A_{Ra}>0$, $a\in\mathcal{A}$. Consequently, if this is true for $p>1$ as well, then Proposition 13 remains true also for those p. We were only able to answer this problem for special measures or, equivalently, for special matrices S (cf. final remark for the case $0<p<2$).

Proposition 14. Let $(X_1,\dots,X_n)$ be a sequence of independent symmetric Gaussian random variables. Define $\sigma_i>0$ by

$$\sigma_i := \{\mathcal{E}|X_i|^2\}^{1/2}, \qquad 1\le i\le n,$$

and assume that

$$\sigma^r := \sum_{i=1}^n \sigma_i^r = 1$$

where $0<p<2$ and $1/r := 1/p-1/2$. Then

$$\mathbb{P}\{(\sum_{i=1}^n |X_i-\xi_i|^p)^{1/p}>u\} \sim (2/\sqrt{2-p})^{n-1}(2/\pi)^{1/2}u^{-1}e^{-u^2/2}$$

iff $x_o = (\xi_1,\dots,\xi_n) = 0$ and it behaves like

$$(2\pi)^{-1/2}2^{\ell}(2-p)^{-(n-1)/2}u^{-1}e^{-(u-\alpha)^2/2}\exp(\tfrac{p-1}{2-p}(\beta^2-\alpha^2)/2)$$

for $x_o \ne 0$. Here

$$\ell := \mathrm{card}\{i:\ \xi_i=0\},$$

$$\alpha := \sum_{i=1}^n \sigma_i^{r/2\ -1}|\xi_i| \quad \text{and}$$

$$\beta^2 := \sum_{i=1}^{n} \sigma_i^{-2} \xi_i^2 .$$

Proof. We have $R=(r_{ij})_{i,j=1}^n$ with $r_{ii} = \sigma_i^2$ and $r_{ij}=0$ for $i\neq j$ in this case. Hence, the matrix S may be chosen as diagonal matrix with diagonal $(\sigma_i)_{i=1}^n$ and

$$C = \{z \in S^{n-1}; \|Sz\|_p=1\} = \{z_e=(\epsilon_i\sigma_i^{r/2})_{i=1}^n,\ e=(\epsilon_1,\dots,\epsilon_n)\in E_n\}$$

possesses 2^n different elements. It is easy to see that

$$S^* D_{Sz_e} S = I$$

for all $e \in E_n$, i.e. we have

$$|I-B_{z_e}|_{\{z_e\}^\perp}|^{½} = |(2-p)I|_{\{z_e\}^\perp}|^{½} = (2-p)^{(n-1)/2}$$

which completes the proof for $x_o = 0$. Observe that

$$\alpha = \sup_{e\in E_n} \langle S^{-1}x_o, z_e\rangle = \sup_{e\in E_n} \sum_{i=1}^{n} \sigma_i^{r/2\ -1}\epsilon_i\xi_i$$

$$= \sum_{i=1}^{n} \sigma_i^{r/2\ -1}|\xi_i|$$

is zero iff $x_o = 0$. Moreoever, there are exactly 2^ℓ such $e\in E_n$ for which $\alpha=\langle S^{-1}x_o,z_e\rangle$ iff $\ell=\mathrm{card}\{i;\ \xi_i=0\}$. To calculate

$$\langle B_{z_e} S^{-1}x_o, (I-B_{z_e})^{-1}S^{-1}x_o\rangle/2$$

we write $S^{-1}x_o$ as

$$S^{-1}x_o = v_o + \langle z_e, S^{-1}x_o\rangle z_e$$

with $v_o \in \{z_e\}^\perp$. Then we have

$$B_{z_e}(v_o) = (p-1)v_o$$

and

$$(I-B_{z_e})^{-1}v_o = (2-p)^{-1}v_o \quad ,$$

i.e. by $B_{z_e}(z_e) = 0$ (19) equals

$$\frac{p-1}{2-p}|v_o|^2/2 = \frac{p-1}{2-p}\left(|S^{-1}x_o|^2-|\langle S^{-1}x_o,z_e\rangle|^2\right)/2$$

$$= \frac{p-1}{2-p}(\beta^2-\alpha^2)/2$$

and the proof is finished.

Remark. For $x_o = 0$ this result has been proved in [14] by different methods.

Finally we treat the case $2<p<\infty$.

Proposition 15. ([14]) Let $(X_1,\dots,X_n)$ be a sequence of independent symmetric Gaussian random variables and assume

$$1 = \sigma_1 = \dots = \sigma_k > \sigma_{k+1} \geq \dots > 0$$

where σ_i is defined as in Proposition 14. If $2<p<\infty$, then

$$\mathbb{P}\{(\sum_{i=1}^{n} |X_i-\xi_i|^p)^{1/p} > u\} \sim cu^{-1}e^{-(u-\alpha)^2/2}$$

where

$$\alpha = \sup_{1\leq i\leq k} |\xi_i|,$$

$$c = (2/\pi)^{1/2}k \quad \text{for } \alpha = 0$$

and

$$c = (2\pi)^{-1/2}\text{card } \{i;\ |\xi_i| = \alpha\} \quad \text{for } \alpha \neq 0.$$

Proof. The diagonal matrix S with diagonal $(\sigma_i)_{i=1}^n$ attains its norm at the unit vectors $\pm e_1,\dots,\pm e_k$. Moreover, if $z \in \{\pm e_1,\dots,\pm e_k\}$, then

$$S^*D_{Sz}S|_{\{z\}^\perp} = 0,$$

that is $B_z = 0$ and the result easily follows from Theorem 1 (Remark 3).

Final Remark. After this paper was finished Prof. W. Lusky (Paderborn/Germany) showed us a paper of R. Grzaslewicz (Lin. Multilin. Alg. 24(1989), 117-125) which can be used to improve some results of our paper.

Proposition 16. Let S be a regular $n\times n$ -matrix with

$$\|S\| = \sup_{|x|\leq 1} \|Sx\|_p = 1.$$

If $1<p<2$, then card (C) $< \infty$ where

$$C := \{z \in \mathbb{R}^n,\ |z| = \|Sz\|_p = 1\}.$$

Moreover, if $z \in C$, then

$$\langle S^*D_{Sz}Sx,x\rangle \leq |x|^2$$

for all $x\in\mathbb{R}^n$ with $x \perp z$. Here D_{Sz} is defined as in Lemma 9.

Proof. R. Grzaslewicz proved card $(C) < \infty$ (cf. Lemma 2 of his paper). Following his ideas we can also prove the second part of the proposition. To do so choose $z \in C$ and $x \perp z$ with $|x| = 1$. Let us denote Sz by $y = (\eta_1, \ldots, \eta_n)$ and Sx by $v = (\upsilon_1, \ldots, \upsilon_n)$. With this notation we define a function f on $\mathbb{R}$ by

$$f(\lambda) := \|y + \lambda v\|_p^p .$$

Then we have

$$f(\lambda) = \|y+\lambda v\|_p^p = \|S(z+\lambda x)\|_p^p \le |z+\lambda x|^p = (1+\lambda^2)^{p/2} \le 1 + \frac{p\lambda^2}{2}$$

because of $\|S\| = 1$, $x \perp z$, $|z| = |x| = 1$ and $p < 2$. Recall that $\eta_i \neq 0$, $1 \le i \le n$, in view of Proposition 10. Thus f may also be written as

$$f(\lambda) = \sum_{i=1}^{n} |\eta_i + \lambda \upsilon_i|^p = \sum_{i=1}^{n} |\eta_i|^p [(1+\lambda \frac{\upsilon_i}{\eta_i})^2]^{p/2}$$

$$= \sum_{i=1}^{n} |\eta_i|^p \left[1 + \frac{2\lambda\upsilon_i}{\eta_i} + \frac{\lambda^2\upsilon_i^2}{\eta_i^2}\right]^{p/2} .$$

Next choose $\rho > 0$ with $\rho < p/2$. Then there exists a $\delta = \delta(\rho, y, v)$ such that

$$\left[1 + \frac{2\lambda\upsilon_i}{\eta_i} + \frac{\lambda^2\upsilon_i^2}{\eta_i^2}\right]^{p/2} \ge 1 + \frac{2\rho\lambda\upsilon_i}{\eta_i} + \frac{\rho\lambda^2\upsilon_i^2}{\eta_i^2}$$

for all $\lambda \in \mathbb{R}$ with $|\lambda| < \delta$ and $1 \le i \le n$. Using this together with $\sum_{i=1}^n |\eta_i|^p = 1$ we obtain the estimate

$$f(\lambda) \ge 1 + 2\rho\lambda\langle v, d_y\rangle + \rho\lambda^2\langle D_y v, v\rangle$$

as long as $|\lambda| < \delta$.

Recall that $d_y = (|\eta_i|^{p-1}\operatorname{sgn}(\eta_i))_{i=1}^n$ and

$$D_y v = (|\eta_i|^{p-2}\upsilon_i)_{i=1}^n .$$

Since

$$\langle v, d_y\rangle = \langle Sx, d_{Sz}\rangle = \langle x, S^* d_{Sz}\rangle = \langle x, z\rangle = 0$$

(Proposition 10) we finally get

$$f(\lambda) \ge 1 + \rho\lambda^2\langle S^* D_{Sz} Sx, x\rangle, \quad |\lambda| < \delta.$$

This being true for any $\rho < p/2$ proves

$$\langle S^* D_{Sz} Sx, x\rangle \le 1 = |x|^2$$

as asserted.

Corollary 17. If B_z is defined as in Proposition 10, then

$$I - B_z > 0$$

for each $z \in C$.

Proof. Since $B_z(z) = 1$, we only have to show

$$\langle B_z x, x\rangle < |x|^2$$

for all $x \neq 0$, $x \perp z$. But for those $x \in \mathbb{R}^n$ we have

$$B_z(x) = (p-1)S^* D_{Sz} Sx.$$

Consequently, Proposition 16 and $p < 2$, i.e. $p-1 < 1$ prove the Corollary.

Theorem 18. Proposition 13 is also valid for $0<p<2$. Especially for any Gaussian symmetric vector $(X_1,\dots,X_n)$ with full support and $\sigma=1$ (defined by (17)) we have

$$\mathbb{P}\{(\sum_{i=1}^{n} |X_i|^p)^{1/p} > u\} \sim cu^{-1}e^{-u^2/2}$$

as $u \longrightarrow \infty$. Here

$$c = (2\pi)^{-1/2} \sum_{z\in C} |I-B_z|^{-1/2}$$

and

$$c \leq (2\pi)^{-1/2} \mathrm{card}(C)\ (2-p)^{-\frac{n-1}{2}}.$$

Proof. We only show the last-written estimate. All other assertions follow easily from Proposition 13 combined with Corollary 17. We have

$$|I-B_z| = |(I-(p-1)S^* D_{Sz} S)|_{\{z\}^\perp}| \geq |(2-p)I|_{\{z\}^\perp}| = (2-p)^{n-1}$$

completing the proof.

Remark 1: It would be interesting to know a formula for card(C) in dependence of S or $(X_1,\dots,X_n)$. Also a general estimate for card(C) seems to be unknown. It's very likely that always $\mathrm{card}(C) \leq 2^n$. If this holds, then among all $(X_1,\dots,X_n)$ with the same σ the tail behaviour of $(\Sigma_1^n |X_i|^p)^{1/p}$ becomes maximal in the case of independent random variables (cf. Proposition 14).

Remark 2: Theorem 18 tells us that for $p<2$ the sum $(\Sigma_1^n |X_i|^p)^{1/p}$ always possesses up to a constant the smallest possible tail behaviour. We do not know whether or not this remains true for $p>2$.

Final Remark. G. Samorodnitsky (Cornell, Operations Research) gave us a preprint of his paper "Probability tails of Gaussian extrema". There he gives two-sided estimates for $\mathbb{P}\{\sup_t X_t > u\}$, $(X_t)_{t\in T}$ Gaussian process, in terms of entropy conditions for the process. It is very likely that these results also give two-sided estimates for the behaviour of $\mu\{\|x\|_p > u\}$ for ℓ_p-norms $\|\cdot\|_p$.

Acknowledgement: The author is grateful to B. Lopez-Melero (Madrid) and to W.-D. Richter (Rostock) for several helpful discussions about the subject of this paper.

REFERENCES

[1] H. Birndt, W.-D. Richter, "Vergleichende Betrachtungen zur Bestimmung des asymptotischen Verhaltens mehrdimensionaler Laplace-Gauβ-Integrale", Z. Anal. u. Anw. **4** (1985), 269-276.

[2] C. Borell, "Gaussian Radon measures on locally convex spaces", Math. Scand. **38** (1976), 265-284.

[3] L.L. Campbell, "Asymptotic value of a multidimensional normal probability integral", J. Comp. and Appl. Math. **19** (1987) 287-292.

[4] V. Dobric, M.B. Marcus, M. Weber, "The distribution of large values of the supremum of a Gaussian process", Asterique 157-158 (1988), 95-127.

[5] R.S. Ellis, J.S. Rosen, "Laplace's method for Gaussian integrals with an application to Statistical Mechanics", Ann. Prob. **10** (1982), 46-66.

[6] R.S. Ellis, J.S. Rosen, "Asymptotic analysis of Gaussian integrals. I: Isolated minimum points", Trans. Amer. Math. Soc. 273 (1982), 447-481.

[7] A. Erdelyi, "Asymptotic expansions", Dover, New York, 1956.

[8] M.V. Fedoryuk, "The method of saddle points", Nauka, Moscow 1977.

[9] M.V. Fedoryuk, "Asymptotic of integrals and series", Nauka, Moscow 1987.

[10] X. Fernique, "Intégrabilité des vecteurs gaussiens", C.R. Acad. Sci. Paris **270** (1970), 1698-1699.

[11] A. Hertle, "On the asymptotic behaviour of Gaussian spherical integrals", Lect. Notes Nath. **990** (1983), 221-234.

[12] C.-R. Hweng, "Gaussian measure of large balls in a Hilbert space", Proc. Amer. Math. Soc. **78** (1980), 107-110. (cf. also Erratum ibidem (1985), 188)

[13] H. Landau, L.A. Shepp, "On the supremum of a Gaussian process, Sankhya Ser. A **32** (1970), 369-378.

[14] W. Linde, "Gaussian measure of large balls in ℓ_p", to appear in Ann. Prob.

[15] M.B. Marcus, L.A. Shepp, "Sample behaviour of Gaussian processes", Proc. Sixth Berkeley Symp. on Prob. and Stat. (1972), 423-441.

[16] V.P. Maslov, M.V. Fedoryuk, "Logarithmic asymptotic of Laplace integrals", Mat. Zametki **30** (1981), 763-768.

[17] G. Pap, W.-D. Richter, "Zum asymptotischen Verhalten der Verteilungen und der Dichten gewisser Funktionale Gaußscher Zufallsvektoren", Math. Nachr. **135** (1988), 119-124.

[18] W.D. Richter, "Laplace integrals and the probability of moderate deviations", In: Prob. distr. and Math. Stat., Tashkent, Fan, 1986, 406-420.

[19] M. Talagrand, "Sur l'integrabilité des vecteurs gaussiens", Z. Wahrsch. verw. Geb. **68** (1984), 1-8.

[20] M. Talagrand, "Small tails for the supremum of a Gaussian process", Ann. Inst. H. Poincaré (B) **24** (1988), 307-315.

[21] V.M. Zolotarev, "Concerning a certain probability problem", Theory Probab. Appl. **6** (1961), 201-204.

Werner Linde
Friedrich-Schiller-Universität Jena
DDR-6900 Jena
German Democratic Republic

On A Class of Infinitely Divisible Processes Represented as Mixtures of Gaussian Processes

JAN ROSINSKI

Abstract. Variance mixtures of the normal distribution with infinitely divisible mixing measures and a class G of stochastic processes, which naturally arises from such distributions, are studied.

1. Introduction.

LePage [6] observed that symmetric stable process can be represented as mixtures of Gaussian processes. This property played a crucial role in the study of continuity and boundedness for sample paths of stable processes (see, e.g., Marcus-Pisier [8], Talagrand [15]). Using this property, Sztencel [14] proved the absolute continuity of the distributions of seminorms of stable processes. With the aim to extend metric entropy conditions in order to characterize the continuity of certain infinitely divisible processes, Marcus [7] introduced a new class of processes which he called a class G. Processes of this class can be represented as mixtures of Gaussian processes in a way analogous to the representation of stable processes. Further, processes of type G (i.e., the elements of class G) can be characterized by a type G distribution on the real line and a measure on the path space (see also Rosinski [11], Section 5). Therefore, the understanding of one-dimensional type G distributions is crucial for studying type G processes.

Sections 2 and 3 of this paper are devoted to study one-dimensional distributions of type G which are defined as variance mixtures of the normal distribution with infinitely divisible mixing measures. In Section 2 the relationship between variance mixtures of the normal distribution and the class G is discussed and several examples of type G distributions are provided. Lévy measures of type G distributions are characterized in Section 3 and the relation of this class with the class of selfdecomposable laws is established. In Section 4, processes of type G are defined (notice that our definition is more inclusive than the one given in [7]) and several properties of type G processes, that can be easily obtained from the fact that

[1] Research supported in part by AFOSR Grant No. 90-0168 and the Tennessee Science Alliance

these are mixtures of Gaussian processes (or otherwise), are presented. It seems reasonable to expect that a systematic use of the unique conditionally Gaussian structure of type G processes may lead to a unified theory for this subclass of infinitely divisible processes that includes Gaussian, stable and also a number of other interesting processes that can be defined on the base of examples of one-dimensional type G distributions.

2. Variance mixtures of the normal distribution and type G distributions.

In this section we shall recall some known results on variance mixtures of the normal (VMN) distribution and discuss the relation of this class with the classes of infinitely divisible (i.d.) and of type G distributions.

A r.v. X (or its distribution) is said to be a VMN if the characteristic function ϕ_X of X can be written in the form

$$\phi_X(s) = \int_0^\infty \exp\left\{-s^2u/2\right\} H(du) \tag{1}$$

where H is a probability measure on $\mathbf{R}^+$. In other words, a r.v. X is a VMN if

$$X \stackrel{d}{=} ZS \tag{2}$$

where Z has $\mathcal{N}(0,1)$ distribution, $S \geq 0$, and Z and S are independent. Here "$\stackrel{d}{=}$" means "equal in distribution" and $\mathcal{L}(S^2) = H$. The distribution function of a VMN r.v. is absolutely continuous, apart from a possible jump at zero, and its continuous part has a density given by $f_X(x) = \int_{0+}^\infty (2\pi u)^{-1/2} \exp\left(-x^2/(2u)\right) H(du)$. Using Bernstein Theorem (see Feller [2], Ch. XIII.4, for this and other facts regarding completely monotone functions) we have immediately the following criterion:

PROPOSITION 1. *The following are equivalent*

(i) *A r.v. X is a VMN;*
(ii) *ϕ_X is real even function such that $\phi_X(s^{1/2})$ is completely monotone on $(0,\infty)$.*

A VMN distribution does not have to be i.d. In fact, Kelker [4] showed the following

PROPOSITION 2. *Any r.v. X, satisfying (2) with S being bounded and non-degenerate, is not i.d.*

Kelker's proof relies on a result concerning zeros of characteristic functions on the complex plane. We notice here that the Proposition 2 follows also from the fact that for a non-Gaussian i.d. r.v. X, $-\log P\{|X| > x\} = O(x \log x)$, as $x \to \infty$. Indeed, in our case, X is non-Gaussian and

$$-\log P\{|X| > x\} \geq -\log P\{|Z| > xb^{-1}\} \sim x^2 b^{-2}/2,$$

as $x \to \infty$, where b is an upper bound of X. Thus X is not i.d.

As we have mentioned it in the Introduction, Marcus [**7**] studied certain type of VMN, which he called type G distributions. They can be defined as follows:

A r.v. X (or its distribution) is said to be of type G if X is a VMN and the mixing distribution H in (1) *is i.d.. In other words, X is of type G if* (2) *holds with S^2 being i.d.*

It is well known that a type G distribution is i.d. (see Feller [**2**], page 427). Moreover, such a distribution can also be obtained by considering Brownian motion subordinated to an i.d. process with positive jumps (see Feller [**2**], p. 538). We should notice here that there are i.d. VMN distributions which are not of type G. Indeed, Kelker [**4**] constructed an example of an i.d. VMN distribution for which the mixing measure is not i.d. The following proposition can be used to verify whether a given distribution is of type G (see Kelker [**4**], Lemma 1, for the proof of equivalence of conditions (ii) and (iii)).

PROPOSITION 3. *The following are equivalent*

(i) *X is of type G;*

(ii) *$\phi_X(s)$ can be written in the form*

$$\phi_X(s) = \exp\left[-\psi(s^2/2)\right],$$

where ψ has completely monotone derivative on $(0, \infty)$ and $\psi(0) = 0$.

(iii) *$\phi_X(s)$ is a non-negative even function such that, for every $n \geq 1$, $\left[\phi_X(s^{1/2})\right]^{1/n}$ is a completely monotone function of $s > 0$.*

Now we shall give some examples of distributions of type G.

EXAMPLES:

(a) *The Laplace distribution.*

(b) *Symmetrized gamma distributions.* Since the characteristic function is of the form $\phi(s) = \left[\lambda^2/(\lambda^2 + s^2)\right]^p$, where $p, \lambda > 0$ are parameters, a symmetrized gamma distribution is of type G by Proposition 3 (ii). The Laplace distribution is a special case here, corresponding to $p = 1$.

(c) *The t distribution.* This distribution is of type G since the reciprocal of a χ_n^2 distribution is i.d. A discussion of and references to this last (highly nontrivial) result can be found in Steutel [**13**].

(d) *Symmetric stable distributions.*

(e) *Convolution of stable distributions of different orders.* This is a distribution with characteristic function

$$\phi(s) = \exp\left[-\int_{(0,2)} |s|^\alpha \, P(d\alpha)\right],$$

where P is a measure on $(0, 2)$.

(f) *A VMN given by* (2) *such that S has completely monotone density on* $(0, \infty)$. Indeed, since S has completely monotone g Wsity, so does S^2 and hence S^2 is i.d. (see, [J/g., Steutel [**13**]). This example shows, in particular, that the products of normal and Pareto or normal and exponential r.v.'s are of type G.

We do not know whether odd powers of $\mathcal{N}(0,1)$ r.v.'s are of type G. We suspect thy¿Pthe answer is affirmative.

Now we shall discuss series representations of type G r.v.'s in terms of a conditionally Gaussian Poisson point process (or a Poisson point process with Gaussian sections in the terminology of Professor LePage). Throughout this paper $\Gamma_1 < \Gamma_2 < \ldots$ will denote the arrival times in a Poisson process with unit rate and $Z, Z_1, Z_2, \ldots$ will be i.i.d. $\mathcal{N}(0,1)$ r.v.'s. We assume that $\{\Gamma_n\}$, $\{Z_n\}$ and Z are independent of each other.

Let now X be a r.v. of type G given by (2) and put $\sigma = \sup\{x : P(S \geq x) = 1\}$. Then $S_1^2 = S^2 - \sigma^2$ is a nonnegative i.d. r.v. such that $P(S_1^2 < \delta) > 0$ for r every $\delta > 0$. Hence the characteristic function of S_1^2 is of the form $\phi_{S_1^2}(t) = \exp\{\int_0^\infty (e^{itx} - 1)\, \rho_1(dx)\}$, where $\int_0^\infty (1 \wedge x)\, \rho_1(dx) < \infty$ (see, e.g., Feller [**2**], page 539). Define

$$R_0^2(u) = \inf\{x > 0 : \rho_1((x, \infty)) \leq u\}, \qquad u > 0. \tag{3}$$

Proposition 4. *In the above notation,*

$$X \stackrel{d}{=} \sigma Z + \sum_{n=1}^{\infty} Z_n R_0(\Gamma_n), \tag{4}$$

where the series converges a.s.

PROOF: Since $X \stackrel{d}{=} \sigma Z + Z_1 S_1$, it is enough to show that

$$\sum_{n=1}^{\infty} Z_n R_0(\Gamma_n) \stackrel{d}{=} Z_1 S_1.$$

We have

$$\sum_{n=1}^{\infty} Z_n R_0(\Gamma_n) \stackrel{d}{=} Z_1 \left[\sum_{n=1}^{\infty} R_0^2(\Gamma_n) \right]^{1/2}$$

and $\sum_{n=1}^{\infty} R_0^2(\Gamma_n) \stackrel{d}{=} S_1^2$ by Lemma 2.2 in [7], which ends the proof. ∎

Finally we notice that every r.v. X which admits a representation (4), with a certain measurable function R_0 and $\sigma \geq 0$, is of type G.

3. Lévy measures of type G distributions.

Since every type G r.v. X is symmetric and i.d., we can also write its characteristic function $\phi_X(s)$ in the Lévy's form:

$$\phi_X(s) = \exp\left[-\frac{1}{2}\sigma^2 t^2 + 2\int_0^{\infty} (\cos su - 1)\, \rho(du) \right], \tag{5}$$

where ρ is a (symmetric) Lévy measure on $\mathbf{R}$.

THEOREM 1. *A measure ρ on $\mathbf{R}$ is a Lévy measure of a type G distribution if and only if it is of the form*

$$\rho(A) = \int_A g(x^2)\, dx, \qquad A \in \mathcal{B}_{\mathbf{R}}, \tag{6}$$

where $g : \mathbf{R}^+ \to \mathbf{R}^+$ is completely monotone on $(0, \infty)$ and $\int_0^{\infty} (1 \wedge x^2) g(x^2)\, dx < \infty$.

PROOF: We first observe that the finiteness of the last integral is necessary and sufficient for ρ, given by (4), to be a Lévy measure. Assume now that ρ is Lévy measure corresponding to a type G distribution. Then, by Proposition 2 and (5),

$$\psi(s^2/2) = \frac{1}{2}\sigma^2 s^2 + 2\int_0^{\infty} (1 - \cos su)\, \rho(du),$$

where $\psi(\lambda) = \int_0^{\infty} \left[1 - e^{-\lambda t}\right] t^{-1}\, Q(dt)$, and Q is a Radon measure on $\mathbf{R}^+$ such that $\int_1^{\infty} t^{-1}\, Q(dt) < \infty$ (see Feller [2], p. 426). We observe that $\sigma^2 = Q(\{0\})$ and

$$2\int_0^{\infty} (1 - \cos su)\, \rho(du) = \int_{0+}^{\infty} [1 - \exp(-s^2 t/2)] t^{-1}\, Q(dt).$$

Since the right side of this equality can be written as

$$2\int_{0+}^{\infty}\int_{0}^{\infty}[1-\cos(st^{1/2}u)](2\pi)^{-1/2}e^{-u^2/2}\,du\,t^{-1}Q(dt),$$

by the uniqueness of Lévy measures we infer that

$$\rho(A)=\int_{0+}^{\infty}\int_{0}^{\infty}I_A(t^{1/2}u)(2\pi)^{-1/2}e^{-u^2/2}\,du\,t^{-1}Q(dt),$$

$A\in\mathcal{B}_{\mathbf{R}^+}$. It is now easy to verify that

$$g(x):=\int_{0+}^{\infty}(2\pi)^{-1/2}e^{-x/(2t)}t^{-3/2}\,Q(dt)$$

satisfies the conditions of the theorem.

Conversely, if (4) holds and g is completely monotone, then $g(x) = \int_0^\infty \exp(-xu)\,Q_1(du)$, for some Radon measure Q_1 on $\mathbf{R}^+$ such that $Q_1(\{0\})=0$. Then we have

$$\begin{aligned}\psi(s^2/2) &:= 2\int_{0}^{\infty}(1-\cos sx)g(x^2)\,dx = \\ &= 2\int_{0+}^{\infty}\int_{0}^{\infty}(1-\cos sx)e^{-x^2u}\,dxQ_1(du) \\ &= \int_{0+}^{\infty}(2\pi)^{-1/2}u^{-1/2}[1-e^{-s^2/(2u)}]\,Q_1(du).\end{aligned}$$

Clearly ψ has a completely monotone derivative, hence ρ is the Lévy measure of a type G distribution by Proposition 3. ∎

COROLLARY 1. *Theorem 1 can be stated equivalently as follows: A symmetric Lévy measure ρ on $\mathbf{R}$ determines a type G distribution if and only if it is absolutely continuous with respect to the Lebesgue measure and there is a version f of $d\rho/dLeb$ such that $f(x^{1/2})$ is complete monotone on $(0,\infty)$.*

COROLLARY 2. *If ρ is a Lévy measure of a type G distribution, then the function $x\to\rho((x,\infty))$, $x>0$, is convex and possesses derivatives of all orders.*

REMARK 1: Corollaries 1 and 2 offer useful criteria to verify whether an i.d. distribution, whose Lévy measure is known, is or is not of type G. For example, semistable distributions are not of type G, in general, because their Lévy measures may have atoms.

Another use of Theorem 1 will be a characterization of those type G distributions which are selfdecomposable (belong to the Lévy's class $\mathbf{L}$).

COROLLARY 3. *A type G distribution is selfdecomposable if and only if $x^{1/2}g(x)$ is nonincreasing on $(0,\infty)$, where g is given by (6).*

PROOF: A necessary and sufficient condition for the selfdecomposability is that $\rho((e^t,\infty))$ is convex (see Feller [2], p.555). Using (6) this condition can be written as

$$g(e^{2t}) + 2e^{2t}g'(e^{2t}) \le 0, \qquad -\infty < t < \infty$$

or

$$(x^{1/2}g(x))' \le 0, \qquad x > 0. \quad \blacksquare$$

REMARK 2: There are type G distributions which are not selfdecomposable. For example, take $g(x) = (1+x)^{-1}$ in (6). On the other hand, we may also give many examples of type G selfdecomposable distributions. For instance, consider $g(x) = cx^{-(\alpha+1)/2}$ (α-stable distributions), or $g(x) = c(x+x^2)^{-1}$, or $g(x) = x^{-1}e^{-cx}$, etc. ($c > 0$ is a parameter).

The following property of Lévy measures of type G distributions was used by Marcus [7], as a definition of class G.

PROPOSITION 5. *A measure ρ is a Lévy measure of a type G distribution if and only if there exists a unique Lévy measure ρ_0 concentrated on $(0,\infty)$ such that*

$$\rho(A) = \int_0^\infty P[Zu \in A]\,\rho_0(du), \qquad A \in \mathcal{B}_{\mathbf{R}}, \tag{7}$$

where Z is $\mathcal{N}(0,1)$ r.v. In other words, ρ is a Lévy measure of a type G distribution if and only if ρ is a VMN, where the mixing measure is another Lévy measure.

PROOF: Assume (7). Then we have

$$\begin{aligned}\psi(s^2/2) &:= \int_{-\infty}^\infty (\cos su - 1)\,\rho(du) = E\int_0^\infty [1-\cos sZu]\,\rho_0(du)\\ &= \int_0^\infty [1-\exp(-s^2u^2/2)]\,\rho_0(du)\end{aligned}$$

Since ρ_0 is a Lévy measure, the last integral is finite and clearly ψ' is completely monotone. Therefore the sufficiency follows by Proposition 3.

Conversely, by Proposition 4 and Lemma 2.1 [7] (or Theorem 2.4 [11]), we have

$$\rho(A) = E\int_0^\infty I_A(Zu)\,\rho_0(du),$$

where $\rho_0(A) = Leb(\{u > 0 : R_0(u) \in A - \{0\}\})$, which yields (7) after interchanging the integrals.

To show that ρ_0, given by (7), is a Lévy measure we observe that

$$\infty > \int_0^\infty (1 \wedge x^2)\,\rho(dx) = E\int_0^\infty (1 \wedge x^2 Z^2)\,\rho_0(dx)$$
$$\geq \int_0^\infty (1 \wedge x^2)\,\rho_0(dx)\,P(|Z| \geq 1).$$

Finally, we prove the uniqueness of ρ_0. Taking $A = (a^{1/2}, \infty)$ in (7) we get

$$\rho((a^{1/2}, \infty)) = \int_0^\infty P[Z > u^{-1}a^{1/2}]\,\rho_0(du).$$

Now differentiating both sides with respect to a (see Corollary 2) we obtain

$$D_x\rho((x,\infty))|_{x=a^{1/2}} = -(2\pi)^{1/2}\int_0^\infty e^{-a/(2u^2)}u^{-1}\rho_0(du). \tag{8}$$

Since the Laplace transform determines a measure uniquely, the proof is complete. ∎

In the above proof ρ_0 was found as the image by R_0 of the Lebesgue measure on $(0, \infty)$. Hence we have

COROLLARY 4. *R_0 in (4) and ρ_0 in (7) are related by the following equality*

$$R_0(u) = \inf\{x > 0 : \rho_0((x, \infty)) \leq u\}, \qquad u > 0.$$

In general, it may be difficult to determine the mixing measure ρ_0 from (7). As we can see in (8), this is essentially the problem of inverting the Laplace transform. However, if ρ is Lévy measure of a stable distribution, i.e. $\rho((x,\infty)) = cx^{-\alpha}$, then it is easy to verify that ρ_0 is proportional to ρ (restricted to $\mathcal{B}_{\mathbf{R}+}$), and we have $\rho_0 = 2[E|Z|^\alpha]^{-1}\rho$ on $\mathcal{B}_{\mathbf{R}+}$.

PROPOSITION 6. *Let μ be a type G distribution with the Lévy measure ρ. Suppose that ρ_0 in (7) is proportional to ρ on the positive half-line. Then, either μ is a stable distribution or μ is the convolution of two stable distributions of different orders α_1 and α_2 such that $E|Z|^{\alpha_1} = E|Z|^{\alpha_2}$.*

PROOF: Interchanging integrals in (7) we get

$$\rho((a,\infty)) = 2^{-1}E\rho_0((a|Z|^{-1},\infty)), \qquad a > 0. \tag{9}$$

Suppose that $\rho_0 = 2k\rho$ on $\mathcal{B}_{\mathbf{R}^+}$, where $k > 0$ is a constant. Define $f(x) = \rho((e^{-x}, \infty))$, $-\infty < x < \infty$. Then (9) yields

$$f(x) = kE\rho((e^{-x}|Z|^{-1}, \infty)) \\ = kEf(x + \log|Z|)$$

or

$$f(x) = \int_{-\infty}^{\infty} f(x+y)\,\nu(dy), \qquad -\infty < x < \infty, \tag{10}$$

where $\nu = k\mathcal{L}\{\log|Z|\}$. It now follows by Lau-Rao Theorem [5] that $f(x) = p_1 e^{\alpha_1 x} + p_2 e^{\alpha_2 x}$, where $p_1, p_2 \geq 0$ are constants and α_i satisfies the equation

$$\int_{-\infty}^{\infty} e^{\alpha_i x}\,\nu(dx) = 1, \qquad i = 1, 2. \tag{11}$$

Hence we get $\rho((u, \infty)) = f(\log u^{-1}) = p_1 u^{-\alpha_1} + p_2 u^{-\alpha_2}$, $u > 0$. The integrability condition for Lévy measures yields $\alpha_i \in (0, 2)$ and (11) gives the relation between α_i and k:

$$E|Z|^{\alpha_i} = k^{-1}, \qquad i = 1, 2, \tag{12}$$

which completes the proof. ∎

REMARKS 3. Since $E|Z|^{\alpha} = \pi^{-1/2} 2^{\alpha/2} \Gamma((\alpha+1)/2)$ is a convex function in $\alpha \in [0, 2]$ which takes on the same values at the end points of the interval, equation (12) has always two different solutions, provided k^{-1} is not the extreme (minimal) value of $E|Z|^{\alpha}$. The minimum occurs for α_0 determined by $\psi((\alpha_0 + 1)/2) = -\log 2$, where ψ is the logarithmic derivative of the gamma function, and one has $\alpha_0 \approx 0.865$. We may also notice that (12) yields the restriction on k: $1 < k \leq (E|Z|^{\alpha_0})^{-1}$.

4. The idea of reducing the problem to equation (10), which is the key point in proving both Proposition 6 and stated below Proposition 7, was suggested to the author by Professor Stanisław Kwapień, whose contribution in this respect is kindly acknowledged.

PROPOSITION 7. *Let μ be a type G distribution with Lévy measure ρ. Suppose that there exists a constant $\sigma > 0$ such that*

$$\rho_0((a, \infty)) = 2\rho((\sigma^{-1}a, \infty)),$$

for all $a > 0$. Then μ is an α-stable distribution, where α is determined by $(E|Z|^{\alpha})^{1/\alpha} = \sigma^{-1}$.

PROOF: Similarly as in the proof of Proposition 6, we obtain equation (10), where $\nu = \mathcal{L}\{log(|\sigma Z|)\}$. Since now ν is a probability measure, one of the α_i's in (11) equals zero. Then the corresponding p_i must be equal to zero, which concludes the proof. ∎

4. Processes of type G.

Roughly speaking, type G processes are those infinitely divisible processes, which can be represented as (special kind) mixtures of Gaussian processes. Before we give a formal definition of type G processes we will discuss some natural examples of such processes and the corresponding series representations.

Let $\{X(s) : s \in [0,1]\}$ be an independent stationary symmetric increment process without Gaussian component such that $\mathcal{L}\{X(1)\}$ is of type G. Let $\{U_n\}$ be a sequence of i.i.d. uniform on $(0,1)$ r.v.'s independent of $\{\Gamma_n\}$ and $\{Z_n\}$.

Then X can be represented as follows

$$\{X(s) : s \in [0,1]\} \stackrel{d}{=} \left\{ \sum_{n=1}^{\infty} Z_n R_0(\Gamma_n) I_{[0,s]}(U_n) : \quad s \in [0,1] \right\},$$

where R_0 is given by (3) with $X = X(1)$ (this representation can be proved in a similar way as Proposition 4, since it is enough to show the equality in distribution of the corresponding finite linear combinations). Moreover, applying a result of Kallenberg [**3**], we infer that the series on the right converges a.s. uniformly in $s \in [0,1]$.

Consider now a stochastic integral process

$$Y(t) = \int_0^1 f(t,s)\, dX(s), \qquad t \in T,$$

where T is an arbitrary parameter set and $f(t,s)$ is such that, for each t, the stochastic integral exists (see Urbanik and Woyczynski [**16**]). Then we have

$$\{Y(t) : t \in T\} \stackrel{d}{=} \left\{ \sum_{n=1}^{\infty} Z_n R_0(\Gamma_n) f(t, U_n) : t \in T \right\}.$$

Therefore process $\{Y(t) : t \in T\}$ is represented as a mixture of Gaussian processes. The underlying Poisson point process $\{Z_n R_0(\Gamma_n) f(.\,, U_n)\}$, with values in $\mathbf{R}^T$, is obtained from a conditionally Gaussian Poisson point process $\{Z_n R_0(\Gamma_n)\}$ by marking the latter by random samples $f(.\,, U_n)$ of the sections of f. Using Proposition 3 we can write the characteristic function of the process Y in the form

$$E \exp\left[i \sum_{k=1}^{n} a_k Y(t_k) \right] = \exp\left[- \int_0^1 \psi\left(2^{-1} |\sum_{k=1}^{n} a_k f(t_k, s)|^2 \right) ds \right]$$

$$= \exp\left[- \int_{\mathbf{R}^T} \psi\left(2^{-1} |\sum_{k=1}^{n} a_k y(t_k)|^2 \right) \nu(dy) \right],$$

where $\psi(2^{-1}s^2) = -\log \phi_{X(1)}(s)$ and ν is the image by $s \to f(.,s)$ of the Lebesgue measure on $[0,1]$. Now we are ready to define:

A process $\{X(t) : t \in T\}$ (T is arbitrary parameter set) is said to be of type G if there exist a function ψ with completely monotone derivative on $(0,\infty)$, $\psi(0) = 0$ and a σ-finite measure ν on the cylindrical σ-field of $\mathbf{R}^T$ such that

(13)

$$E\exp\left[i\sum_{k=1}^{n} a_k X(t_k)\right] = \exp\left[-\int_{\mathbf{R}^T} \psi\left(2^{-1}|\sum_{k=1}^{n} a_k x(t_k)|^2\right)\nu(dx)\right],$$

for every $a_1,\dots,a_n \in \mathbf{R}$, $t_1,\dots,t_n \in T$ and $n \geq 1$.

We notice that X does not have Gaussian component if and only if $\psi'(\infty) = 0$. To avoid trivial complications, for the rest of this section we shall assume that considered processes of type G do not have Gaussian components.

In order to obtain a conditionally Gaussian representation of a type G process X, given by (13), consider an arbitrary probability measure λ on $\mathbf{R}^T$ such that ν is absolutely continuous with respect λ. Let $h(x) = (d\nu/d\lambda)(x)$, $x \in \mathbf{R}^T$ be a Radon-Nikodym derivative of ν with respect to λ. Let now $\{V_n(t) : t \in T\}$ be i.i.d. stochastic processes with the common distribution λ, defined on the same probability space as and independent of $\{\Gamma_n\}$ and $\{Z_n\}$. Then we have

$$\{X(t) : t \in T\} \stackrel{d}{=} \left\{\sum_{n=1}^{\infty} Z_n R_0(\Gamma_n[h(V_n)]^{-1})V_n(t) : t \in T\right\}, \tag{14}$$

where R_0 is given by (3) and is determined by ψ via Proposition 3 (ii) (to verify (14) apply, e.g., Theorem 2.4 in Rosinski [**11**]). Combining Proposition 3 (ii) and Theorem 1 we infer that the Lévy measure of the process $\{X(t) : t \in T\}$ (see Maruyama [**9**] for a construction of the Lévy measure of a general i.d. process) is of the form

$$F(A) = \int_{\mathbf{R}^T}\int_{-\infty}^{\infty} I_{(A-\{0\})}(ux)\, g(u^2)du\, \nu(dx), \tag{15}$$

where g is completely monotone on $(0,\infty)$ and $\int_0^\infty (1 \wedge u^2)g(u^2)\, du < \infty$. Using the relation between ψ and g, established in the proof of Theorem 1, we notice that $\psi(\infty) = \rho(\mathbf{R}) = \int_{-\infty}^{\infty} g(u^2)\, du$.

REMARK 5: It may help to clarify this subject if we consider, for a moment, a finite set T of cardinality d, so that the process X can be viewed as a random vector in $\mathbf{R}^d$. Then X is of type G if and only if its characteristic function is of the form

$$E\exp(i<y,X>) = \exp\left[-\int_{\mathbf{R}^d}\psi(<y,x>^2/2)\,\nu(dx)\right], \qquad y \in \mathbf{R}^d,$$

where ψ has completely monotone derivative with $\psi(0) = 0$ and ν is a σ-finite Borel measure on $\mathbf{R}^d$. In view of Theorem 1, the Lévy measure F of X admits the decomposition (15) with T replaced by d; and conversely, if Lévy measure F of a symmetric with no Gaussian component random vector X is of the form (15), then X is of type G. Finally, in representation (14), V_n are i.i.d. random vectors taking values in $\mathbf{R}^d$ such that measure ν is absolutely continuous with respect to their common distribution λ (except for this condition there are no other restrictions on λ).

We now return to a general T and observe that (15) implies that processes of type G satisfy the zero-one law for subgroups.

THEOREM 2. *Let $\{X(t) : t \in T\}$ be a process of type G given by* (13) *such that $\psi(\infty) = \infty$. Then, for every $\mathcal{L}\{X\}$-measurable subgroup H of $\mathbf{R}^T$ and $x \in \mathbf{R}^T$, $P[X \in H + x] = 0$ or 1.*

PROOF: Since the "radial component" ρ of F in (15) is absolutely continuous, $\rho(du) = g(u^2)du$ and $\rho(\mathbf{R}) = \infty$, Theorem 2 follows by Corollary 3 in Rosinski [**12**]. ∎

The fact that F is symmetric and has absolutely continuous non-vanishing "radial part" also implies, under some regularity conditions on sample paths (e.g., that sample paths belong to a separable metrizable linear subspace of $\mathbf{R}^T$), that (properly defined) support of $\mathcal{L}\{X\}$ is a linear subspace. Indeed, Theorem 1 in Rajput [**10**] gives that the support is a subgroup which is generated by the support of F. Since g is positive, (15) yields that the support of F is closed under multiplication by scalars, which completes the argument.

Now we notice that type G processes satisfy *Anderson's inequality*. Namely, if $\{X(t) : t \in T\}$ is of type G, then for every symmetric convex sets A and B in the cylindrical σ-field of $\mathbf{R}^T$, $x_0 \in \mathbf{R}^T$ and $|\lambda| \le 1$,

$$P[X \in (A + \lambda x_0) \cap B] \ge P[X \in (A + x_0) \cap B].$$

Indeed, this follows immediately by (14) and Anderson's inequality for Gaussian measures (see [**1**], Sect. 6), by conditioning.

As the last property of type G processes we shall establish the continuity of the distributions of subadditive functionals of sample paths. Let $\Phi : \mathbf{R}^T \to \mathbf{R}$ be a subadditive positively homogeneous mapping such that $\{x : \Phi(x) \le 1\}$ is $\mathcal{L}\{X\}$-measurable. Let $K_\Phi(u) = P[\Phi(X) \le u]$ and $u_0 = \inf\{u : K_\Phi(u) > 0\}$.

THEOREM 3. *Let $\{X(t) : t \in T\}$ be a process of type G given by (13) such that $\psi(\infty) = \infty$ and let Φ satisfies above assumptions. Then K_Φ is absolutely continuous on (u_0, ∞).*

PROOF: Let λ be a probability measure on $\mathbf{R}^T$ equivalent to ν, so that $h = d\nu/d\lambda$ can be taken positive everywhere. By (7) and our assumption, we have $\rho_0((0,\infty)) = \infty$ which, in view of Corollary 4, is equivalent to that $R_0(u) > 0$, for every $u > 0$. Now we can adapt (and slighty modify) the idea of proof given by Sztencel [**14**] in the case of stable distributions.

Without loss of generality we may assume that X is defined by the a.s. convergent series (14). Since the conditional distribution of X, given that $\Gamma_n - \Gamma_{n-1} = s_n$, $V_n = v_n$, is Gaussian, the conditional distribution function

$$P[\Phi(X) \leq u \,|\, \Gamma_n - \Gamma_{n-1} = s_n, V_n = v_n, n \geq 1]$$

is absolutely continuous on the half-line $(c(\{s_n, v_n\}), \infty)$, where $c(\{s_n, v_n\})$ is the point of the first increase of this conditional distribution function (equals to $-\infty$ if the distribution function is positive everywhere). This fact on the absolute continuity follows from the convexity of Gaussian measures (see [**1**]) and the properties of Φ (usually it is also assumed that Φ is non-negative, cf. [**14**] and references therein, but this is unnecessary restriction on Φ). To conclude our proof, it is enough to show that $c(\{\Gamma_n - \Gamma_{n-1}, V_n\})$ is constant a.s. This will follow by Hewitt-Savage zero-one law, if we show that $c(\{s_{i(n)}, v_{i(n)}\}) = c(\{s_n, v_n\})$, for every finite permutation $\{i(n)\}$ of natural numbers. Let $\{i(n)\}$ be such a permutation. Since R_0 and h do not vanish, the conditional distribution of X, given that $\Gamma_n - \Gamma_{n-1} = s_{i(n)}$, $V_n = v_{i(n)}$, is equivalent to the conditional distribution of X, given that $\Gamma_n - \Gamma_{n-1} = s_n$, $V_n = v_n$. Therefore, the points of first increase for boths conditional distribution function of $\Phi(X)$ must coincide, i.e., $c(\{s_{i(n)}, v_{i(n)}\}) = c(\{s_n, v_n\})$, which ends the proof. ∎

Finally we notice that examples of Φ, for which the above theorem applies, include $\Phi(x) = \sup\{x(t) : t \in T\}$, $\Phi(x) = \sup\{|x(t)| : t \in T\}$ (under obvious separability assumptions), and other interesting functionals of sample paths.

REFERENCES

[1] Borell, C.(1974). Convex measures on locally convex spaces. Arkiv Mat. 12, 239-252.

[2] Feller, W. An introduction to probability theory and its applications. Volume II. First ed. J. Wiley & Sons, New York, 1966.

[3] Kallenberg, O. (1973). Series of random processes without discontinuities of the second kind. Ann. Probab. 2, 729- 737.

[4] Kelker, D. (1971). Infinite divisibility and variance mixtures of the normal distribution. Ann. Math. Statist. 42, 802-808.

[5] Lau, Ka-Sing and Rao, C. Radhakrishna (1984). Solution to the integrated Cauchy functional equation on the whole line. Sankhyā A 46, 311-318.

[6] LePage, R. (1980). Multidimensional infinitely divisible variables and processes. Part I: Stable case. Technical report no. 292. Dept. of Statistics, Stanford University.

[7] Marcus, M. B. (1987). ξ-radial processes and random Fourier series. Memoirs Amer. Math. Soc. 368.

[8] Marcus, M. B. and Pisier, G. (1984). Characterizations of almost surely continuous p-stable random Fourier series and strongly stationary processes. Acta Math. 152, 245-301.

[9] Maruyama, G. (1970). Infinitely divisible processes. Theor. Prob. Appl. 15, 3-23.

[10] Rajput, B. S. (1977). On the support of symmetric infinitely divisible and stable probability measures on LCTVS. Proc. Amer. Math. Soc. 66, 331-334.

[11] Rosinski, J. (1990). On series representations of infinitely divisible random vectors. Ann. Probab. 18, 405-430.

[12] Rosinski, J. (1990). An application of series representations for zero-one laws for infinitely divisible random vectors. Probability in Banach Spaces 7, Progress in Probability 21, Birkhauser, 189-199.

[13] Steutel, F. W. (1979). Infinite divisibility in theory and practice. Scand. J. Statist. 6, 57-64.

[14] Sztencel, R. (1986). Absolute continuity of the lower tail of stable seminorms. Bull. Pol. Acad. Sci. Math. 34, 231-234.

[15] Talagrand, M. (1988). Necessary conditions for sample boundedness of p-stable processes. Ann. Probab. 16, 1584-1595.

[16] Urbanik, K. and Woyczynski, W. A. (1967). Random integrals and Orlicz spaces. Bull. Acad. Polon. Sci. 15, 161- 169.

Keywords: Variance mixtures of the normal distribution, marked Poisson point processes, infinitely divisible processes
1980 *Mathematics subject classifications:* 60E07, 60G15

Department of Mathematics
University of Tennessee
Knoxville, TN 37996-1300

Capacities, Large Deviations and Loglog Laws

GEORGE L. O'BRIEN* and WIM VERVAAT**

Abstract

Spaces of capacities are considered with their natural subspaces and two topologies, the vague and the narrow. Large deviation principles are identified as a class of limit relations of capacities. Narrow large deviation principles occasionally can be tied to loglog laws, and this relationship is studied. Specific narrow large deviation principles and loglog laws are presented (without proof) for the Poisson process on the positive quadrant that is the natural foundation for extremal processes and spectrally positive stable motions. Related loglog laws for extremal processes and stable motions are discussed.

0. Introduction

Capacities are increasing functions from the power set of a topological space E into $[0, \infty]$ that satisfy certain regularity conditions. The space of all capacities contains as subspaces the Radon measures, tight bounded measures, the upper semicontinuous functions on E, the closed subsets of E and the compact subsets of E, the last three spaces only after identification with (subspaces of) the sup measures, capacities c such that $c(K_1 \cup K_2) \leq c(K_1) \vee c(K_2)$ for all compact $K_1, K_2 \subset E$.

The usual topologies on all these subspaces can be identified as the traces of just two topologies on the space of all capacities $\mathcal{C}(E)$. We say that a sequence (or net) (c_n) converges vaguely (narrowly) to c in $\mathcal{C}(E)$ if

$$\begin{aligned} \liminf c_n(G) &\geq c(G) \qquad \text{for open } G \subset E, \\ \limsup c_n(B) &\leq c(B) \qquad \text{for compact (closed) } B \subset E. \end{aligned}$$

* Supported by the Natural Sciences and Engineering Research Council of Canada.

** Collaboration between the two authors supported by a NATO grant for international collaboration in research.

The vague convergence mimics vague convergence of Radon measures, and so does the narrow convergence with narrow (= weak) convergence of bounded measures, in particular probability measures. Capacities and these topologies are examined in Sections 1–4.

Let (c_n) be a sequence of subadditive capacities (i.e., $c_n(K_1 \cup K_2) \leq c_n(K_1) + c_n(K_2)$ for all compact $K_1, K_2 \subset E$), and let $a_n \to \infty$ in $(0, \infty)$. If $c_n^{1/a_n} \to c$ vaguely (narrowly) for some $c \in \mathcal{C}(E)$, then we say that a vague (narrow) large deviation principle holds. Necessarily, the limit c is a sup measure (this is Section 5). In the literature, the c_n are mostly probability measures, and the limit relation is often presented in a logarithmic form. Narrow large deviation principles in which the c_n are probability distributions of random variables X_n in E on one probability space can be transformed into a *preloglog law* after a change of time:

$$\mathsf{P}[X_{n(m)} \in \cdot\,] = m^{-J(\,\cdot\,)+o(1)} \quad \text{narrowly},$$

where $c =: e^{-J}$, so J is an 'inf measure'. If the $X_{n(m)}$ are sufficiently independent to apply the Borel-Cantelli lemma in both directions, we obtain a *loglog law:* the sequence $(X_{n(m)})$ is wp1 relatively compact in E with set of limit points $\{x \in E\colon J(\{x\}) \leq 1\}$ (Section 6).

Here are a few more details. The general assumption about E is that E is a separable metric space. In the regularity conditions on the capacities the compact and open sets play a role (Section 1). For general E, $\mathcal{C}(E)$ is vaguely compact, but things become nice, $\mathcal{C}(E)$ is vaguely Hausdorff for instance, only if E is locally compact (Section 2). This restriction on E can be dropped if we consider $\mathcal{C}(E)$ with the narrow topology, but then we have to restrict $\mathcal{C}(E)$ to those capacities for which we can replace at will the compact sets by the closed in the regularity conditions on capacities. This is guaranteed by tightness, a technical condition which for subadditive capacities translates into the more familiar $c(E \backslash K_m) \to 0$ for some sequence of compacts (K_m) (Section 3). The tight capacities with the narrow topology (in this combination only) behave well under continuous mappings from E into another space E', as we are used to with probability measures (Section 4).

In the final sections (7–10) we formulate and discuss specific large deviation principles and loglog laws as they will be handled in forthcoming work by the authors. A particular aspect is that the space E on which the capacities are defined and in which the random variables take their values is itself a space of capacities (or a subspace of it) on another space, a 'quadrant' $\square_\ell := [0, \infty) \times (\ell, \infty]$ for some $\ell \in [-\infty, \infty)$.

Let (ξ_k) be a sequence of iid random variables in R. As has become common in the context of extremal processes and stable motions, we study

(ξ_k) by the *observation process,* the random measure Ξ in $\square_{-\infty}$ that counts the points (k, ξ_k) of the graph of (ξ_k). Weak convergence to extremal processes is attained by transformations $(k, \xi_k) \mapsto (\frac{k}{n}, a_n\xi_k + b_n)$ of Ξ for suitable choices of $a_n > 0$ and b_n, and the limiting random measure on $\square_\ell$ (for some minimal ℓ) turns out to be a Poisson process. Up to simple transformations, the limiting process can be represented by the *self-affine* Poisson process Π on $\square_0$ with mean measure $\pi(dt, dx) = dt\, x^{-2}dx$ (Section 7). In the present paper we restrict our attention to Π, and do not consider the observation process Ξ. Narrow large deviation principles and loglog laws are presented for two different sequences of transformations of Π (Section 8). By different functionals they are transformed into results for extremal processes (Section 9) and spectrally positive stable motions (Section 10). In the former case the application is straightforward, but in the latter much additional work is needed, and even the presentation of the results is restricted here to the simplest case of increasing stable motions.

Capacities with the vague topology on locally compact spaces were studied by Norberg (1986). The first try to embed the probability measures with the narrow topology into larger spaces is Salinetti & Wets (1987). The present setup with the two topologies was announced in Vervaat (1988a). A detailed study of sup measures and spaces of upper semicontinuous functions with the two topologies is Vervaat(1988b). Norberg & Vervaat (1989) investigate the relations between capacities and Radon measures in the generality of non-Hausdorff E.

The first study of relations between large deviation principles in the two topologies is Lynch & Sethuraman (1987).

Loglog laws for extremal processes have been obtained by Wichura (1974a) and Mori & Oodaira (1976), and loglog laws for stable motions by Wichura (1974b), Mijnheer (1975) and Pakshirajan & Vasudeva (1981). Mori & Oodaira (1976) obtained one loglog law directly in terms of the self-affine Poisson process. Deheuvels & Mason (1989a) obtained loglog laws for extremal processes and stable motions with as starting point loglog laws for quantile processes (more papers on loglog laws for tail empirical processes and quantile processes are Deheuvels & Mason (1989b, 1990) and Mason (1988)). Only in the last papers are related preloglog laws dealt with explicitly.

1. Capacities

Throughout the paper E is a separable metric space. In some places, especially in Section 2, it will be assumed in addition that E is locally compact. This condition is always stated explicitly. The families of open, closed and compact sets in E are denoted by $\mathcal{G}(E)$, $\mathcal{F}(E)$ and $\mathcal{K}(E)$, or $\mathcal{G}$,

$\mathcal{F}$ and $\mathcal{K}$ for short.

We will make regular use of the following technical conditions involving interaction between two classes $\mathcal{A}$ and $\mathcal{B}$ of subsets of E:

(i) if $A \in \mathcal{A}$, $B \in \mathcal{B} \cup \{E\}$ and $A \subset B$, then there exist sets $A_1, A_2, \ldots \in \mathcal{A}$ and $B_1, B_2, \ldots \in \mathcal{B}$ such that

$$B \supset A_1 \supset B_1 \supset A_2 \supset \cdots \supset A_m \supset B_m \supset A_{m+1} \supset \cdots \downarrow A;$$

(ii) if $A \in \mathcal{A} \cup \{\emptyset\}$, $B \in \mathcal{B}$ and $A \subset B$, then there exist sets $A_1, A_2, \ldots \in \mathcal{A}$ and $B_1, B_2, \ldots \in \mathcal{B}$ such that

$$A \subset B_1 \subset A_1 \subset B_2 \subset \cdots \subset B_m \subset A_m \subset B_{m+1} \subset \cdots \uparrow B;$$

(iii) if $A \in \mathcal{A}$, $B \in \mathcal{B}$ and $A \subset B$, then there exist sets $A_1 \in \mathcal{A}$ and $B_1 \in \mathcal{B}$ such that $A \subset B_1 \subset A_1 \subset B$.

We say that $\mathcal{A}$ *nestles* in $\mathcal{B}$ if all three of these conditions hold. Note that (i) and (ii) each imply (iii). The next lemma gives two important examples.

1.1. Lemma. (a) *$\mathcal{F}$ nestles in $\mathcal{G}$.*
(b) *$\mathcal{K}$ nestles in $\mathcal{G}$ iff E is locally compact.*

The proofs of (a) and (b) involve straightforward arguments using the metric separable nature of E. Note that each of (i), (ii) and (iii) by itself implies local compactness in case (b).

Many of the spaces considered in the paper are subspaces of a master space $\mathcal{C}$ to be defined below. This space has been studied for locally compact E by Norberg (1986) and for more general topological spaces E by Vervaat (1988a) and Norberg & Vervaat (1989). Many of the assertions of this section and the next are proved in these papers, but the reader should be able to obtain them directly for separable metric E; hints are given in some cases.

1.2. Definition. A *capacity* on E is a function c from the power set $\mathcal{P}(E)$ of E into $\bar{\mathbb{R}}_+ := [0, \infty]$ such that

$$c(\emptyset) = 0, \tag{1.1a}$$

$$c(A) \le c(B) \quad \text{if } A \subset B \subset E, \tag{1.1b}$$

$$c(A) = \sup_{K \in \mathcal{K}, K \subset A} c(K) \quad \text{for } A \subset E, \tag{1.1c}$$

$$c(K) = \inf_{G \in \mathcal{G}, G \supset K} c(G) \quad \text{for } K \in \mathcal{K}. \tag{1.1d}$$

The set of capacities on E is indicated by $\mathcal{C}(E)$ or $\mathcal{C}$. The relationship between capacities in $\mathcal{C}$ and Choquet capacities is discussed at the end of the section.

Here are a few basic properties of capacities. First, note that if $G_n \uparrow G$ in $\mathcal{G}$, $K \in \mathcal{K}$ and $K \subset G$ then $K \subset G_n$ eventually. It follows from (1.1b,c) that

$$c(G_n) \uparrow c(G) \quad \text{if } G_n \uparrow G \text{ in } \mathcal{G}. \tag{1.1e}$$

Similarly, if $K_n \downarrow K$ in $\mathcal{K}$, $G \in \mathcal{G}$ and $K \subset G$ then $K_n \subset G$ eventually. It follows from (1.1b,d) that

$$c(K_n) \downarrow c(K) \quad \text{if } K_n \downarrow K \text{ in } \mathcal{K}. \tag{1.1f}$$

If E is locally compact so that $\mathcal{K}$ nestles in $\mathcal{G}$, then (1.1f) and (1.1b) together imply (1.1d). Thus any function $c\colon \mathcal{K} \to \bar{\mathsf{R}}_+$ that satisfies (1.1b) restricted to $\mathcal{K}$ and (1.1a) can be extended to a unique capacity by means of (1.1c) iff (1.1f) holds, in case E is locally compact. The following sequential versions of (1.1d,c) are useful: we can find for each $K \in \mathcal{K}$ a sequence (G_n) in $\mathcal{G}$ such that

$$G_n \downarrow K, \qquad c(G_n) \downarrow c(K) \quad \text{for } all\ c \in \mathcal{C} \tag{1.1g}$$

(for example $G_n := \{x \in E\colon d(x, K) < n^{-1}\}$, where d is the metric on E), and, if $\mathcal{K}$ nestles in $\mathcal{G}$, then by (1.1e) we can find for each $G \in \mathcal{G}$ a sequence (K_n) in $\mathcal{K}$ such that

$$K_n \uparrow G, \qquad c(K_n) \uparrow c(G) \quad \text{for } all\ c \in \mathcal{C}. \tag{1.1h}$$

We now review some important subspaces of $\mathcal{C}$. By SA(E) or SA we denote the space of all *subadditive capacities,* capacities c such that

$$c(K_1 \cup K_2) \leq c(K_1) + c(K_2) \qquad \text{for } K_1, K_2 \in \mathcal{K}. \tag{1.2}$$

Using the following lemma and (1.1c,d) we have that (1.2) is equivalent to

$$c\left(\textstyle\bigcup_\alpha G_\alpha\right) \leq \sum_\alpha c(G_\alpha) \tag{1.3}$$

for arbitrary collections $(G_\alpha)_\alpha$ in $\mathcal{G}$.

1.3. Lemma. *If $K \in \mathcal{K}$, $G_i \in \mathcal{G}$ for $i = 1, 2, \ldots, n$ and $K \subset \bigcup_{i=1}^n G_i$, then there are compact K_i such that $K_i \subset G_i$ for $i = 1, 2, \ldots, n$ and $K \subset \bigcup_{i=1}^n K_i$.*

To prove this lemma we may take

$$K_i = \{x \in K : d(x, G_i^c) = \max_{j=1}^n d(x, G_j^c)\}.$$

For proofs in more general contexts, see Berg, Christensen & Ressel (1984, Lemma 2.1.17) and Vervaat (1988c).

All other subspaces of interest are in fact subspaces of SA. In particular, we denote by AD(E) or AD the space of *additive* capacities, namely those which are subadditive *and* satisfy

$$c(K_1 \cup K_2) = c(K_1) + c(K_2) \qquad \text{for disjoint } K_1, K_2 \in \mathcal{K}. \tag{1.4}$$

By Lemma 1.3 and (1.1c,d) it follows that (1.4) and (1.2) are together equivalent to (1.3) with $\leq$ replaced by $=$ in case the G_α's are disjoint (only countably many can then be nonempty).

The additive capacities that are finite on $\mathcal{K}$ are called *Radon measures*, and the space of all of them is denoted by $\mathcal{M}(E)$ or $\mathcal{M}$. Note that the restrictions of the Radon measures to Bor E, the Borel field of E, coincide with (one version of) the classical Radon measures (Bourbaki (1969)), countably additive measures c on Bor E that are finite on $\mathcal{K}$ and satisfy (1.1c,d). By $\mathcal{N}(E)$ or $\mathcal{N}$ we denote the space of $(\mathsf{N} \cup \{\infty\})$-valued Radon measures, the *point measures.*

Let c be a capacity. By (1.1d) applied to singletons, the set

$$\operatorname{supp}_\infty c := \{x \in E : c(\{x\}) = \infty\} \tag{1.5}$$

is closed. We call $\operatorname{supp}_\infty c$ the *infinity support* of c. If c is subadditive, then $c(K) < \infty$ for all $K \in \mathcal{K}$ for which

$$K \cap \operatorname{supp}_\infty c = \emptyset. \tag{1.6}$$

It follows that if $c \in$ AD, the restriction of c to $E \backslash \operatorname{supp}_\infty c$ is a Radon measure. We now consider the question of when a measure c on Bor E can be extended to a capacity.

1.4. Lemma. *Suppose E is locally compact and c is a measure on* Bor E. *Then c can be extended to a capacity via (1.1c) iff* $\operatorname{supp}_\infty c$ *(defined in (1.5)) is closed and $c(K) < \infty$ for all compact K for which (1.6) holds. In particular, if c is σ-finite, so that* $\operatorname{supp}_\infty c = \emptyset$, *then c is extendable to a capacity iff c is Radon.*

Proof. The necessity was proved above; we show the sufficiency. Suppose $K_n \downarrow K$ in $\mathcal{K}$ and K satisfies (1.6). Then eventually K_n satisfies (1.6) so (1.1f) holds. If $K \cap \operatorname{supp}_\infty c \neq \emptyset$, then $c(K) = \infty$ so again (1.1f) holds. Since E is locally compact, c restricted to $\mathcal{K}$ can be extended uniquely by (1.1c) to an additive capacity c_0. It turns out that c_0 is countably additive on Bor E and so $c_0 = c$ on Bor E. QED

Next, we denote by SM(E) or SM the space of *sup measures*, capacities c such that

$$c(K_1 \cup K_2) = c(K_1) \vee c(K_2) \qquad \text{for } K_1, K_2 \in \mathcal{K} \tag{1.7}$$

(note that (1.7) with $\leq$ instead of $=$ is equivalent, by (1.1b)). By Lemma 1.3, (1.7) is equivalent to

$$c\left(\bigcup_\alpha G_\alpha\right) = \bigvee_\alpha c(G_\alpha) \tag{1.8}$$

for arbitrary collections $(G_\alpha)_\alpha$ in $\mathcal{G}$. Any function $c: \mathcal{G} \to \bar{\mathbb{R}}_+$ satisfying (1.1a) and (1.8) can be extended uniquely to a sup measure via

$$c(A) = \inf_{G \in \mathcal{G}: G \supset A} c(G) \qquad \text{for } A \subset E. \tag{1.9}$$

To verify (1.1.c) one can easily check that in this case we even have

$$c(A) = \sup_{x \in A} c(\{x\}). \tag{1.10}$$

It follows that the definition of sup measure used here is consistent with the one used in Vervaat (1988b) and O'Brien, Torfs & Vervaat (1989). Moreover, from (1.10) we also see that (1.8) holds for arbitrary $G_\alpha \subset E$. From (1.1b,d) we see that f: $E \ni x \mapsto c(\{x\})$ is upper semicontinuous (usc). Then we write $f = d^\vee c$ (the *sup derivative* of c) and $c = f^\vee$ (the *sup integral* of f). The space of usc functions $E \to \bar{\mathbb{R}}_+$ is denoted by US(E) or US, and is regarded as a subspace of SA $\subset \mathcal{C}$ after identification with SM. By identifying closed sets $F \in \mathcal{F}(E)$ with their indicator functions $1_F \in$ US we make $\mathcal{F}$ a subspace of US and next, after identification with the sup integral $1_F^\vee \in$ SM, a subspace of SA $\subset \mathcal{C}$.

Although $\mathcal{C}$ will be made a measurable space only at the end of the next section, we remark now that we will encounter random variables with values in $\mathcal{M}$, $\mathcal{N}$, SM, US and $\mathcal{F}$. They are called *random measures, point processes, extremal processes, usc processes* and *random closed sets,* respectively.

Note that capacities in $\mathcal{C}$ are not necessarily Choquet capacities as defined by Dellacherie & Meyer (1978, p.51). Specifically, $A_n \uparrow A$ does not imply $c(A_n) \uparrow c(A)$, even in AD. Nor does the dual relation for decreasing sequences always hold, even in SM. Nevertheless, the definitions are similar and the classes AD and SM are strongly subadditive. Also, $c(A) = \inf\{c(G) : G \in \mathcal{G},\ G \supset A\}$ for $c \in$ SM and, if $A \in \operatorname{Bor} E$, for $c \in$ AD, so we do have regularity properties similar to those of Choquet. Inner regular Choquet capacities on locally compact E with paving $\mathcal{K}$ are capacities in our sense. Capacities of different types on non-Hausdorff spaces are studied in Norberg & Vervaat (1989).

2. Capacities with the vague topology

The space $\mathcal{C}$ and its subspaces are made topological by providing $\mathcal{C}$ with the *vague* topology, generated by the subbase consisting of the sets $\{c\colon c(G) > x\}$ and $\{c\colon c(K) < x\}$ for $x \in (0,\infty)$, $G \in \mathcal{G}$ and $K \in \mathcal{K}$. Equivalently, the vague topology on $\mathcal{C}$ is the coarsest that makes the evaluations $c \mapsto c(A)$ lower semicontinuous (lsc) for $A \in \mathcal{G}$ and upper semicontinuous (usc) for $A \in \mathcal{K}$. Convergence of sequences in $\mathcal{C}$ has the 'portmanteau' characterization

$$c_n \to c \text{ vaguely} \iff \begin{cases} \liminf c_n(G) \geq c(G) & \text{for } G \in \mathcal{G}, \\ \limsup c_n(K) \leq c(K) & \text{for} K \in \mathcal{K}. \end{cases} \tag{2.1}$$

We will see below that $\mathcal{C}$ with the vague topology is compact and, if E is locally compact, then $\mathcal{C}$ is Hausdorff with a countable base, which implies $\mathcal{C}$ is a compact metric space. Actually, $\mathcal{C}$ will turn out to be vaguely compact for general topological spaces E.

So far we have not obtained a direct recipe for obtaining limits given a convergent sequence of capacities. Moreover, limits are unique if $\mathcal{C}$ is vaguely Hausdorff, but need not be otherwise. To deal with these gaps, the following construction is useful.

2.1. Construction. For any function $\gamma\colon \mathcal{P}(E) \to \bar{\mathbb{R}}_+$ that satisfies (1.1a,b), set

$$\gamma_*(G) := \sup_{K\in\mathcal{K}, K\subset G} \gamma(K) \qquad \text{for } G \in \mathcal{G},$$

$$\gamma^*(K) := \inf_{G\in\mathcal{G}, G\supset K} \gamma(G) \qquad \text{for } K \in \mathcal{K}.$$

Then extend γ_* to $\mathcal{K}$ by (1.1d) and γ^* to $\mathcal{G}$ by (1.1c).

It is clear that $\gamma_* \geq \gamma$ on $\mathcal{K}$ so γ_* satisfies (1.1c) on $\mathcal{G}$; similarly, $\gamma^* \leq \gamma$ on $\mathcal{G}$ so γ^* satisfies (1.1d). Thus γ_* and γ^* can be extended to capacities by (1.1c). Then $\gamma_* \leq \gamma^*$ on $\mathcal{K}$ and hence on $\mathcal{P}(E)$, with equality if E is locally compact, by Lemma 1.1(b). By the definition of γ^* on $\mathcal{K}$, γ^* is the largest capacity such that $\gamma^* \leq \gamma$ on $\mathcal{G}$; likewise, γ_* is the smallest capacity such that $\gamma_* \geq \gamma$ on $\mathcal{K}$.

If (c_n) is a sequence in $\mathcal{C}$, let $\liminf c_n$ and $\limsup c_n$ denote the 'pointwise' lower and upper limits (so $(\liminf c_n)(A) := \liminf c_n(A)$). Then $\liminf c_n$ and $\limsup c_n$ need not be capacities, but $(\liminf c_n)^*$ and $(\limsup c_n)_*$ are. From (2.1) it is clear that $(\liminf c_n)^*$ is the largest capacity c that satisfies the upper right-hand side of (2.1), and $(\limsup c_n)_*$ is the smallest capacity satisfying the lower right-hand side. Consequently, c is a limit of (c_n) iff $(\limsup c_n)_* \leq c \leq (\liminf c_n)^*$. If E is locally compact, $(\liminf c_n)^* = (\liminf c_n)_* \leq (\limsup c_n)_*$, so in that case c_n converges vaguely to c iff $c = (\limsup c_n)_* = (\liminf c_n)^*$.

All characterizations of vague convergence ending with (2.1) hold verbatim for the subspaces SA, AD, $\mathcal{M}$, $\mathcal{N}$, SM, US with $c = f^\vee$ and $\mathcal{F}$ with $c = 1_F^\vee$. We next consider the topological structure of $\mathcal{C}$ and various subspaces.

2.2. Theorem. *Let L be compact in $\bar{\mathbb{R}}_+$ and $0 \in L$. Then*
(a) *the space of L-valued capacities is vaguely compact;*
(b) *the spaces of L-valued sup measures and L-valued usc functions are vaguely compact;*
(c) *$\mathcal{F}$ is vaguely compact.*

Proof. (a) (due to H. Holwerda). Denote the L-valued capacities by $\mathcal{C}_L$. By Alexander's subbase theorem we must indicate a finite subcover for each instance of

$$\mathcal{C}_L \subset \bigcup_{i\in I}\{c: c(G_i) > x_i\} \cup \bigcup_{j\in J}\{c: c(K_j) < y_j\}, \tag{2.2}$$

where $G_i \in \mathcal{G}$, $K_j \in \mathcal{K}$ and $x_i, y_j \in \bar{\mathbb{R}}_+$. We may assume that $x_i = 0$ for at least one i, otherwise extend the cover by $\emptyset = \{c: c(\emptyset) > 0\}$. Replacing each x_i by $\sup\{x \in L: x \leq x_i\}$ we may assume each $x_i \in L$. For $K \in \mathcal{K}$, set

$$c_0(K) := \bigwedge_{i: G_i \supset K} x_i \qquad (\text{here } \textstyle\bigwedge\emptyset := \sup L).$$

Then c_0 is L-valued, satisfies (1.1a,b), and after extension by (1.1c) also (1.1d), so the extension results in a capacity. Since $c_0(G_i) \leq x_i$ for all i, there must by (2.2) be a $j_0 \in J$ such that $c_0(K_{j_0}) < y_{j_0}$. By definition of $c_0(K_{j_0})$ then there is an $i_0 \in I$ such that $G_{i_0} \supset K_{j_0}$ and $x_{i_0} < y_{j_0}$. So the right-hand side of (2.2) with I restricted to $\{i_0\}$ and J to $\{j_0\}$ already covers $\mathcal{C}_L$.
(b) (the proof in Vervaat (1988b)). As (a), but now

$$c_0 := \left(\bigwedge_{i\in I}(x_i 1_{G_i} \vee (\sup L)\cdot 1_{G_i^c})\right)^\vee$$

and I is restricted to a finite subset of the indices of the G_i such that $x_i < y_{j_0}$, which G_i cover K_{j_0}.
(c) We may identify $\mathcal{F}$ with the $\{0,1\}$-valued sup measures. Apply (b) with $L = \{0,1\}$. QED

Theorem 2.2 actually holds for general topological spaces E (even non-Hausdorff or without countable base). In Vervaat (1988b, Section 15) it is shown that $\mathcal{C}(E) \simeq \mathrm{US}(\mathcal{K}(E)\backslash\{\emptyset\})$ with a certain non-Hausdorff topology on $\mathcal{K}(E)\backslash\{\emptyset\}$. In this setting (a) follows from (b), and that is the way (a) is obtained in Vervaat (1988b).

The vaguely compact spaces in Theorem 2.2 need not be vaguely Hausdorff. A sufficient condition for Hausdorffness (in fact also necessary, cf. Vervaat (1988b)) is given by the next theorem.

2.3. Theorem. *Let E be locally compact. Then*
(a) *the space $\mathcal{C}$ is vaguely Hausdorff;*
(b) *the spaces* SA, AD, SM, US *and $\mathcal{F}$ are vaguely closed in $\mathcal{C}$, hence vaguely compact.*

Proof. (a) To show that $\mathcal{C}$ is Hausdorff, consider $c_1 \neq c_2$ in $\mathcal{C}$. Then $c_1(K) \neq c_2(K)$ for some $K \in \mathcal{K}$, say $c_1(K) < c_2(K)$. Select $x \in \mathsf{R}_+$, $G \in \mathcal{G}$ and $K' \in \mathcal{K}$ such that $c_1(K) < x < c_2(K)$, $K \subset G \subset K'$ and $c_1(K') < x$ (combine (1.1d) with consequence (iii) of Lemma 1.1(b)). Then $\{c: c(K') < x\}$ and $\{c: c(G) > x\}$ are disjoint neighborhoods of c_1 and c_2.
(b) Compact sets in Hausdorff spaces are closed. So the statement for SM, US and $\mathcal{F}$ follows from Theorem 2.2(b,c). To show that SA is vaguely closed in $\mathcal{C}$, let $c_0 \in \mathrm{SA}^c$. Then there exist $K_1, K_2 \in \mathcal{K}$ and $x_1, x_2 \in (0,\infty)$ such that $c_0(K_1 \cup K_2) > x_1 + x_2$, $c_0(K_1) < x_1$ and $c_0(K_2) < x_2$. By Lemma 1.1 and (1.1d), there exist $K_1', K_2' \in \mathcal{K}$ and $G_1, G_2 \in \mathcal{G}$ such that $K_j' \supset G_j \supset K_j$ and $c_0(K_j') < x_j$ for $j = 1, 2$. Then c_0 is in the open set

$$\{c: c(G_1 \cup G_2) > x_1 + x_2,\ c(K_1') < x_1,\ c(K_2') < x_2\},$$

which is disjoint from SA. To show AD is closed in SA, we apply a similar argument for $c_0 \in \mathrm{SA}\backslash\mathrm{AD}$. QED

We continue to assume that E is locally compact. The space $\mathcal{M}$ of Radon measures is not vaguely closed in $\mathcal{C}$, because Radon measures are required to be finite on $\mathcal{K}$. Hence, subsets Π of $\mathcal{M}$ are relatively vaguely compact in $\mathcal{M}$ iff they are *uniformly locally finite,* i.e., $\sup_{c\in\Pi} c(K) < \infty$ for $K \in \mathcal{K}$. In particular, the subprobability measures $\{c \in \mathcal{M}: c(E) \leq 1\}$ form a vaguely compact space. The space $\mathcal{N}$ is a vaguely closed subspace of $\mathcal{M}$, since by Theorem 2.2(a) with $L = \mathsf{N} \cup \{\infty\}$ and Theorem 2.3(a) the L-valued capacities are a closed subspace of all capacities. Intersect with $\mathcal{M}$.

The portmanteau characterization (2.1) applied to the space $\mathcal{M}$ of Radon measures re-establishes 'the' vague convergence of Radon measures in the literature. A more familiar functional-analytic characterization is

$$c_n \to c \text{ vaguely} \iff \int_E f\,dc_n \to \int_E f\,dc$$

for continuous $f\colon E \to \mathsf{R}$ with compact support.

The portmanteau characterization (2.1) applied to the space US of usc functions re-establishes what is known in the literature as hypo convergence (cf. Vervaat (1988b), Salinetti & Wets (1981)). The characterization applied to $c_n = 1^{\vee}_{F_n}$ for $F_n \in \mathcal{F}$ results in

$$F_n \to F \text{ in } \mathcal{F}(E) \iff \begin{cases} F \cap G \neq \emptyset \Rightarrow F_n \cap G \neq \emptyset \text{ eventually} & \text{for } G \in \mathcal{G}, \\ F \cap K = \emptyset \Rightarrow F_n \cap K = \emptyset \text{ eventually} & \text{for } K \in \mathcal{K} \end{cases} \tag{2.3}$$

and re-establishes the convergence considered in Matheron (1975) for locally compact E. The same topology for general E is due to Fell (1962).

For applications it is convenient to replace $\mathcal{G}$ and $\mathcal{K}$ in the criteria for vague convergence by smaller subclasses. In general, if $\mathcal{S}$ is a family of subsets of E, then $\mathcal{S}^{\cup}$ will denote the family of finite unions from $\mathcal{S}$. We say that (the topology on) E is *locally* $\mathcal{S}$ if for each instance of $x \in G \in \mathcal{G}$ there is an $S \in \mathcal{S}$ such that $x \in \operatorname{int} S \subset S \subset G$. Note that E is locally $\mathcal{F}$ since E is assumed to be metric, and that E is locally compact iff E is locally $\mathcal{K}$, which holds in particular if E is locally $\mathcal{K}_0$ for some $\mathcal{K}_0 \subset \mathcal{K}$. The statement "$E$ is locally $\mathcal{G}_0$" for a $\mathcal{G}_0 \subset \mathcal{G}$ is equivalent to saying that $\mathcal{G}_0$ is a base of $\mathcal{G}$.

The notion E is locally $\mathcal{S}$ is our regularity condition that replaces what Kallenberg (1983) expresses in terms of DC-(semi)rings and separating classes in the context of locally compact E.

2.4. Lemma. *Let $\mathcal{G}_0 \subset \mathcal{G}$ and $\mathcal{K}_0 \subset \mathcal{K}$ be such that E is locally $\mathcal{G}_0$ and locally $\mathcal{K}_0$ (hence E is locally compact). Then the vague topology on $\mathcal{C}$ is already the coarsest that makes the evaluations $c \mapsto c(A)$ lsc for $A \in \mathcal{G}_0^{\cup}$ and usc for $A \in \mathcal{K}_0^{\cup}$.*

2.5. Corollary. *If E is locally compact (with countable base since E is assumed to be separable), then the vague topology on $\mathcal{C}$ has a countable base (so makes $\mathcal{C}$ a compact metric space).*

Proof of Lemma 2.4. Since E is separable by general assumption, each open G is a union of countably many sets in $\mathcal{G}_0$, so is the union of an increasing sequence in $\mathcal{G}_0^{\cup}$. By (1.1e) it follows that $c \mapsto c(G)$ is lsc for $G \in \mathcal{G}$ if it is for $G \in \mathcal{G}_0^{\cup}$, since suprema of lsc functions are lsc.

We now show that

$$c(K) = \bigwedge\{c(K_0)\colon K_0 \in \mathcal{K}_0^{\cup},\ K_0 \supset K\} \qquad \text{for } K \in \mathcal{K}, \tag{2.4}$$

so that $c \mapsto c(K)$ is usc as infimum of usc functions. Only the inequality $c(K) \geq$ RHS in (2.4) needs a proof. So let $y > c(K)$. By (1.1d) there is an open $G \supset K$ such that $y > c(G)$. Since E is locally $\mathcal{K}_0$, we have

$$G = \bigcup_{K_0 \in \mathcal{K}_0, K_0 \subset G} K_0 = \bigcup_{K_0 \in \mathcal{K}_0, K_0 \subset G} \operatorname{int} K_0.$$

Since K is compact and $K \subset G$, there is a finite set $\mathcal{I}$ of K_0's as above such that $K \subset \bigcup_{K_0 \in \mathcal{I}} \operatorname{int} K_0 \subset \bigcup_{K_0 \in \mathcal{I}} K_0 \subset G$, where $\bigcup_{K_0 \in \mathcal{I}} K_0 \in \mathcal{K}_0^{\cup}$. So the right-hand side of (2.4) is at most $c(G) < y$. QED

For the subspaces of interest the following Lemma 2.6 is a more convenient version of Lemma 2.4. Lemma 2.6 and its counterpart Lemma 3.10 in the next section are not going to be used in the present paper, but they will be in subsequent papers, for instance in the proofs of the results stated (but not proved) in the last sections of this paper.

2.6. Lemma. *Let $\mathcal{G}_0$ and $\mathcal{K}_0$ be as in Lemma 2.4.*
(a) *For the subspaces* SM, US *and $\mathcal{F}$ of $\mathcal{C}$ the conclusion of Lemma 2.4 holds with $\mathcal{G}_0$ and $\mathcal{K}_0$ instead of $\mathcal{G}_0^{\cup}$ and $\mathcal{K}_0^{\cup}$.*
(b) *Suppose in addition that $\mathcal{G}_0$ and $\mathcal{K}_0$ are closed for finite intersections and that $\mathcal{K}_0$ nestles in $\mathcal{G}_0$. Then the conclusion of Lemma 2.4 holds for the subspaces* AD, *$\mathcal{M}$ and $\mathcal{N}$ with $\mathcal{G}_0$ and $\mathcal{K}_0$ instead of $\mathcal{G}_0^{\cup}$ and $\mathcal{K}_0^{\cup}$.*

If E is a block (product of intervals) in Euclidean space, then the conditions of Lemma 2.6(b) are satisfied by $\mathcal{K}_0$ consisting of the compact blocks in E and $\mathcal{G}_0$ consisting of the relatively compact open blocks in E. More generally, these conditions are satisfied if we take for $\mathcal{G}_0$ the finite intersections from a base of open relatively compact balls in E, and for $\mathcal{K}_0$ their closures (cf. Billingsley (1968, Th.2.2)).

Proof of Lemma 2.6. (a) Suprema of lsc functions are lsc and suprema of finitely many usc functions are usc. So the appropriate semicontinuities of $c \mapsto c(A)$ extend from $A \in \mathcal{G}_0$ to $A \in \mathcal{G}_0^{\cup}$ and from $A \in \mathcal{K}_0$ to $A \in \mathcal{K}_0^{\cup}$ by (1.7) and (1.8).
(b) Let $\mathcal{G}_n$ be the collection of unions of at most $n+1$ sets from G_0, and $\mathcal{K}_n$ the collection of unions of at most $n+1$ sets from K_0. We will prove by induction on n that the evaluations $c \mapsto c(A)$ are lsc for $A \in \mathcal{G}_n$ and usc for $A \in \mathcal{K}_n$ (for $n = 0$ part of the hypotheses of the lemma), so that they are lsc for $A \in \bigcup_{n=0}^{\infty} \mathcal{G}_n = \mathcal{G}_0^{\cup}$ and usc for $A \in \bigcup_{n=0}^{\infty} \mathcal{K}_n = \mathcal{K}_0^{\cup}$. Then apply Lemma 2.4.

Ingredients in the proof of the induction step are that $\mathcal{K}_n$ nestles in $\mathcal{G}_n$, that $G_n G_0 \in \mathcal{G}_n$ in case $G_n \in \mathcal{G}_n$ and $G_0 \in \mathcal{G}_0$, and that $K_n K_0 \in \mathcal{K}_n$ in case $K_n \in \mathcal{K}_n$ and $K_0 \in \mathcal{K}_0$, because $\mathcal{G}_0$ and $\mathcal{K}_0$ are closed for finite intersections.

Other ingredients are the following generalities about semicontinuous functions in relation to their top values. For functions f from some topological space T to $\bar{\mathbb{R}}$, let f° denote its restriction to the complement of its infinity support, to $\{t \in T: f(t) < \infty\}$, with the relative topology on its restricted domain. Then f is usc iff f° is usc and $\{t \in T: f(t) = \infty\}$ is closed, and f is lsc iff f° is lsc and f is lsc at each t such that $f(t) = \infty$. Furthermore, f is lsc at a t such that $f(t) = \infty$ iff there is a $g: T \to \bar{\mathbb{R}}$ such that $g \leq f$, $g(t) = \infty$ and g is lsc at t.

Now assume that $c \mapsto c(A)$ is lsc for $A \in \mathcal{G}_n$ and usc for $A \in \mathcal{K}_n$. The generic element of $\mathcal{G}_{n+1}$ is $G_n \cup G_0$ and that of $\mathcal{K}_{n+1}$ is $K_n \cup K_0$, where $G_j \in \mathcal{G}_j$ and $K_j \in \mathcal{K}_j$ for $j = n, 0$. We first consider $K_n \cup K_0$. Since $\mathcal{K}_j$ nestles in $\mathcal{G}_j$ for $j = n, 0$, we can find

$$G_{j,1} \supset K_{j,2} \supset G_{j,2} \supset \cdots \supset K_{j,m} \supset G_{j,m} \supset \cdots \downarrow K_j,$$

where every $G_{j,m} \in \mathcal{G}_j$ and $K_{j,m} \in \mathcal{K}_j$. For c such that $c(K_n \cup K_0) < \infty$ we find, for $m \geq m_0 = m_0(c)$ such that $c(G_{n,m_0} \cup G_{0,m_0}) < \infty$,

$$\begin{aligned} c(K_n \cup K_0) &\leq c(G_{n,m} \cup G_{0,m}) = c(G_{n,m}) + c(G_{0,m}) - c(G_{n,m}G_{0,m}) \\ &\leq c(K_{n,m}) + c(K_{0,m}) - c(G_{n,m}G_{0,m}) \\ &\to c(K_n) + c(K_0) - c(K_nK_0) = c(K_n \cup K_0) \qquad \text{as } m \to \infty. \end{aligned}$$

The second line is usc on the open set of all c such that $c(G_{j,m}) < \infty$ for $j = n, 0$. Since $c(K_n \cup K_0)$ is the infimum of the second line over m, we find that it is usc in c on the union of all these open sets, the set $\{c: c(K_n \cup K_0) < \infty\}$. The complement $\{c: c(K_n \cup K_0) = \infty\}$ being closed, we conclude that $c \mapsto c(K_n \cup K_0)$ is usc.

A similar argument proves that $c \mapsto c(G_n \cup G_0)$ is lsc when restricted to $\{c: c(G_n \cup G_0) < \infty\}$. Things are even simpler here, because the analogous inequalities hold on all of this set. From $c(G_n \cup G_0) \geq c(G_n) \vee c(G_0)$ we see that $c \mapsto c(G_n \cup G_0)$ is lsc at all c for which $c(G_n \cup G_0) = \infty$, so the whole map is lsc. QED

We conclude this section by characterizing the Borel field Bor $\mathcal{C}$ generated by the vague topology. We only get nice results in case E is locally compact (with countable base by general assumption). Then $\mathcal{C}$ has a countable base for the vague topology (Corollary 2.5), so Bor $\mathcal{C}$ is already generated by any base. Hence Bor $\mathcal{C}$ is generated by the evaluations $c \mapsto c(A)$ for $A \in \mathcal{G} \cup \mathcal{K}$. By Lemma 1.1(b) in combination with (1.1g,h) either of $\mathcal{G}$ and $\mathcal{K}$ already suffices. By Lemma 2.4 we can restrict A further to either $\mathcal{G}_0^\cup$ or $\mathcal{K}_0^\cup$, or even to $\mathcal{G}_0$ or $\mathcal{K}_0$ for the subspaces AD, $\mathcal{M}$, $\mathcal{N}$, SM, US and $\mathcal{F}$ in case the conditions of Lemma 2.6 are satisfied.

3. Tight capacities with the narrow topology

In all of this section, E is a separable metric space, so $\mathcal{F}$ nestles in $\mathcal{G}$. Specifically we do not assume that E is locally compact, so that many results of the previous section do not hold. In particular, $\mathcal{C}$ need not be vaguely Hausdorff.

In the present section we provide $\mathcal{C}$ with the *narrow topology*, generated by the subbase consisting of the sets $\{c: c(G) > x\}$ and $\{c: c(F) < x\}$ for $x \in (0, \infty)$, $G \in \mathcal{G}$ and $F \in \mathcal{F}$. Equivalently, the narrow topology on $\mathcal{C}$ is the coarsest that makes the evaluations $c \mapsto c(A)$ lsc for $A \in \mathcal{G}$ and usc for $A \in \mathcal{F}$. Narrow convergence in $\mathcal{C}$ has the portmanteau characterization

$$c_n \to c \text{ narrowly} \iff \begin{cases} \liminf c_n(G) \geq c(G) & \text{for } G \in \mathcal{G}, \\ \limsup c_n(F) \leq c(F) & \text{for } F \in \mathcal{F}. \end{cases} \tag{3.1}$$

Note that the characterizations of narrow convergence are obtained from those of vague convergence by giving $\mathcal{F}$ the place of $\mathcal{K}$. As a consequence, narrow convergence implies vague convergence. A smoother theory might be obtained if we shifted our attention to capacities such that $\mathcal{F}$ takes over the role of $\mathcal{K}$ in (1.1c,d,f). Such a theory does not seem to have been developed so far (the theory in this section was announced in Vervaat (1988a); for a development in another direction, see Salinetti & Wets (1987)). Here we escape this by restricting our attention mostly to capacities that turn out to satisfy (1.1) in both the $\mathcal{K}$ and $\mathcal{F}$ versions, the *tight* capacities.

3.1. Definition. A capacity c on E is said to be *tight* if there is a sequence $(K_m)_{m=1}^{\infty}$ in $\mathcal{K}$ such that $c(KK_m) \to c(K)$ as $m \to \infty$ uniformly for $K \in \mathcal{K}$. Such a sequence (K_m) is called *tightening*.

By uniform convergence of $[0, \infty]$-valued functions f_m to f we mean uniform convergence of $\arctan f_m$ to $\arctan f$ in $[0, \frac{\pi}{2}]$. In particular, $\mathcal{C}$ and SA are mapped by arctan into bounded subcollections. Therefore we may assume sets of capacities to be bounded in most proofs where uniform convergence plays a role. It is always possible to take increasing tightening sequences. If E is compact, then we may take $K_m \equiv E$. If E is locally compact there exists by Lemma 1.1(b) a sequence (K_m) in $\mathcal{K}$ with $K_m \uparrow E$ such that eventually $K \subset K_m$ for all $K \in \mathcal{K}$; it follows that (K_m) is a tightening sequence for *all* tight c.

3.2. Definition. A capacity c on E is *classically tight* if there is a sequence (K_m) in $\mathcal{K}$ such that $c(K_m^c) \to 0$. The sequence (K_m) is called *classically tightening*.

For the subspaces of interest to us, tightness and classical tightness turn out to be equivalent.

3.3. Lemma. (a) *If a capacity c on E is tight, then we have $c(K_m^c) \to 0$ for tightening sequences (K_m), so c is classically tight.*
(b) *If $c \in$ SA and c is classically tight, then c is tight with the same tightening sequences.*

Proof. We may assume c to be bounded by the remark after Definition 3.1.
(a) Set $\varepsilon_m := \sup_{K\in\mathcal{K}}(c(K) - c(KK_m))$. Then we have $\varepsilon_m \to 0$ and $c(K) \leq \varepsilon_m$ if $KK_m = \emptyset$. By (1.1c) it follows that $c(K_m^c) \leq \varepsilon_m$.
(b) For $K \in \mathcal{K}$ and subadditive c we have

$$c(K) - c(KK_m) \leq c(KK_m \cup K_m^c) - c(KK_m) \leq c(K_m^c). \qquad \text{QED}$$

The uniform convergence in Definition 3.1 is not restricted to $\mathcal{K}$, as the next lemma states.

3.4. Lemma. *If a capacity c on E is tight and (K_m) is the tightening sequence, then $c(AK_m) \to c(A)$ as $m \to \infty$ uniformly for $A \subset E$.*

Proof. We may assume c to be finite. By (1.1c)

$$\begin{aligned} c(A) - c(AK_m) &= \sup_{K\in\mathcal{K}, K\subset A}(c(K) - c(AK_m)) \\ &\leq \sup_{K\in\mathcal{K}, K\subset A}(c(K) - c(KK_m)) \\ &\leq \sup_{K\in\mathcal{K}}(c(K) - c(KK_m)). \end{aligned}$$

The last upper bound is independent of A and vanishes as $m \to \infty$. QED

We now derive the $\mathcal{F}$ analogue of (1.1).

3.5. Lemma. *If c is a tight capacity on E, then we have besides (1.1a,b,e)*

$$c(A) = \sup_{F\in\mathcal{F}, F\subset A} c(F) \quad \text{for } A \subset E, \tag{3.2c}$$

$$c(F) = \inf_{G\in\mathcal{G}, G\supset F} c(G) \quad \text{for } F \in \mathcal{F}, \tag{3.2d}$$

$$c(F_n) \downarrow c(F) \quad \text{for } F_n \downarrow F; \tag{3.2f}$$

for each $F \in \mathcal{F}$ there is a sequence (G_n) in $\mathcal{G}$ such that

$$G_n \downarrow F, \qquad c(G_n) \downarrow c(F) \quad \text{for all tight } c \in \mathcal{C}, \tag{3.2g}$$

and for each $G \in \mathcal{G}$ there is a sequence (F_n) in $\mathcal{F}$ such that

$$F_n \uparrow G, \qquad c(F_n) \uparrow c(G) \quad \textit{for all tight } c \in \mathcal{C}. \tag{3.2h}$$

Proof. (c) Follows from (1.1c) and $\mathcal{K} \subset \mathcal{F}$. Tightness is not needed.
(f) Let (K_m) be a tightening sequence. By (1.1f) we have $c(F_n K_m) \downarrow c(F K_m)$ as $n \to \infty$ for each m. By the uniform convergence in Lemma 3.4 we obtain (3.2f) as $m \to \infty$.
(d) Follows from (3.2f) since $\mathcal{F}$ nestles in $\mathcal{G}$.
(g,h) Follows from (3.2f,e) and Lemma 1.1(a). QED

By $\mathcal{C}_t$ we denote the collection of tight capacities. The corresponding subspaces are denoted by SA_t, AD_t, $\mathcal{M}_t$, $\mathcal{N}_t$, SM_t, US_t and $\mathcal{F}_t$. We now investigate these subspaces and the relative narrow topologies on them.

3.6. Lemma. *The space of tight capacities $\mathcal{C}_t$ is narrowly Hausdorff, and its subspaces SA_t, AD_t, SM_t, US_t and $\mathcal{F}_t$ are narrowly closed.*

Proof. By Lemma 3.5 we may repeat the proof of Theorem 2.3(a) with F and $\mathcal{F}$ instead of K and $\mathcal{K}$, which proves $\mathcal{C}_t$ to be narrowly Hausdorff. By Lemma 3.4, formulae (1.2), (1.3) and (1.4) extend to closed sets for tight capacities. So Lemma 3.5 allows us to repeat the proof of Theorem 2.3(b), with compact sets replaced by closed sets and sequences replaced by nets. QED

The space $\mathcal{M}_t$ consists of the *finite* (classically) tight measures on Bor E, which follows from Lemma 3.3 and the fact that Radon measures are finite on $\mathcal{K}$. Narrow convergence in $\mathcal{M}_t$ is just narrow or weak convergence of finite measures as in Billingsley (1968). Note that the two lines on the right-hand side of (3.1) are equivalent by complementation if $c_n(E) \to c(E)$ is given, in particular if all the capacities are probability measures. It is well-known that all finite measures on E are tight in case E is Polish (Billingsley (1968)).

The space US_t consists of the *upper compact* functions, i.e., those functions $f\colon E \to \bar{\mathbb{R}}_+$ such that $f^{\leftarrow}[x, \infty]$ is compact for $x > 0$ (not $x = 0$), where $f^{\leftarrow}$ denotes the inverse of f as a set function. (Recall that $f^{\leftarrow}[x, \infty]$ being closed for $x > 0$ is one characterization of upper semicontinuity.) Consequently, we have $\mathcal{F}_t = \mathcal{K}$. The narrow topology on $\mathcal{K}$ is known in the literature as the finite or Vietoris topology. For metric E it is metrized by the Hausdorff distance (Matheron (1975)).

For the remainder of this section we need a uniform variant of tightness.

3.7. Definition. Let $\Pi \subset \mathcal{C}$. Then Π is *equitight* if there is one sequence (K_m) in $\mathcal{K}$ such that the condition in Definition 3.1 holds uniformly with this sequence (so uniformly for both $K \in \mathcal{K}$ and $c \in \Pi$). We call (K_m) *equitightening.*

The definition of classical equitightness is now obvious, and Lemmas 3.3 and 3.4 have straightforward generalizations to equitightness.

It is well-known that the narrow topology on $\mathcal{M}_t$ has a countable base (because E is assumed to be separable), cf. Billingsley (1968). The situation for all of $\mathcal{C}_t$ has not yet been well explored, but the following analogue of Lemma 2.4 is adequate for our purposes.

3.8. Lemma. *Let $\Pi \subset \mathcal{C}$ be equitight, and let $\mathcal{G}_0 \subset \mathcal{G}$ and $\mathcal{F}_0 \subset \mathcal{F}$ be such that E is locally $\mathcal{G}_0$ and locally $\mathcal{F}_0$. Then the relative narrow topology on Π is already the coarsest that makes the evaluations $c \mapsto c(A)$ lsc for $A \in \mathcal{G}_0^{\cup}$ and usc for $A \in \mathcal{F}_0^{\cup}$.*

Proof. The first part of the proof of Lemma 2.4 serves here as well. The following replaces the second part.

The second part of the proof of Lemma 2.4 starting with $\mathcal{F}_0$ and $\mathcal{F}_0^{\cup}$ instead of $\mathcal{K}_0$ and $\mathcal{K}_0^{\cup}$, but with compact K as it stands, results in $\mathcal{C} \ni c \mapsto c(K)$ being usc for compact K, with the narrow topology in the domain $\mathcal{C}$. If (K_m) is an equitightening sequence for Π, then $c(AK_m)$ converges uniformly on Π to $c(A)$. If $A = F$ is closed, then $c \mapsto c(FK_m)$ is usc on $\mathcal{C}$, and so is $c \mapsto c(F)$ on Π as a uniform limit. QED

3.9. Corollary. *The relative narrow topology on equitight sets of capacities is countably generated (take countable $\mathcal{G}_0$ and $\mathcal{F}_0$ in Lemma 3.8).*

The following stronger version of Lemma 3.8 for subspaces of $\mathcal{C}$ is more useful. Its proof is similar to that of Lemma 2.6.

3.10. Lemma. (a) *If, in addition to the hypotheses of Lemma 3.8,* $\Pi \subset$ SM, *then the conclusion of the lemma holds with $\mathcal{G}_0$ and $\mathcal{F}_0$ instead of $\mathcal{G}_0^{\cup}$ and $\mathcal{F}_0^{\cup}$.*
(b) *If, in addition to the hypotheses of Lemma 3.8,* $\Pi \subset$ AD, *$\mathcal{G}_0$ and $\mathcal{F}_0$ are closed for finite intersections and $\mathcal{F}_0$ nestles in $\mathcal{G}_0$, then the conclusion of the lemma holds with $\mathcal{G}_0$ and $\mathcal{F}_0$ instead of $\mathcal{G}_0^{\cup}$ and $\mathcal{F}_0^{\cup}$.*

In contrast to the situation with the vague topology, $\mathcal{C}$ need not be narrowly compact. So it is worthwhile to characterize narrow relative compactness of subsets Π of $\mathcal{C}$. The following can be regarded as the generalization to $\mathcal{C}$ of Prokhorov's tightness theorem for finite measures. Note that E is in particular Polish if E is locally compact (in addition to our general assumption of separability). By $\mathcal{C}_0$ and SA_0 we denote the subspaces of $\mathcal{C}$ and SA consisting of those c that satisfy (3.2f). From Lemma 3.5 it follows that $\mathcal{C}_t \subset \mathcal{C}_0$ and that $\mathrm{SA}_t \subset \mathrm{SA}_0$. Note that SA_0 contains all finite measures.

3.11. Theorem. (a) *If* $\Pi \subset \mathcal{C}$ *is equitight, then* Π *is narrowly relatively compact and (narrow)* $\operatorname{clos}\Pi$ *is classically equitight.*
(b) *If* $\Pi \subset \mathrm{SA}$ *is narrowly relatively compact in* $\mathcal{C}_0$ *and* E *is Polish, then (narrow)* $\operatorname{clos}\Pi$ *is equitight and contained in* SA_t.
(c) *If* E *is Polish, then* $\mathrm{SA}_0 = \mathrm{SA}_t$.

Proof. (a) If Π is equitight, then narrow relative compactness can be checked in sequential form, because of Corollary 3.9. Since $\mathcal{C}$ is vaguely compact (cf. Theorem 2.2(a)), Π is relatively vaguely compact, so each sequence in Π has a vaguely convergent subsequence. Hence it suffices to prove that a vaguely convergent equitight sequence in $\mathcal{C}$ actually converges narrowly to the same limit and that this limit is classically tight with the same tightening sequence.

So let $c_n \to c$ vaguely in $\mathcal{C}$, and let $\{c_n\}$ be equitight with equitightening sequence (K_m). Let $F \in \mathcal{F}$. By (2.1) we have $\limsup_n c_n(FK_m) \le c(FK_m)$ for each m. Lemma 3.4 in a uniform version on $\{c_n\}$ now implies that $\limsup_n c_n(F) \le \sup_m c(FK_m) \le c(F)$. We have obtained the $\mathcal{F}$ part of (3.1), so $c_n \to c$ narrowly. By the uniform version of Lemma 3.3(a), $\{c_n\}$ is classically equitight with the same tightening sequence (K_m). By the $\mathcal{G}$ part of (2.1) or (3.1) we now obtain $c(K_m^c) \le \liminf_n c_n(K_m^c) \le \sup_n c_n(K_m^c) \to 0$ as $m \to \infty$. We see that c is classically tight with the same tightening sequence.
(b) This part of the proof follows closely p.40 in Billingsley (1968). Recall that equitightness and classical equitightness are equivalent in SA. Fix a complete metric on the Polish space E. A sufficient condition for classical equitightness is that for each positive ε and δ there exist finite collections of δ-spheres $A_1, A_2, \ldots, A_n$ such that $c\left(\left(\bigcup_{i=1}^n A_i\right)^c\right) < \varepsilon$ for all $c \in \Pi$. To see this, choose for given $\varepsilon > 0$ and for $k = 1, 2, \ldots$ finitely many $\frac{1}{k}$-spheres $A_{k,1}, \ldots, A_{k,n_k}$ such that $c\left(\left(\bigcup_{i=1}^{n_k} A_{k,i}\right)^c\right) < \varepsilon 2^{-k}$ for all $c \in \Pi$. The closure K of the totally bounded set $\bigcap_{k=1}^\infty \bigcup_{i=1}^{n_k} A_{k,i}$ is totally bounded, hence compact since E is complete. Furthermore, as $\Pi \subset \mathrm{SA}$,

$$c(K^c) \le c\left(\bigcup_{k=1}^\infty \left(\bigcup_{i=1}^{n_k} A_{k,i}\right)^c\right) \le \sum_{k=1}^\infty c\left(\left(\bigcup_{i=1}^{n_k} A_{k,i}\right)^c\right) < \varepsilon.$$

So Π is indeed classically equitight.

Now suppose, in addition to the hypotheses, that Π is not classically equitight. Then the sufficient condition fails for some ε and δ. So there are

positive ε and δ such that every finite collection $A_1, \ldots, A_n$ of δ-spheres satisfies $c\left(\left(\bigcup_{i=1}^n A_i\right)^c\right) \geq \varepsilon$ for some c in Π. Since E is separable, it is the union of a sequence of open spheres A_i of radius δ. Let $B_n := \bigcup_{i=1}^n A_i$ and choose c_n in Π such that $c_n(B_n^c) \geq \varepsilon$. By the hypotheses, some subsequence $(c_{n'})$ converges narrowly to some $c \in \mathcal{C}_0$. Since B_m is open, we would have $c(B_m^c) \geq \limsup_{n'} c_{n'}(B_m^c)$ for fixed m, by (3.1). But since $B_m \subset B_{n'}$ for large n', it follows that $c(B_m^c) \geq \varepsilon$. Since $B_m^c \downarrow \emptyset$ in $\mathcal{F}$ and $c \in \mathcal{C}_0$, we have $c(B_m^c) \downarrow 0$, a contradiction. So Π must be (classically) equitight, and $\operatorname{clos} \Pi$ is by (a). Finally, $\operatorname{clos} \Pi \subset \mathrm{SA}_t$ by Lemma 3.6.
(c) Consider singleton $\Pi \subset \mathrm{SA}_0$ in (b) and recall that $\mathrm{SA}_t \subset \mathrm{SA}_0$. QED

For ease of reference we summarize some consequences for sequences of capacities.

3.12 Corollary. (a) *If $(c_n) \to c$ vaguely in $\mathcal{C}$ and $\{c_n\}$ is equitight, then $c_n \to c$ narrowly and c is classically tight.*
(b) *If $c_n \to c$ narrowly in $\mathcal{C}$, then $c_n \to c$ vaguely in $\mathcal{C}$.*
(c) *If E is Polish and (c_n) is a sequence in SA_0 such that $c_n \to c$ narrowly in $\mathcal{C}_0$, then $c \in \mathrm{SA}_t$ and $\{c_n\}$ is equitight.*

In Theorem 3.11(b) and Corollary 3.12(c) we imposed an additional condition on the limit (points). We cannot do without it, as the following example shows. Let $E = \mathsf{R}$, $c_n = 1^\vee_{[-n,n]}$, $c = 1^\vee_{\mathsf{R}}$. Then $c_n \to c$ narrowly in $\mathrm{SM}(\mathsf{R})$, the c_n's are tight, but c is not.

We conclude this section with examining the Borel field $\mathcal{B}$ of $\mathcal{C}$ generated by the narrow topology. So far our results are rather unsatisfactory. We only can conclude that the trace of $\mathcal{B}$ on an equitight subset Π of $\mathcal{C}$ is generated by the evaluations $c \mapsto c(A)$ for $A \in \mathcal{G}$ or $A \in \mathcal{F}$ by proofs and with further reductions anologous to those in the final paragraph of Section 2. Moreover, $\mathcal{F}$ may be replaced by $\mathcal{K}$, so the traces of the Borel fields of the vague and narrow topologies coincide. The situation for all of $\mathcal{C}_t$ is not yet clear. If E is compact, as sometimes happens in Section 8, $\mathcal{C}$ is equitight, so these statements all hold for $\mathcal{C} = \mathcal{C}_t$.

4. Transformations of capacities

In this section we investigate the (semi)continuity of several transformations of capacities.

Let T be a topological space and φ a mapping from T into $\mathcal{C}(E)$. We continue to assume that E is separable and metric. From (2.1) it follows that φ is vaguely continuous iff

$$T \ni t \mapsto \varphi(t)(G) \text{ is lsc for } G \in \mathcal{G}, \tag{4.1a}$$

$$T \ni t \mapsto \varphi(t)(K) \text{ is usc for } K \in \mathcal{K}, \tag{4.1b}$$

and from (3.1) that φ is narrowly continuous iff (4.1a) holds and

$$T \ni t \mapsto \varphi(t)(F) \text{ is usc for } F \in \mathcal{F}. \tag{4.1c}$$

4.1. Definition. A mapping $\varphi: T \to \mathcal{C}(E)$ is *lower semicontinuous* (*lsc*) if (4.1a) holds, *vaguely upper semicontinuous* (*usc*) if (4.1b) holds, and *narrowly usc* if (4.1c) holds.

If also T is of the form $\mathcal{C}(E^*)$ for some separable metric space E^*, then $\mathcal{C}(E^*)$ is silently understood to have the same type of topology as $\mathcal{C}(E)$: the vague, the narrow, or in case of lsc maps φ even only the $\mathcal{G}$ half of the vague or narrow topology, with subbase consisting of $\{c: c(G) > x\}$ for $G \in \mathcal{G}(E^*)$ and $x \in \mathsf{R}_+$.

First we explore $c \mapsto fc$ for nondecreasing functions $f: \mathsf{R}_+ \to \mathsf{R}_+$ such that $f(0) = 0$. More specifically, if $c \in \mathcal{C}$ and f is lsc (so left-continuous), then fc is defined by

$$fc(G) = (fc)(G) := f(c(G)) \qquad \text{for } G \in \mathcal{G} \tag{4.2a}$$

and (1.1d). Indeed, this determines a capacity, as follows by verifying (1.1c) for $A \in \mathcal{G}$. The mapping $\mathcal{C} \ni c \mapsto fc(G)$ is lsc, so $\mathcal{C} \ni c \mapsto fc \in \mathcal{C}$ is lsc.

Similarly, let f be a nondecreasing usc (so right-continuous) function $\mathsf{R}_+ \to \mathsf{R}_+$ such that $f(0) = 0$. Then

$$fc(K) := f(c(K)) \qquad \text{for } K \in \mathcal{K} \tag{4.2b}$$

together with (1.1c) determines a capacity fc, and the map $c \mapsto fc$ is vaguely usc. The map need not be narrowly usc, nor must it map tight capacities to tight capacities. (Since f is continuous at 0, $f \mapsto fc$ maps classically tight capacities to classically tight capacities, but this does not help since f need not map SA into SA.)

However, if f is continuous (in addition to being nondecreasing and fixing 0), then (4.2a) and (4.2b) are equivalent starts for defining fc, $c \mapsto fc$ is vaguely and narrowly continuous and it maps $\mathcal{C}_t$ continuously into $\mathcal{C}_t$ (note that f, with its compact domain, is uniformly continuous). The most important example we will encounter is $f(x) = x^\alpha$ ($\alpha > 0$, especially $0 < \alpha \leq 1$).

Here is an application of the above results.

4.2. Theorem. (a) *Let the* **support** *of* $c \in \mathcal{C}$ *be defined by*

$$\operatorname{supp} c := \Big(\bigcup\{G \in \mathcal{G} : c(G) = 0\}\Big)^c$$

(so $x \in \operatorname{supp} c$ *iff* $c(G) > 0$ *for all open* $G \ni x$*). Then* supp *is a lsc map from* $\mathcal{C}$ *provided with the* $\mathcal{G}$ *half of the topologies into* $\mathcal{F}$*. Furthermore,* supp *is vaguely continuous on the integer-valued capacities, and narrowly continuous on the tight integer-valued capacities.*
(b) *The map* $c \mapsto \operatorname{supp}_\infty c \in \mathcal{F}$*, cf. (1.5), is vaguely usc.*

Proof. (a) Let $f := 1_{(0,\infty]}$. Then f is lsc and $fc = 1^\vee_{\operatorname{supp} c}$. We see that $c \mapsto 1^\vee_{\operatorname{supp} c}$ is lsc, and so is $c \mapsto \operatorname{supp} c$ since $\mathcal{F}$ is regarded as a subspace of $\mathcal{C}$ via $F \mapsto 1^\vee_F$.

Let $g := x \wedge 1$ for $x \in \bar{\mathsf{R}}_+$. Then g is continuous, and $gc = 1^\vee_{\operatorname{supp} c}$ in case c is integer-valued. So all statements in the theorem follow for $c \mapsto gc$ restricted to integer-valued c.
(b) Let $f := 1_{\{\infty\}}$. Then f is usc and $fc = 1^\wedge_{\operatorname{supp}_\infty c}$, so $c \mapsto \operatorname{supp}_\infty c$ is vaguely usc. QED

Next we consider maps $\varphi\colon E \to E'$, where both E and E' are separable and metric. We write $\varphi^\leftarrow$ for the inverse of φ regarded as set function. Such φ induce maps $c \mapsto c\varphi^\leftarrow$ from $\mathcal{C}(E)$ into functions on the power set of E' that satisfy (1.1a,b), by $(c\varphi^\leftarrow)(A') := c(\varphi^\leftarrow A')$ for $A' \subset E'$. Recall that φ is continuous iff $\varphi^\leftarrow$ maps $\mathcal{G}(E')$ into $\mathcal{G}(E)$ iff $\varphi^\leftarrow$ maps $\mathcal{F}(E')$ into $\mathcal{F}(E)$, in which case φ maps $\mathcal{K}(E)$ into $\mathcal{K}(E')$. We call φ *cocompactly continuous* if $\varphi^\leftarrow$ maps $\mathcal{K}(E')$ into $\mathcal{K}(E)$ (then φ is continuous as a mapping between E and E' provided with the cocompact topologies).

4.3. Theorem. (a) *If* $\varphi : E \to E'$ *is continuous, then* $c \mapsto c\varphi^\leftarrow$ *is a narrowly continuous map from* $\mathcal{C}_t(E)$ *into* $\mathcal{C}_t(E')$*.*
(b) *If* E *and* E' *are locally compact and* $\varphi : E \to E'$ *is continuous and cocompactly continuous, then* $c \mapsto c\varphi^\leftarrow$ *is a vaguely continuous map from* $\mathcal{C}(E)$ *into* $\mathcal{C}(E')$*.*

Proof. Once it has been proved that $c\varphi^\leftarrow$ is a capacity of the right type, then the continuity of $c \mapsto c\varphi^\leftarrow$ in either case follows directly from Definition 4.1. The harder point is to prove that $c' := c\varphi^\leftarrow$ is indeed a capacity.

For (1.1c) it suffices to show that

$$c'(A') \le \bigvee\{c'(K') : K' \in \mathcal{K}(E'),\ K' \subset A'\} \qquad \text{for } A' \subset E'. \tag{4.3}$$

If $x < c'(A') = c(\varphi^\leftarrow A')$, then there is a compact $K \subset \varphi^\leftarrow A'$ such that $x < c(K)$. Hence $x < c(K) \le c(\varphi^\leftarrow(\varphi K)) = c'(\varphi K)$. Now $\varphi K \subset A'$ and $\varphi K \in \mathcal{K}(E')$ if φ is continuous, so (4.3) follows in cases (a) and (b).

In case (b), both E and E' are locally compact, so (1.1d) is equivalent to (1.1f) for monotone set functions. If $K'_n \downarrow K'$ in $\mathcal{K}(E')$, then $\varphi^{\leftarrow}K'_n \downarrow \varphi^{\leftarrow}K'$ in $\mathcal{K}(E)$ since φ is cocompactly continuous. So $c'(K'_n) = c(\varphi^{\leftarrow}K'_n) \downarrow c(\varphi^{\leftarrow}K') = c'(K')$, and (1.1f) follows for c'.

In case (a) we know that c satisfies (3.2f) as a tight capacity. If $F'_n \downarrow F'$ in $\mathcal{F}(E')$, then $\varphi^{\leftarrow}F'_n \downarrow \varphi^{\leftarrow}F'$ in $\mathcal{F}(E)$ since φ is continuous. So $c'(F'_n) = c(\varphi^{\leftarrow}F'_n) \downarrow c(\varphi^{\leftarrow}F') = c'(F')$, and (3.2f) follows for c'. This implies (1.1f) for c'.

Finally, if (K_m) is a tightening sequence for c, then (φK_m) is one for c'. Here Lemma 3.4 must be used. QED

Let $E_0 \subset E$. Applying Theorem 4.3 for φ the identity map from E_0 into E, we find the following result for the extension e of capacities on E_0 to capacities on E.

4.4. Theorem. *Let $E_0 \subset E$. Then*

$$(ec)(A) := c(AE_0) \qquad \text{for } c \in \mathcal{C} \text{ and } A \subset E$$

defines
(a) *a narrowly continuous map e from $\mathcal{C}_t(E_0)$ into $\mathcal{C}_t(E)$,*
(b) *a vaguely continuous map e from $\mathcal{C}(E_0)$ into $\mathcal{C}(E)$ in case E is locally compact and $E_0 \in \mathcal{F}(E)$.*

Next we investigate the restriction operator r of capacities on E to capacities on a subspace E_0 of E. For relevant facts about SA_0, see Theorem 3.11 and the lines preceding it.

4.5. Theorem. *Let $E_0 \subset E$ and set*

$$(rc)(A) := c(A) \qquad \text{for } c \in \mathcal{C}(E) \text{ and } A \subset E_0.$$

Then the following hold.
(a) *rc is a capacity on E_0, i.e., $rc \in \mathcal{C}(E_0)$.*
(b) *$c \mapsto rc$ is vaguely usc, and vaguely continuous in case E_0 is open (but not in general).*
(c) *$c \mapsto rc$ is narrowly continuous when restricted to $\{c \in \mathrm{SA} : c(E_0^c) = 0\}$.*
(d) *$c \mapsto rc$ maps $\{c \in \mathrm{SA}_0 : c(E_0^c) = 0\}$ into $\mathrm{SA}_0(E_0)$.*

Proof. (a) Condition (1.1c) for rc follows from $\mathcal{K}(E_0) = \{K \in \mathcal{K}(E) : K \subset E_0\}$. Condition (1.1d) for rc follows from

$$rc(K) \le \bigwedge_{G \in \mathcal{G} : G \supset K} rc(GE_0) \le \bigwedge_{G \in \mathcal{G} : G \supset K} c(G) = c(K) = rc(K)$$

for $K \in \mathcal{K}(E_0)$.
(b) The first statement follows from Definition 4.1 and $\mathcal{K}(E_0) \subset \mathcal{K}(E)$, the second from Definition 4.1 and $\mathcal{G}(E_0) \subset \mathcal{G}(E)$ in case E_0 is open in E.
(c) For $c \in \mathrm{SA}$ with $c(E_0^c) = 0$ we have

$$c(AE_0) \leq c(A) \leq c(AE_0) + c(AE_0^c) = c(AE_0), \tag{4.4}$$

so $c(A) = c(AE_0)$ for all $A \subset E$. Now verify the conditions of Definition 4.1 on $c \mapsto rc(A) = c(A)$ for $A \in \mathcal{G}(E)$ and $A \in \mathcal{F}(E)$.
(d) If $F'_m \downarrow \emptyset$ in $\mathcal{F}(E_0)$, then $F'_m = F_m E_0$ for some $F_m \in \mathcal{F}(E)$, which can be chosen to be decreasing in m (replace F_m by $\bigcap_{k=1}^m F_k$ if necessary). By (4.4) we have for $c \in \mathrm{SA}_0$ such that $c(E_0^c) = 0$:

$$c(F'_m) = c(F_m) \downarrow c\left(\bigcap\nolimits_{k=1}^{\infty} F_k\right) = c\left(\bigcap\nolimits_{k=1}^{\infty} F_k E_0\right) = c(\emptyset) = 0. \qquad \square$$

4.6. Remark. There are some variations on Theorem 4.5(b,c). For instance, $c \mapsto rc$ is vaguely continuous on $\{c \in \mathrm{SA}: c(E_0 \cap \partial E_0) = 0\}$ (since $c(AE_0) = c(A \cap \operatorname{int} E_0)$ for such c), $c \mapsto rc$ is narrowly usc if E_0 is closed (since then $\mathcal{F}(E_0) \subset \mathcal{F}(E)$), and $c \mapsto rc$ is narrowly continuous on $\{c \in \mathrm{SA}: c(\partial E_0) = 0\}$ (since $c(AE_0) = c(A \cap \operatorname{int} E_0) = c(A \cap \operatorname{clos} E_0)$ for such c).

5. Large deviations

This section generalizes results of Lynch & Sethuraman (1987). Recall that E is a separable metric space. Powers of capacities are taken 'pointwise': $c^\alpha(A) := (c(A))^\alpha$.

5.1. Definition. (a) Let E be locally compact, (c_n) a sequence in SA and $c \in \mathcal{C}$. We say that (c_n) satisfies a *vague large-deviation principle* (VLDP) with limit c if there is a sequence (a_n) in $(0, \infty)$ such that $a_n \to \infty$ and $c_n^{1/a_n} \to c$ vaguely.
(b) Let (c_n) be a sequence in SA_t and $c \in \mathcal{C}_t$. We say that (c_n) satisfies a *narrow large-deviation principle* (NLDP) with limit c if there is a sequence (a_n) in $(0, \infty)$ such that $a_n \to \infty$ and $c_n^{1/a_n} \to c$ narrowly.

5.2. Lemma. (a) *If c is a limit in a* VLDP, *then $c \in \mathrm{SM}$: c is a sup measure.*
(b) *If c is tight and a limit in an* NLDP, *then $c \in \mathrm{SM}_t$: c is a tight sup measure.*

Proof. (a) We must prove (1.7), where $\leq$ at the place of $=$ suffices. For all instances of $K_j \subset G_j \subset K'_j$ with $j = 1, 2$, $K_j, K'_j \in \mathcal{K}$ and $G_j \in \mathcal{G}$ we have by (2.1) and (1.2):

$$\begin{aligned} c(K_1 \cup K_2) &\leq c(G_1 \cup G_2) \leq \liminf c_n^{1/a_n}(G_1 \cup G_2) \\ &\leq \limsup c_n^{1/a_n}(K'_1 \cup K'_2) \leq \limsup (c_n(K'_1) + c_n(K'_2))^{1/a_n} \\ &\leq \limsup 2^{1/a_n} \left(c_n^{1/a_n}(K'_1) \vee c_n^{1/a_n}(K'_2)\right) \leq c(K'_1) \vee c(K'_2). \end{aligned}$$

Since $\mathcal{K}$ nestles in $\mathcal{G}$ (E is assumed to be locally compact), (1.7) follows by (1.1d).
(b) Tight capacities satisfy (3.2) besides (1.1) and in narrow convergence $\mathcal{F}$ assumes the role of $\mathcal{K}$ in vague convergence. So the proof of (a) applies with F's instead of K's. QED

Considering sup derivatives of sup measures, we see that $d^\vee c$ is an usc function ($d^\vee c \in \mathrm{US}$) in case of a VLDP, and that $d^\vee c$ is upper compact ($d^\vee c \in \mathrm{US}_t$) in case of an NLDP. In the probability and statistical physics literature one usually considers $J := -\log c$, so J is an $\bar{\mathbb{R}}$-valued inf measure (tight in the obvious meaning, with $\infty = -\log 0$ taking over the role of 0, in case of an NLDP), and its inf derivative $d^\wedge J = -\log d^\vee c$ is an $\bar{\mathbb{R}}$-valued lsc function (lower compact in case of an NLDP). Names for J and $d^\wedge J$ in the literature are *free energy, information* and *rate function* (Lynch & Sethuraman (1987)). We adopt the last term.

In the literature one mostly considers probability measures c_n. Then the limit c is $[0, 1]$-valued, so J and $d^\wedge J$ are $[0, \infty]$-valued. Lower compactness of $d^\wedge J$ now means that $(d^\wedge J)^{\leftarrow}[0, y] = \{x \in E : d^\wedge J(x) \leq y\}$ is compact for $y < \infty$ (not $y = \infty$).

Applying Theorem 3.11 and Corollary 3.12 combined with Lemma 3.3, we find

5.3. Lemma. *Let (c_n) be a sequence in SA_t. If a* VLDP *holds for (c_n) with limit c and $\{c_n : n = 1, 2, \ldots\}$ is (classically) equitight, then c is tight and an* NLDP *holds for (c_n) with limit c. Conversely, if E is Polish and an* NLDP *holds for (c_n) with limit c, then c is tight, a* VLDP *holds for (c_n) with limit c and $\{c_n : n = 1, 2, \ldots\}$ is equitight.*

Lemma 5.3 suggests that in practice an NLDP should be obtained by proving the corresponding VLDP first and then the additional tightness conditions. This procedure can often be avoided by embedding the space E of interest into a larger compact space. Note that NLDP's and VLDP's are equivalent in case E is compact, the most important case of application of the next lemma.

5.4. Lemma. *Let E be Polish and let E_0 be a G_δ set in E. If an* NLDP *holds for (c_n) with limit c and if $c_n(E_0^c) = c(E_0^c) = 0$, then an* NLDP *holds for (rc_n) with limit rc, where r denotes the restriction of capacities on E to E_0.*

Proof. By Theorem 4.5(c) we have the right convergence for rc_n to rc. It remains to show that $rc_n, rc \in \mathrm{SA}_t(E_0)$. From $c_n, c \in \mathrm{SA}_t(E) \subset \mathrm{SA}_0(E)$ and Theorem 4.5(d) it follows that $rc_n, rc \in \mathrm{SA}_0(E_0)$. We have $\mathrm{SA}_0(E_0) = \mathrm{SA}_t(E_0)$ by Theorem 3.11(c), because E_0 is Polish as a G_δ subset of a Polish space (Dugundji (1966)). QED

Interpreting the LDP's by (2.1) and (3.1) we see that their convergence contents hold iff

$$\liminf c_n^{1/a_n}(G) \geq c(G) \quad \text{for } G \in \mathcal{G}, \tag{5.1a}$$

$$\limsup c_n^{1/a_n}(B) \leq c(B) \quad \text{for } B \in \mathcal{K} \text{ (VLDP) or } B \in \mathcal{F} \text{ (NLDP)}. \tag{5.1b}$$

For applications it is convenient to replace $\mathcal{G}$ and $\mathcal{K}$ or $\mathcal{F}$ by smaller collections, so the following is useful.

5.5. Lemma. (a) *Formula (5.1a) (with $c \in \mathrm{SM}$ and $c_n \in \mathrm{SA}$) is implied by its restriction to $\mathcal{G}_0$ in case E is locally $\mathcal{G}_0$.*
(b) *Formula (5.1b) (with $c \in \mathrm{SM}$ (SM_t) and $c_n \in \mathrm{SA}$ (SA_t)) is implied by its restriction to $\mathcal{K}_0$ ($\mathcal{F}_0$) in case E is locally $\mathcal{K}_0$ ($\mathcal{F}_0$).*

Proof. (a) By Lemma 2.4 it suffices to show that (5.1a) for $G \in \mathcal{G}_0$ implies (5.1a) for $G \in \mathcal{G}_0^\cup$. If (5.1a) holds for $G = G_1, G_2$, then it also holds for $G = G_1 \cup G_2$ because

$$\begin{aligned}\liminf c_n^{1/a_n}(G_1 \cup G_2) &\geq \liminf (c_n(G_1) \vee c_n(G_2))^{1/a_n} \\ &\geq \liminf c_n^{1/a_n}(G_1) \vee \liminf c_n^{1/a_n}(G_2) \\ &\geq c(G_1) \vee c(G_2) = c(G_1 \cup G_2).\end{aligned}$$

(b) Similar considerations do the job, with the chain of inequalities replaced by

$$\begin{aligned}\limsup c_n^{1/a_n}(B_1 \cup B_2) &\leq \limsup (c_n(B_1) + c_n(B_2))^{1/a_n} \\ &\leq \limsup 2^{1/a_n} \left(c_n^{1/a_n}(B_1) \vee c_n^{1/a_n}(B_2)\right) \\ &\leq c(B_1) \vee c(B_2) = c(B_1 \cup B_2)\end{aligned}$$

QED

Applying Theorem 4.3 to LDP's we find

5.6. Theorem. (a) *A* VLDP*:* $c_n^{1/a_n} \to c$ *in a locally compact space* E *is transformed into a* VLDP*:* $(c_n\varphi^{\leftarrow})^{1/a_n} \to c\varphi^{\leftarrow}$ *in a locally compact space* E' *by a continuous and cocompactly continuous mapping* $\varphi\colon E \to E'$*.*
(b) *An* NLDP*:* $c_n^{1/a_n} \to c$ *is transformed into an* NLDP*:* $(c_n\varphi^{\leftarrow})^{1/a_n} \to c\varphi^{\leftarrow}$ *in* E' *by a continuous mapping* $\varphi\colon E \to E'$*.*

The limit $c\varphi^{\leftarrow}$ is a sup measure (tight in case (b)) with sup derivative

$$d^{\vee}(c\varphi^{\leftarrow})(x') = c\varphi^{\leftarrow}(\{x'\}) = c(\varphi^{\leftarrow}\{x'\}) = \sup\{d^{\vee}(x)\colon \varphi(x) = x'\}.$$

For the rate function $J = -\log c$ this takes the form

$$d^{\wedge}(J\varphi^{\leftarrow})(x') = \inf\{d^{\wedge}J(x)\colon \varphi(x) = x'\}. \tag{5.2}$$

6. From large deviations to loglog laws

In this section we consider NLDP's in which the tight subadditive capacities are in fact tight probability measures. We write μ_n instead of c_n and e^{-J} instead of c:

$$\mu_n^{1/a_n} \to e^{-J} \quad \text{narrowly, where } a_n \to \infty.$$

Recall that J is a tight $\bar{\mathsf{R}}_+$-valued inf measure, so that $d^{\wedge}J(x) = J(\{x\})$ is a lower compact function of x.

Since $a_n \to \infty$, we can find a function $m \mapsto n(m)$ on N or on an infinite subset of N such that $a_{n(m)} \sim \log m$. We then have

$$\mu_{n(m)}^{1/\log m} \to e^{-J} \quad \text{narrowly.} \tag{6.1}$$

In a suggestive but sloppy notation (6.1) reads

$$\mu_{n(m)} = m^{-J+o(1)} \quad \text{narrowly.}$$

An NLDP for tight probability measures $\mu_{n(m)}$ with the special choice $a_{n(m)} = \log m$ is a *preloglog law,* and the function $m \mapsto n(m)$ the *loglog time.*

Suppose that the μ_n are in fact the probability distributions of E-valued random variables X_n on one probability space. Then we have by (3.1)

$$\mathsf{P}[X_{n(m)} \in G] \geq m^{-J(G)+o(1)} \quad \text{for } G \in \mathcal{G}, \tag{6.2a}$$

$$\mathsf{P}[X_{n(m)} \in F] \leq m^{-J(F)+o(1)} \quad \text{for } F \in \mathcal{F} \text{ with } J(F) < \infty, \tag{6.2b}$$

and $\mathsf{P}[X_{n(m)} \in F] \leq m^{-a}$ eventually, for all $a > 0$ in case $F \in \mathcal{F}$ and $J(F) = \infty$.

For sequences (x_m) in E we introduce the notation

$$x_m \rightsquigarrow K,$$

which means that the sequence (x_m) is relatively compact and has set of limit points K. Note that K must be compact. We write "$x_m \rightsquigarrow$" if (x_m) is relatively compact, and $x_m \rightsquigarrow\supset A$ ($\subset A$) if in addition its set of limit points contains (is contained in) A.

6.1. Lemma. *If (6.1) holds for the probability distributions $\mu_{n(m)}$ of $X_{n(m)}$, then $X_{n(m)} \rightsquigarrow\subset [d^\wedge J \leq 1]$ wp1 in E.*

Proof. Note that $K := [d^\wedge J \leq 1] = \{x \colon J(\{x\}) \leq 1\}$ is compact since $d^\wedge J$ is lower compact. Let G be any open set such that $K \cap \operatorname{clos} G = \emptyset$. Then

$$J(\operatorname{clos} G) = J((\operatorname{clos} G) \cap [d^\wedge J \leq 2]) \wedge J((\operatorname{clos} G) \cap [d^\wedge J > 2]).$$

The first term on the right is J of a compact set so equals $\min d^\wedge J$ over the set, which is > 1; the second term is ≥ 2. Therefore $\mathsf{P}[X_{n(m)} \in \operatorname{clos} G \text{ i.o.}] = 0$ by (6.2b) and the Borel-Cantelli lemma. Hence, for any open set G containing K, $X_{n(m)}$ is eventually in G wp1. Choosing $G_n \downarrow K$ as near (1.1g), we get the desired conclusion. QED

In several specific cases the $X_{n(m)}$ happen to be sufficiently independent for applying versions of the other half of the Borel-Cantelli lemma. If so, then it follows by (6.2a) that $\mathsf{P}[X_{n(m)} \in G \text{ i.o.}] = 1$ for $G \in \mathcal{G}$ such that $J(G) < 1$. Combining this for a countable base of $[d^\wedge J < 1]$ we conclude that

$$X_{n(m)} \rightsquigarrow\supset \operatorname{clos}\,[d^\wedge J < 1] \quad \text{wp1}. \tag{6.3}$$

If $\operatorname{clos}\,[d^\wedge J < 1] = [d^\wedge J \leq 1]$, which is obvious in virtually all specific cases, then Lemma 6.1 and (6.3) combine into the *loglog law:*

$$X_{n(m)} \rightsquigarrow [d^\wedge J \leq 1] \quad \text{wp1}. \tag{6.4}$$

We have established the following terminology.

6.2. Definition. Let (X_n) be a sequence of random variables in E. We say that (X_n) satisfies a preloglog law with rate function J and loglog time $n(m)$ if $d^\wedge J$ is lower compact and

$$\left(\mathsf{P}[X_{n(m)} \in \cdot\,]\right)^{1/\log m} \to e^{-J} \quad \text{narrowly as } m \to \infty. \tag{6.5}$$

If (6.4) holds in addition, then we say that the corresponding loglog law holds.

In many specific cases we may replace $X_{n(m)}$ in (6.4) by X_n. The general pattern behind this and the connection with regular variation is explained in Vervaat (1987).

6.3. Example (Strassen's loglog law, Stroock (1984), Freedman (1972)). Let B be Brownian motion regarded as random variable in $E = C[0, \infty)$ with the topology of locally uniform convergence, and set

$$X_n := (2n \log\log n)^{-1/2} B(n\,\cdot) \qquad \text{for real } n > e.$$

Then X_n satisfies the preloglog law and the corresponding loglog law with loglog time $m \mapsto b^m$ $(b > 1)$ and rate function determined by

$$d^\wedge J(f) = \begin{cases} \int_0^\infty (f'(t))^2\,dt & \text{if } f(0) = 0 \text{ and } f \text{ is absolutely continuous,} \\ \infty & \text{else.} \end{cases}$$

We have seen that preloglog laws (6.5) may give rise to loglog laws (6.4). However, it is not true that loglog laws in the form of (6.4) always occur in conjunction with the preloglog law (6.5) (as was brought to our attention by David M. Mason). If the $X_{n(m)}$ satisfy both (6.4) and (6.5), then we can construct Y_m that satisfy (6.4) but not (6.5). Fix a y such that $d^\wedge J(y) > 1$. Let (ζ_m) be a sequence of independent random variables, independent of (X_n) and such that $\mathsf{P}[\zeta_m = 1] =: p_m = 1 - \mathsf{P}[\zeta_m = 0]$. Then set $Y_m := y$ if $\zeta_m = 1$, $Y_m := X_{n(m)}$ if $\zeta_m = 0$. If $\sum_m p_m < \infty$, then wp1 $Y_m = X_{n(m)}$ eventually, so (Y_m) satisfies (6.4). On the other hand, (p_m) can be selected such that $\sum_m p_m < \infty$ and

$$\mathsf{P}[Y_m = y] \geq p_m > m^{-d^\wedge J(y) + \varepsilon}$$

for some $\varepsilon > 0$ and infinitely many m. This contradicts (6.2b) for $F = \{y\}$, so (6.5) fails.

We may expect violation of the the corresponding preloglog law in instances of the Strassen loglog law with B replaced by the affine interpolation of partial sums of iid random variables in R with zero mean and unit variance. As a rule of thumb, one may expect the conjunction of preloglog law and loglog law when the X_n are transformations of one fixed limiting process in a weak limit theorem, and possible violation of the preloglog law if the limiting process is replaced by approximating processes in the weak limit theorem.

Nor is it true that each NLDP gives rise to a loglog law via its preloglog version. For instance, in the classical Cramér-Chernoff NLDP (Deuschel & Stroock (1989), Varadhan (1984)) we have $E = \mathsf{R}$, $X_n = \frac{1}{n}\sum_{k=1}^n \xi_k$

with iid ξ_k and $a_n = n$. Then $n(m) \sim \log m$, and there is no way to expect sufficient independence for the Borel-Cantelli lemma in the practically constant sequence $(X_{n(m)})$. So preloglog laws must be carefully arranged for obtaining loglog laws.

In the literature many preloglog laws are hidden in the technical details of the proofs of the corresponding loglog laws. They warrant a separate statement.

7. Observation processes and the self-affine Poisson process

In the final few sections we state and explain results that are being obtained by O'Brien & Vervaat (1990). Capacities play two different roles here. On the one hand they form the appropriate context for loglog laws in general, as explained in the previous section. Then they must be provided with the narrow topology. On the other hand, our generic separable metric space E, in which the random variables take their values, will turn out to be a subspace of the capacities on some other space E'. Here they are provided with the vague topology.

In fact, we consider random variables in $E = \mathrm{AD}([0,\infty) \times (\ell,\infty])$, where $\ell \in [-\infty,\infty)$. We write $\square_\ell := [0,\infty) \times (\ell,\infty]$ for short. In a canonical case we will have $\ell = 0$. Note that Radon measures are finite on northwest corners of $\square_\ell$, because they are required to be finite on $\mathcal{K}(\square_\ell)$. For general E' and $y \in E'$ we denote by ι_y the Dirac measure at y. For point measures $\mu \in \mathcal{N}(E')$ the support is a locally finite set, and $\mu = \sum_{b \in E} \mu\{b\}\iota_b$, where $\mu\{b\} \in \mathsf{N}$.

Let $\xi = (\xi_k)_{k=1}^\infty$ be a sequence of iid random variables in R. The *observation process* constructed from ξ is the point process

$$\Xi := \sum_{k=1}^{\infty} \iota_{(k,\xi_k)}$$

in $\square_{-\infty}$, a random variable in $\mathcal{N}(\square_{-\infty})$.

Resnick (1986, 1987) codified the following approach to weak functional limit theorems for partial maxima of ξ as a consequence of weak limit theorems for transformations of the observation process Ξ. Let $(\gamma_n)_{n=1}^\infty$ be a sequence of affine transformations of R: $\gamma_n(x) = a_n x + b_n$, $a_n > 0$, and set

$$W_n\Xi := \sum_{k=1}^{\infty} \iota_{(k/n,\gamma_n(\xi_k))},$$

the image of Ξ under the transformation W_n: $(t,x) \mapsto (t/n, \gamma_n(\xi_k))$ of $\square_{-\infty}$ (here $\gamma_n(\infty) := \infty$). We suppose there is a level ℓ (chosen to be minimal) such that $W_n\Xi$ restricted to Bor $\square_\ell$ converges in distribution to a nonzero point process Π in $\square_\ell$ with support in $\mathsf{R}_+ \times (\ell, \infty)$ (note that the level ∞ is left out here).

For $\mu \in \mathcal{N}(\square_\ell)$ and $A \in \text{Bor}\,\mathsf{R}_+$, set

$$(M\mu)(A) = M\mu(A) := \sup\{x\colon (t,x) \in \operatorname{supp}\mu,\, t \in A\},$$

where $\sup\emptyset := \ell$. Then M maps $\mathcal{N}(\square_\ell)$ with the vague topology continuously into $\mathrm{SM}(\mathsf{R}_+)$, the sup measures on R_+, with the vague topology. So the extremal processes $MW_n\Xi$ converge in distribution to the extremal process $M\Pi$. Note that

$$MW_n\Xi(0,t] = \gamma_n(\sup_{k=1}^{\lfloor nt \rfloor} \xi_k).$$

It turns out that Π is always a Poisson process in $\square_\ell$ with mean measure $\mathsf{E}\Pi = \text{Leb} \times \nu$ for specific $\nu \in \mathcal{M}((\ell,\infty])$. Furthermore, the convergence of $W_n\Xi$ to a nonconstant limit takes place iff $MW_n\Xi(0,1] = \gamma_n(\sup_{k=1}^n \xi_k)$ converges in distribution in R to a nondegenerate limit. So the limiting processes Π are closely related to the classical limit distributions in extreme-value theory. All these limit distributions and the convergence to them can be transformed into each other by compositions of affine transformations and the mappings $x \mapsto x^H$ $(H > 0)$, $x \mapsto -1/x$, log and exp applied to ξ or the vertical coordinate in $\square_\ell$. So we select one possible Π as standard case, to be considered exclusively from now on. It is the Poisson process with intensity $\mathsf{E}\Pi(dt,dx) = dt\,x^{-2}dx =: \pi(dt,dx)$ on $\square_0$), corresponding to the extreme-value distribution function $\exp(-1/x)$ on R_+, whose domain of attraction consists of those ξ such that $\mathsf{P}[\xi_1 > x]$ varies regularly at ∞ with exponent -1. We call this Poisson process the *self-affine Poisson process*, for reasons explained in O'Brien, Torfs & Vervaat (1989).

Applying functionals that add up functions of the heights of the points in the support of $\mu \in \mathcal{N}(\square_\ell)$, LePage, Woodroofe & Zinn (1981) obtain convergence in distribution of partial sum processes to stable motions (= Lévy's stable processes). The general case is complicated, and the functional applied to $\mathcal{N}(\square_\ell)$ not continuous, so we restrict ourselves at this point to Π and ξ as above, the ξ_k being nonnegative and the γ_n being pure multiplications. For $\mu \in \mathcal{N}(\square_0)$, $H > 0$ and $A \in \text{Bor}\,\mathsf{R}_+$ set

$$(S_H\mu)(A) = S_H\mu(A) := \int_0^\infty x^H \mu(A \times dx).$$

Then $S_HW_n\Xi \to_d S_H\Pi$ in $\mathcal{M}(\mathsf{R}_+)$ for $H > 1$, where $S_H\Pi[0,\cdot]$ is increasing stable motion with characteristic exponent $\alpha = 1/H$, and

$$S_HW_n\Xi[0,t] = \sum_{k=1}^{\lfloor nt \rfloor} (\gamma_n(\xi_k))^H.$$

In order to set out our strategy for deriving corresponding strong laws, let us review some features of the classical functional limit laws for partial sums $X_n = \sum_{k=1}^n \xi_k$ of iid random variables ξ_k with zero mean and unit variance. Set $X(t) := X_{\lfloor t \rfloor}$ for $t \geq 0$. Donsker's theorem tells us that $n^{-1/2}X(n\,\cdot) \to B$ (:= standard Brownian motion) in $D[0,\infty)$. Strassen's theorem tells us that wp1 the sequence with elements $(2n \log\log n)^{-1/2}X(n\,\cdot)$ is relatively compact in $D[0,\infty)$ with set of limit points $[d^\wedge J \leq 1]$, as in Example 6.3. Note that the latter normalization differs from the former by a slightly stronger compression, by a factor $(2\log\log n)^{1/2}$. Furthermore, one way to derive Strassen's theorem (cf. Freedman (1972)) is to replace X first by Donsker's weak limit B, get the result for $(2n\log\log n)^{-1/2}B(n\,\cdot)$, and extend it afterwards to X by embedding techniques.

In analogy with this O'Brien & Vervaat (1990) first derive loglog laws for transformations of the self-affine Poisson process Π, and make afterwards the transition from transformations of Π to transformations of the observation process Ξ. In this paper we restrict ourselves to quoting the results for Π.

8. Loglog laws for the self-affine Poisson process

In this section Π is the self-affine Poisson process, the Poisson process on $\square := \square_0 = [0,\infty) \times (0,\infty]$ with mean measure

$$\mathsf{E}\Pi(dt,dx) = dt\, x^{-2}dx =: \pi(dt,dx).$$

It is understood to be 0 on $\mathsf{R}_+ \times \{\infty\}$, so that $\Pi(\mathsf{R}_+ \times \{\infty\}) = 0$ wp1 and

$$\pi(A \times (x,\infty]) = (\operatorname{Leb} A)\cdot x^{-1}. \tag{8.1}$$

We now list some transformations that play a role in the remainder of this paper; we consider them for all *real* $n \geq 3$:

$$\varrho_n(t) := \tfrac{t}{n} \qquad \text{for } t \in [0,\infty]; \tag{8.2a}$$

$$\tau_n(x) := \left(\tfrac{x}{n}\right)^{1/\log\log n} \qquad \text{for } x \in (0,\infty]; \tag{8.2b}$$

$$\upsilon_n(x) := \tfrac{x}{n}\log\log n \qquad \text{for } x \in (0,\infty]; \tag{8.2c}$$

$$R_n := \varrho_n \otimes \varrho_n\colon \qquad (t,x) \mapsto \left(\tfrac{t}{n},\tfrac{x}{n}\right); \tag{8.2d}$$

$$T_n := \varrho_n \otimes \tau_n\colon \qquad (t,x) \mapsto \left(\tfrac{t}{n},\left(\tfrac{x}{n}\right)^{1/\log\log n}\right); \tag{8.2e}$$

$$U_n := \varrho_n \otimes \upsilon_n\colon \qquad (t,x) \mapsto \left(\tfrac{t}{n},\tfrac{x}{n}\log\log n\right). \tag{8.2f}$$

Note that $R_n\pi = \pi$, so $R_n\Pi =_d \Pi$. To compare τ_n and υ_n, note that

$$\begin{aligned}
\tau_n(x_n) \to 1,\ \upsilon_n(x_n) \to \infty \quad & \text{if } \varrho_n(x_n) \to x \text{ in } (0,\infty);\\
\varrho_n(x_n) \to 0,\ \upsilon_n(x_n) \to 0 \quad & \text{if } \tau_n(x_n) \to x \text{ in } (0,1);\\
\varrho_n(x_n) \to \infty,\ \upsilon_n(x_n) \to \infty \quad & \text{if } \tau_n(x_n) \to x \text{ in } (1,\infty);\\
\varrho_n(x_n) \to 0,\ \tau_n(x_n) \to 1 \quad & \text{if } \upsilon_n(x_n) \to x \text{ in } (0,\infty).
\end{aligned}$$

So υ_n serves as a microscope for what τ_n compresses to 1. We will see that there is a completely different limiting behavior for what τ_n brings above or below level 1 eventually.

The invariance in distribution of Π under R_n is related to Π being the limit in weak convergence results. Here is the strong limit result for $T_n = \varrho_n \otimes \tau_n$. We write $\mathrm{AD}_{\mathsf{N}}(\square)$ for the additive $(\mathsf{N} \cup \{\infty\})$-valued capacities on $\square$. In contrast to the point measures from $\mathcal{N}(\square)$ they need not be finite on compact sets. Recall Lemma 1.4.

8.1. Theorem. *Let* Π *be the self-affine Poisson process in* $\square$, *regarded as a random variable in* $\mathrm{AD}_{\mathsf{N}}(\square)$ *with the vague topology. Then (the distribution of)* $T_n\Pi$ *satisfies the* NLDP *with* $a_n = \log\log n$ *and with sup limit* e^{-J_T} *determined by*

$$d^\wedge J_T(\mu) = \begin{cases} \int_\square \log^+ x\, \mu(dt \times dx) & \text{if } \operatorname{supp}_\infty \mu \supset \square \backslash \square_1, \\ \infty & \text{for other } \mu \in \mathrm{AD}_{\mathsf{N}}(\square), \end{cases} \tag{8.3}$$

so $T_{b^m}\Pi$ *satisfies a preloglog law with the same* J_T *for each* $b > 1$. *Also the corresponding loglog laws hold.*

In other words, we have

$$(\mathsf{P}[T_n\Pi \in \cdot\,])^{1/\log\log n} \to e^{-J_T} \quad \text{narrowly,}$$

so, for each $b > 1$,

$$(\mathsf{P}[T_{b^m}\Pi \in \cdot\,])^{1/\log m} \to e^{-J_T} \quad \text{narrowly}$$

(with the consequences described in (6.2)), and finally

$$T_{b_m}\Pi \rightsquigarrow [d^\wedge J_T \le 1] \quad \text{wp1 vaguely in } \mathrm{AD}_{\mathsf{N}}(\square). \tag{8.4}$$

By general considerations as in Vervaat (1987) it follows that (8.4) also holds for $T_n\Pi$ as $n \to \infty$ through N or R. The key ingredients are that T_n varies regularly:

$$T_{an}T_n^{-1} \to \varrho_a \otimes \mathrm{id} \qquad \text{as } n \to \infty, \text{ for all } a > 0,$$

and that J_T and hence also the set of limit points $[d^\wedge J_T \leq 1]$ are invariant under $\varrho_a \otimes \mathrm{id}$.

By Theorem 4.2 the mapping $\mathrm{supp}\colon \mathrm{AD}_{\mathbf{N}}(\square) \to \mathcal{F}(\square)$ is vaguely continuous. So Theorem 8.1 implies (note that $\mathrm{supp}\, T_n\Pi = T_n \,\mathrm{supp}\,\Pi$):

8.2. Theorem. *Let Π be the self-affine Poisson process in $\square$. Then $T_n \,\mathrm{supp}\,\Pi$, regarded as a random variable in $\mathcal{F}(\square)$ with the vague topology, satisfies the* NLDP *with $a_n = \log\log n$ and with sup limit $e^{-J_{T,\mathcal{F}}}$ determined by*

$$d^\wedge J_{T,\mathcal{F}}(F) = \begin{cases} \sum_{(t,x)\in F} \log^+ x & \text{if } F \supset \square\backslash\square_1, \\ \infty & \text{for other } F \in \mathcal{F}(\square), \end{cases} \tag{8.5}$$

so $T_{b^m} \mathrm{supp}\,\Pi$ satisfies a preloglog law with the same $J_{T,\mathcal{F}}$ for each $b > 1$. Also the corresponding loglog laws hold.

Note that each F in the set of limit points $[d^\wedge J_{T,\mathcal{F}} \leq 1]$ contains the whole strip $\square\backslash\square_1 = [0,\infty) \times (0,1]$, but contains at most $\lfloor a \rfloor$ points at or above the level e^a, in particular none above the level e. More particularly, we see that the integer-valued measures in $[d^\wedge J_T \leq 1]$ are finite on compact sets in $\square_1$. From Theorem 4.5(b) it follows that Theorem 8.1 implies another version with the random variables taking values in $\mathcal{N}(\square_1)$. In this form the loglog law has been obtained by Mori & Oodaira (1976).

In the results for $T_n\Pi$ only the locations of the atoms of Π were moved around by T_n, but their masses remained unchanged. In the corresponding result for $U_n = \varrho_n \otimes v_n$ the masses are also transformed, by a factor $1/\log\log n$.

8.3. Theorem. *Let Π be the self-affine Poisson process on $\square$, regarded as a random variable in $\mathcal{M}(\square)$ with the vague topology. Then $\frac{1}{\log\log n} U_n\Pi$ satisfies the* NLDP *with $a_n = \log\log n$ and with sup limit e^{-J_U} determined by*

$$d^\wedge J_U = \begin{cases} \int_\square \left(\int_1^{\frac{d\mu}{d\pi}} \log y \, dy \right) d\pi & \text{if } \mu \ll \pi, \\ \infty & \text{for other } \mu \in \mathcal{M}(\square), \end{cases} \tag{8.6}$$

so T_{b^m} satisfies a preloglog law with the same J_U for each $b > 1$. Also the corresponding loglog laws hold.

The loglog law in Theorem 8.3 holds also for the full sequence (cf. Vervaat (1987)), because $\frac{1}{\log\log n} U_n$ varies regularly:

$$\frac{1}{\log\log na} U_{na} \left(\frac{1}{\log\log n} U_n \right)^{-1} \to \varrho_a \otimes \varrho_a = R_a$$

as $n \to \infty$, for each $a > 0$, and π, hence J_U, hence $[d^\wedge J_U \leq 1]$ are invariant under R_a.

Here is the result that follows from Theorem 8.3 for the supports. It is not an immediate consequence because supp is only lsc (Theorem 4.2) on $\mathcal{M}(\square)$. In contrast to the result of Theorem 8.2, supports of limit points may extend to the level $x = \infty$. Note that $\operatorname{supp}\left(\frac{1}{\log\log n} U_n \Pi\right) = U_n \operatorname{supp} \Pi$.

8.4. Theorem. *Let Π be the self-affine Poisson process in $\square$. Then $U_n \operatorname{supp} \Pi$, regarded as a random variable in $\mathcal{F}(\square)$ with the vague topology satisfies the* NLDP *with $a_n = \log\log n$ and with sup limit $e^{-J_{U,\mathcal{F}}}$ determined by*

$$d^\wedge J_{U,\mathcal{F}}(F) = \pi(\square \backslash F) \qquad \text{for } F \in \mathcal{F}(\square), \tag{8.7}$$

so U_{b^m} satisfies a preloglog law with the same $J_{U,\mathcal{F}}$ for each $b > 1$. Also the corresponding loglog laws hold.

We see that the complements in $\square$ of F in the set of limit points $[d^\wedge J_{U,\mathcal{F}} \leq 1]$ are bounded in π-measure. In particular, $([0,t) \times (x,\infty]) \cap F$ can be empty iff $x \geq t$ (cf. (8.1)).

9. Applications to extremal processes

In this section Π is the self-affine Poisson process. We present results about the classical extremal processes as corollaries of the results in the previous section. If we restrict our attention to extremal processes in the strict sense and ignore second and further largest observations, then the extremal processes are simple continuous functionals of the supports of the random measures in Theorems 8.1 and 8.3, so Theorems 8.2 and 8.4 are relevant here.

Let $\downarrow$ be the function from the subsets of $\square$ to the subsets of $\square$ defined by

$$\downarrow\{(t,x)\} := \{t\} \times (0,x] \qquad \text{for } (t,x) \in \square,$$

$$\downarrow A := \bigcup_{z \in A} \downarrow\{z\} \qquad \text{for } A \subset \square.$$

Then $\downarrow$ is a vaguely continuous map from $\mathcal{F}(\square)$ into $\mathcal{F}(\square)$, and the images are the closed sets F such that $F = \downarrow F$. They can be identified as the *hypographs* of $f \in \mathrm{US}(\mathsf{R}_+)$, defined by $\operatorname{hypo} f := \downarrow(\square \cap \operatorname{graph} f)$. It is well-known (Vervaat (1988b), Salinetti & Wets (1981)) that hypo is a vague homeomorphism from $\mathrm{US}(\mathsf{R}_+)$ into $\mathcal{F}(\square)$.

Let us define the *extremal functional* $M : \mathrm{AD}(\square) \to \mathrm{SM}(\mathsf{R}_+)$ by

$$\operatorname{hypo} d^\vee(M\mu) = \downarrow \operatorname{supp} \mu \qquad \text{for } \mu \in \mathrm{AD}(\square), \tag{9.1a}$$

which is equivalent to

$$(M\mu)(B) = \sup\{x : \mu(B \times [x, \infty]) > 0\} \qquad \text{for } B \in \operatorname{Bor} \mathsf{R}_+. \tag{9.1b}$$

Then all classical limit extremal processes are simple transformations of $M\Pi$. So the loglog laws for the classical limit extremal processes amount to the following.

9.1. Theorem. *The conclusions of Theorems 8.2 and 8.4 hold verbatim with* $\operatorname{supp}\Pi$ *replaced by* $\downarrow\operatorname{supp}\Pi = \operatorname{hypo} d^\vee M\Pi$, *and with the sets* F *restricted to* $\{F \in \mathcal{F}(\square) : F = \downarrow F\}$.

The loglog part of Theorem 9.1 has been obtained directly by Wichura (1974a), and the half referring to Theorem 8.2 in a context comparable to the present by Mori & Oodaira (1976). In a by now old-fashioned setting, extremal processes are regarded as random nondecreasing right-continuous functions on R_+ with the topology of pointwise convergence at continuity points of the limit (which is the trace of the vague topology on $\mathrm{US}(\mathsf{R}_+)$). The following corollary presents Theorem 9.1 in this setting. By $\mathrm{US}\!\uparrow\!(\mathsf{R}_+)$ we denote the subspace of nondecreasing functions in $\mathrm{US}(\mathsf{R}_+)$.

9.2. Corollary. *Let* Π *be the self-affine Poisson process in* $\square$. *Then*

$$M(T_n\Pi)\,[0,\cdot] = \left(\frac{1}{n} M\Pi[0, n\,\cdot]\right)^{1/\log\log n}$$

regarded as a random variable in $\mathrm{US}\!\uparrow\!(\mathsf{R}_+)$ *with the vague topology satisfies the* NLDP *with* $a_n = \log\log n$ *and with sup limit* $e^{J_{T,\uparrow}}$ *determined by*

$$d^\wedge J_{T,\uparrow}(f) = \begin{cases} \sum_{x\in \mathrm{Range}\, f} \log^+ x & \text{if } f(0) \geq 1, \\ \infty & \text{for other } f \in \mathrm{US}\!\uparrow\!(\mathsf{R}_+), \end{cases}$$

so T_{b^m} *satisfies a preloglog law with the same* $J_{T,\uparrow}$ *for each* $b > 1$. *Also the corresponding loglog laws hold.*

If one is interested in second and further largest observations, then obvious functionals of Π and the observation process Ξ rather than of their supports should be considered. In this way one can obtain NLDP's and loglog laws for tails of quantile processes. At this point it should be mentioned that Deheuvels & Mason (1989a) have obtained NLDP's and loglog laws for extremal processes and stable motions (cf. next section) with the quantile processes as starting point.

10. Applications to stable motions

In this section we discuss the counterparts of the theorems in Section 8 for stable motions. Only loglog laws are considered, because the situation around the related preloglog laws has not been settled yet. We only aim at giving an impression of the results, and refer to the literature for full statements and proofs. The results are only stated for the simplest case of increasing stable motions and made plausible by heuristics. The literature covers also the other spectrally positive stable motions and their attraction domains, but not more than that.

Let Π be the self-affine Poisson process, the Poisson process on $\square$ with mean measure $\pi(dt, dx) = dt\, x^{-2}dx$. Apart from a scaling constant and a drift term, all spectrally positive non-Brownian stable motions can be obtained for $H > \frac{1}{2}$ as random functions on R_+ by

$$S_H(t) := \int_0^\infty x^H \left(\Pi([0,t] \times dx) - 1_{B_H}(x)\pi([0,t] \times dx)\right), \tag{10.1a}$$

where

$$B_H := \begin{cases} \emptyset & \text{if } H > 1, \\ (0,1] & \text{if } H = 1, \\ (0,\infty) & \text{if } \frac{1}{2} < H < 1. \end{cases} \tag{10.1b}$$

Here H is the self-similarity exponent, which equals $1/\alpha$ where α is the stability index (the exponent in the characteristic function). From this representation it is clear that the simplest results can be expected for $H > 1$, to which case we restrict ourselves.

For $\mu \in \mathrm{AD}(\square)$, $B \in \mathrm{Bor}\,\mathsf{R}_+$ and $H > 1$ we define

$$(S_H\mu)(B) := \int_0^\infty x^H \mu(B \times dx), \tag{10.2}$$

so $S_H\mu \in \mathrm{AD}(\mathsf{R}_+)$. There is no guarantee that $S_H\mu$ is Radon, although $S_H\Pi$ is, wp1. We recognize $S_H\Pi$ as the random Radon measure on R_+ whose (random) distribution function is the random function S_H on R_+ in (10.1).

The result for $H > 1$ corresponding to Theorem 8.1 is suggested by the following heuristic arguments. Consider the random capacity on R_+:

$$X_{H,n} := \left(n^{-H} S_H\Pi(n\,\cdot)\right)^{1/\log\log n}. \tag{10.3}$$

Substituting (10.1) on the right-hand side and forcing it by T_n (cf. (8.2)) into a form appropriate for applying Theorem 8.1, we obtain

$$X_{H,n}(B) = \left(\int_0^\infty \left(\frac{x}{n}\right)^H \Pi(B \times dx)\right)^{1/\log\log n}$$

$$= \left(\int_0^\infty \left(\frac{T_n^{-1}(x)}{n} \right)^H T_n \Pi(B \times dx) \right)^{1/\log\log n}$$

$$= \left(\int_0^\infty x^{H \log\log n} \, T_n \Pi(B \times dx) \right)^{1/\log\log n} . \qquad (10.4)$$

Writing $\|\cdot\|_p$ for the p-norm of (infinite) sequences, we find

$$X_{H,n}(B) = \left\| (x^H)\colon (u,x) \in \operatorname{supp} T_n \Pi \text{ for some } u \in B \right\|_{\log\log n} . \qquad (10.5)$$

In many instances the p-norm converges to the ∞-norm, so we are led to guess

$$\begin{aligned} X_{H,n}(B) &\approx \sup\{(x^H)\colon (u,x) \in \operatorname{supp} T_n \Pi \text{ for some } u \in B\} \\ &= (M T_n \Pi(B))^H , \end{aligned} \qquad (10.6)$$

with M as in (9.1). The following definition is useful here.

10.1. Definition. Let E be a metric space, and let $(x_n)_{n=1}^\infty$ and $(y_n)_{n=1}^\infty$ be two sequences in E. We say that (x_n) and (y_n) are convergence equivalent, notation $x_n \approx y_n$ as $n \to \infty$, if for all sequences $n_k \to \infty$ in N and all $z \in E$ we have $x_{n_k} \to z$ iff $y_{n_k} \to z$.

If $x_n \approx y_n$, then $x_n \rightsquigarrow K$ iff $y_n \rightsquigarrow K$. If d is the metric in E and $d(x_n, y_n) \to 0$, then $x_n \approx y_n$. If E is locally compact and (x_n) relatively compact, then $x_n \approx y_n$ iff $d(x_n, y_n) \to 0$. If both (x_n) and (y_n) have no limit points, then $x_n \approx y_n$.

The following theorem states that (10.6) can be made rigorous.

10.2. Theorem. *Let Π be the self-affine Poisson process on $\square$, and let $S_H \Pi$ for $H > 1$ be defined by (10.2), regarded as a random variable in $\mathcal{C}(\mathsf{R}_+)$ with the vague topology. Then*

$$\left(n^{-H} S_H \Pi(n\,\cdot)\right)^{1/\log\log n} \approx (M T_n \Pi)^H = \left(\frac{1}{n} M \Pi(n\,\cdot) \right)^{H/\log\log n} \qquad \textit{wp1}.$$

Combining this with Theorems 9.1 and 8.1 we obtain

$$\left(n^{-H} S_H \Pi(n\,\cdot)\right)^{1/\log\log n} \rightsquigarrow \{M\mu\colon \mu \in \mathrm{AD}(\square),\ d^\wedge J_T(\mu) \le H\} \qquad (10.7)$$

wp1 in $\mathcal{C}(\mathsf{R}_+)$ as $n \to \infty$ through N, R or integer powers of $b > 1$. Similar results have been obtained in a more classical setting by Pakshirajan &

Vasudeva (1981). Note that each convergent subsequence on the left-hand side of (10.7) satisfies a VLDP.

The loglog laws corresponding to Theorem 8.3 have been derived by Wichura (1974b) and Mijnheer (1975). Again we discuss only the easiest case $H > 1$. If one would proceed from Theorem 8.3 with the (false) assumption that S_H in (10.2) maps $\mathcal{M}(\square)$ vaguely continuously into $\mathrm{AD}(\mathsf{R}_+)$, then we would obtain

$$\begin{aligned}
\frac{1}{\log\log n}&\int_0^\infty x^H U_n \Pi(\cdot \times dx) = \frac{1}{\log\log n}\int_0^\infty (v_n(x))^H\, \Pi((n\,\cdot)\times dx)\\
&= \frac{(\log\log n)^{H-1}}{n^H}\int_0^\infty x^H \Pi((n\,\cdot)\times dx) = \frac{(\log\log n)^{H-1}}{n^H} S_H \Pi(n\,\cdot)\\
&\rightsquigarrow \Big\{\int_\cdot\int_0^\infty x^H f(t,x)\,\pi(dt,dx):\\
&\qquad \text{Borel measurable } f\geq 0,\ \int_\square\Big(\int_1^f \log y\,dy\Big)\,d\pi \leq 1\Big\}\\
&=: K.
\end{aligned} \tag{10.8}$$

Let $\kappa \in K$. Then κ may have nonempty infinity support (cf. Lemma 1.4). In particular, we obtain the constant measure ∞ on R_+ by taking $f \equiv 1$. Moreover, κ is absolutely continuous outside $\mathrm{supp}_\infty \kappa$ with density $\int_0^\infty f(\cdot,x)x^{-2}dx$. We see that $A \cap \mathrm{supp}_\infty\kappa = \varnothing$ only if, above A, f is equal to 0 in a neighborhood of $x = \infty$ or vanishes very rapidly. In a crude attempt we could consider only f of the form $f = 1_F$ with $F \in \mathcal{F}(\square)$. Then the condition behind the colon in (10.8) specializes to $\pi(\square\backslash F) \leq 1$ (cf. Theorem 8.4), and then a natural guess is that in fact all limit points κ are represented by F such that $F = \downarrow F$, i.e., $F = \mathrm{hypo}\,\varphi$ for some $\varphi \in \mathrm{US}(\mathsf{R}_+)$. For this particular choice, the density of κ outside its infinity support is, as function of t,

$$\dot\kappa(t) = \int_0^{\varphi(t)} x^{H-2}dx = \frac{(\varphi(t))^{H-1}}{H-1},$$

and the condition $\pi(\square\backslash F) \leq 1$ becomes $\int_0^\infty \frac{dt}{\varphi(t)} \leq 1$ (cf. (8.1)). So our attempt ends with the guess

$$\begin{aligned}
K := \Big\{&\kappa \in \mathrm{AD}(\mathsf{R}_+):\\
&\kappa \ll \mathrm{Leb} \text{ on } \mathsf{R}_+\backslash\mathrm{supp}_\infty\kappa,\quad (H-1)^{\frac{1}{H-1}}\int_{\mathsf{R}_+\backslash\mathrm{supp}_\infty\kappa}(\dot\kappa(t))^{\frac{-1}{H-1}}\,dt \leq 1\Big\}.
\end{aligned}$$

The true result differs in two aspects. The constant before the integral is different, and κ need not be absolutely continuous outside its infinity domain. However, the density $\dot\kappa$ of the absolutely continuous part of κ outside the infinity support must satisfy the same inequality. Here is the result.

10.3. Theorem. *Let* Π *be the self-affine Poisson process in* $\square$, $H > 1$ *and let* $S_H\Pi$ *be the stable motion, regarded as a random variable in* $\mathrm{AD}(\mathsf{R}_+)$. *Then*

$$\frac{(\log\log n)^{H-1}}{n^H} S_H\Pi(n\cdot) \rightsquigarrow$$
$$\left\{\kappa \in \mathrm{AD}(\mathsf{R}_+) \colon \left(\Gamma\left(2-\tfrac{1}{H}\right)\right)^H \frac{1}{H-1}\int_{\mathsf{R}_+\setminus\mathrm{supp}_\infty\kappa} (\dot\kappa(t))^{\frac{-1}{H-1}}\,dt \le 1\right\}$$

wp1 vaguely in $\mathrm{AD}(\mathsf{R}_+)$, *where* $\dot\kappa$ *denotes the density of the absolutely continuous part of* κ *on* $\mathsf{R}_+\setminus\mathrm{supp}_\infty\kappa$.

For $H \le 1$, the calculus and the topology become more difficult, because $t \mapsto S_H(t)$ in (10.1) is no longer monotone. For all $H > \frac{1}{2}$, topological problems with finding appropriate function spaces force us to consider stable motions which are not spectrally one-sided only via their decomposition as difference of two independent spectrally positive stable motions.

Acknowledgement. We would like to thank A. Dow, H. Holwerda, D. Mason, T. Salisbury and a referee for their useful comments.

REFERENCES

[1] C. Berg, J.P.R. Christensen, P. Ressel (1984): *Harmonic Analysis on Semigroups.* Springer.

[2] P. Billingsley (1968): *Convergence of Probability Measures.* Wiley.

[3] N. Bourbaki (1969): *Éléments de Mathématique, Livre VI, Ch. IX, 2e éd.* Hermann.

[4] P. Deheuvels, D.M. Mason (1989a): A tail empirical process approach to some non-standard laws of the iterated logarithm. *J. Theor. Probab.* (to appear).

[5] P. Deheuvels, D.M. Mason (1989b): Functional laws of the iterated logarithm for the increments of empirical and quantile processes. Submitted.

[6] P. Deheuvels, D.M. Mason (1990): Non-standard functional laws of the iterated logarithm for tail empirical and quantile processes. *Ann. Probab.* **18** (to appear).

[7] C. Dellacherie, P.-A. Meyer (1978): *Probabilities and Potential.* Hermann and North-Holland.

[8] J.-D. Deuschel, D.W. Stroock (1989): *Large Deviations.* Academic Press.

[9] J. Dugundji (1966): *Topology.* Allyn and Bacon.

[10] J.M.G. Fell (1962): A Hausdorff topology for the closed subsets of a locally compact non-Hausdorff space. *Proc. Amer. Math. Soc.* **13** 472–476.

[11] D. Freedman (1972): *Brownian Motion and Diffusion.* Holden-Day.

[12] O. Kallenberg (1983): *Random Measures, 2nd Ed.* Akademie-Verlag & Academic Press.

[13] R. LePage, R. Woodroofe, M. Zinn (1981): Convergence to a stable distribution via order statistics. *Ann. Probab.* **9** 624–632.

[14] J. Lynch, J. Sethuraman (1987): Large deviations for processes with independent increments. *Ann. Probab.* **15** 610–627.

[15] D.M. Mason (1988): A strong invariance theorem for the tail empirical process. *Ann. Inst. H. Poincaré. Probab. Statist.* **74** 491–506.

[16] G. Matheron (1975): *Random Sets and Integral Geometry.* Wiley.

[17] J.L. Mijnheer (1975): *Sample Path Properties of Stable Processes.* Mathematical Centre Tracts **133**, Mathematisch Centrum, Amsterdam.

[18] T. Mori, H. Oodaira (1976): A functional law of the iterated logarithm for sample sequences. *Yokohama J. Math.* **24** 35–49.

[19] T. Norberg (1986): Random capacities and their distributions. *Probab. Th. Rel. Fields* **73** 281–297.

[20] T. Norberg, W. Vervaat (1989): Capacities on non-Hausdorff spaces. Report 1989-11, Dept. Math., U. of Gothenburg; to appear in [34].

[21] G.L. O'Brien, P.J.J.F. Torfs, W. Vervaat (1989): Stationary self-similar extremal processes. Report 89-07, Dept. of Math., York U.; to appear in *Probab. Th. Rel. Fields.*

[22] G.L. O'Brien, W. Vervaat (1990): Loglog laws for point processes. In preparation.

[23] R.P. Pakshirajan, Vasudeva (1981): A functional law of the iterated logarithm for a class of subordinators. *Ann. Probab.* **9** 1012–1018.

[24] S.I. Resnick (1986): Point processes, regular variation and weak convergence. *Adv. Appl. Probab.* **18** 66–138.

[25] S.I. Resnick (1987): *Extreme Values, Regular Variation and Point Processes.* Springer.

[26] G. Salinetti, R.J.-B. Wets (1981): On the convergence of closed-valued measurable multifunctions, *Trans. Amer. Math. Soc.* **266** 275–289.

[27] G. Salinetti, R.J.-B. Wets (1987): Weak convergence of probability measures revisited. Working Paper 87-30, IIASA, Laxenburg, Austria.

[28] D.W. Stroock (1984): *An Introduction to the Theory of Large Deviations.* Springer.
[29] S.R.S. Varadhan (1984): *Large Deviations and Applications.* SIAM, Philadelphia.
[30] W. Vervaat (1987): Functional iterated logarithm laws for geometric subsequences and full sequences. Report 8721, Dept. of Math., Catholic U., Nijmegen; to appear in *Probab. Th. Rel. Fields* under the title: Transformations in functional iterated logarithm laws and regular variation.
[31] W. Vervaat (1988a): Narrow and vague convergence of set functions. *Statist. Probab. Letters* **6** 295–298.
[32] W. Vervaat (1988b): Random upper semicontinuous functions and extremal processes. Report MS-8801, Center for Math. and Comp. Sci., Amsterdam; to appear in [34].
[33] W. Vervaat (1988c): Spaces with vaguely upper semicontinuous intersection. Report 88-30, Faculty of Techn. Math and Informatics, Delft U. of Techn.; to appear in [34].
[34] W. Vervaat (1990) ed.: *Probability and Lattices.* CWI Tracts, Center for Math. and Comp. Sci., Amsterdam (to appear).
[35] M.J. Wichura (1974a): On the functional form of the law of the iterated logarithm for the partial maxima of independent identically distributed random variables. *Ann. Probab.* **2** 202–230.
[36] M.J. Wichura (1974b): Functional laws of the iterated logarithm for the partial sums of i.i.d. random variables in the domain of attraction of a completely asymmetric stable law. *Ann. Probab.* **2** 1108–1138.

George L. O'Brien
Department of Mathematics
York University
4700 Keele Street
North York, Ontario
Canada M3J 1P3
obrien@vm1.yorku.ca

Wim Vervaat
Mathematisch Instituut
Katholieke Universiteit
Toernooiveld 1
6525 ED Nijmegen
The Netherlands
vervaa@sci.kun.nl

Conditional variance of symmetric stable variables

WEI WU and STAMATIS CAMBANIS

Abstract.

For two symmetric α-stable random variables with $1 < \alpha < 2$ we find a necessary and sufficient condition for the conditional variance to exist and be finite, we show it has a fixed functional form independent of their joint distribution, we describe its asymptotic behavior and we illustrate its global dependence on the joint distribution.

1. INTRODUCTION

A useful property of multivariate normal distributions is that conditional distributions are also normal with conditional mean linear and conditional variance constant. For multivariate non-Gaussian stable distributions, conditional distributions are stable only under very restrictive conditions (see Proposition 3.1 in Adler et al. [1] for the bivariate stable distributions with $1 < \alpha < 2$), in which case the conditional second moment is of course infinite. In Theorem 1 we describe all bivariate symmetric α-stable distributions with $1 < \alpha < 2$ which have finite conditional second moment (though the unconditional second moment is always infinite). We show in Theorem 2 that the conditional variance has always the same functional form, independent of the joint distribution, and we derive its asymptotic behavior. The joint distribution affects the conditional variance only through a global multiplier and several examples in Section 3 illustrate the various ways this multiplier may depend on the joint distribution. The general case of skewed stable distributions is more complex and will be considered elsewhere.

2. FORM OF THE CONDITIONAL VARIANCE

The joint characteristic function of symmetric α-stable random variables X_1 and X_2 is of the form

$$\phi(t_1, t_2) = Eexp\{i(t_1X_1 + t_2X_2)\} = exp\{-\int_{S_2} |t_1y_1 + t_2y_2|^\alpha d\Gamma(y_1, y_2)\},$$

where Γ is a finite symmetric (spectral) measure on the unit circle S_2 in $\mathbf{R}^2$. We will use bracketed power to denote: $y^{<\alpha>} = |y|^\alpha \operatorname{sgn}(y)$ for y real.

If for some s_1, s_2, $\Gamma\{(y_1, y_2) \in S_2 : s_1y_1 + s_2y_2 \neq 0\} = 0$, i.e. Γ is concentrated at two (opposite) points, then X_1, X_2 are linearly dependent: $s_1X_1 + s_2X_2 = 0$ a.s., and thus $E(X_2|X_1) = cX_1, E(X_2^2|X_1) = c^2X_1^2$ and $Var(X_2|X_1) = 0$. We will therefore assume throughout that X_1 and X_2 are linearly independent, i.e. that for all t_1, t_2,

$$\Gamma\{(y_1, y_2) \in S_2 : t_1y_1 + t_2y_2 \neq 0\} > 0.$$

Then, as in the proof of Lemma 2.1 in Samorodnitsky and Taqqu [5],

$$\gamma = inf\{\int_{S_2} |t_1y_1 + t_2y_2|^\alpha d\Gamma(y_1, y_2);\ (t_1, t_2) \in S_2\} > 0,$$

$\phi(t_1, t_2) \leq exp\{-\gamma(t_1^2 + t_2^2)^{\frac{1}{2}}\}$ and thus $\phi(t_1, t_2)$ is in $L^1(\mathbf{R}^2)$. It follows that X_1, X_2 have a continuous joint probability density function $f(x_1, x_2)$ and the regular conditional second moment of X_2 given $X_1 = x_1$ is given *for all* $x_1 \in \mathbf{R}^1$ by

$$E(X_2^2|X_1 = x_1) = \frac{1}{f(x_1)} \int_{-\infty}^{\infty} x_2^2 f(x_1, x_2) dx_2.$$

where $f(x_1)$ is the density of X_1. Whenever we write $E(X_2^2|X_1 = x_1)$ in the sequel, we will always mean this regular version. The first result provides a necessary and sufficient condition for this regular conditional second moment to be finite.

Theorem 1. *Let X_1, X_2 be linearly independent, jointly symmetric α-stable with $1 < \alpha \leq 2$. Then $E(X_2^2|X_1 = x_1) < \infty$ for all $x_1 \in \mathbf{R}^1$ if and only if*

$$\int_{S_2} |y_1|^{\alpha-2} y_2^2 d\Gamma(y_1, y_2) < \infty.$$

The necessary and sufficient condition is clearly equivalent to $\int_{S_2} |y_1|^{\alpha-2} d\Gamma(y_1, y_2) < \infty$, but in the form written in Theorem 1 it is easier to evaluate. This condition restricts the value of the spectral measure Γ in the neighbourhood of the points $\pm\pi/2$ as follows,

$$\int_{\pi/2-\epsilon}^{\pi/2+\epsilon} \frac{d\Gamma(\theta)}{|\theta - \pi/2|^{2-\alpha}} < \infty.$$

Thus Γ cannot have an atom at $\pm\pi/2$; and if in a neighbourhood of $\pi/2$ it is absolutely continuous with bounded density, then the condition is satisfied. The sufficiency of this condition is also established in Theorem 3.1 of Samorodnitsky and Taqqu [5].

We now specify the form of the conditional variance, whenever it is finite.

Theorem 2. *Let X_1, X_2 be linearly independent, jointly symmetric α-stable with $1 < \alpha \leq 2$ and assume that $\int_{S_2} |y_1|^{\alpha-2} y_2^2 d\Gamma(y_1, y_2) < \infty$. Then the conditional variance $Var(X_2|X_1)$ exists and is a.s. finite and has the form*

$$Var(X_2|X_1) = C_{2|1}^2(\alpha)\, S^2(X_1/\sigma_1; \alpha),$$

where
$$C_{2/1}^2(\alpha) = \alpha(\alpha-1)\sigma_1^{2-2\alpha}[\int_{S_2} |y_1|^\alpha d\Gamma \int_{S_2} |y_1|^{\alpha-2} y_2^2 d\Gamma - (\int_{S_2} y_1^{<\alpha-1>} y_2 d\Gamma)^2],$$

$$S^2(x;\alpha) = \frac{\int_0^\infty \cos(xt) e^{-t^\alpha} t^{\alpha-2} dt}{\int_0^\infty \cos(xt) e^{-t^\alpha} dt} = \frac{1}{\alpha f(x;\alpha)} \int_{|x|}^\infty u f(u;\alpha) du,$$

$$\sigma_1^\alpha = \int_{S_2} |y_1|^\alpha d\Gamma(y_1, y_2),$$

and $f(x;\alpha)$ is the α-stable density with characteristic function $exp\{-|t|^\alpha\}$. Moreover, when $1 < \alpha < 2$, the conditional "standard deviation" function $S(x;\alpha)$ is symmetric and satisfies

$$\begin{aligned} S(x;\alpha) &= [\alpha(\alpha-1)]^{\frac{1}{2}} x + o(x) \text{ as } x \to \infty, \\ S(x;\alpha) &= S(0;\alpha) + A(\alpha)x^2 + o(x^2) \text{ as } x \to 0, \end{aligned}$$

where

$$S(0;\alpha) = [\alpha\Gamma(1-\frac{1}{\alpha})/\Gamma(\frac{1}{\alpha})]^{\frac{1}{2}}, \quad A(\alpha)S(0;\alpha) = \frac{1}{4\alpha}[-1 + \alpha\Gamma(\frac{3}{\alpha})\Gamma(1-\frac{1}{\alpha})/\Gamma^2(\frac{1}{\alpha})] \geq 0.$$

It is seen from Theorem 2 that, whenever it is finite, the conditional variance has a fixed functional form $S^2(\cdot/\sigma_1;\alpha)$, depending only on the index of stability α and scaled by the scale parameter σ_1 of X_1, *independent of the joint distribution of* X_1 *and* X_2.

Whereas in the Gaussian case $\alpha = 2$ the conditional variance is constant, $S^2(x;2) \equiv 1$, in the non-Gaussian stable case with $1 < \alpha < 2$ the conditional variance $Var(X_2|X_1)$ can never be a nonzero constant and indeed as $|x_1| \to \infty$ the value of $Var(X_2|X_1 = x_1)$ tends to infinity. The conditional "standard deviation" function $S(x;\alpha)$ is an even function, approximately quadratic around the origin and approximately linear at infinity with slope $[\alpha(\alpha-1)]^{-\frac{1}{2}}, 1 < \alpha < 2$. Thus the heavy tailed stable distributions with $1 < \alpha < 2$ are such that the larger the observed value of X_1, the larger the uncertainty in the (conditional) value of X_2. Note that the quadratic asymptote $A(x)x^2 + S(0;\alpha)$ around 0 lies always strictly above the linear asymptote $[\alpha(\alpha-1)]^{-\frac{1}{2}}|x|$ at infinity; this follows from the fact that $A(\alpha)S(0;\alpha) \geq [4(\alpha-1)]^{-1}$.

The joint distribution of X_1 and X_2 affects the conditional variance only through the factor $C^2_{2|1}(\alpha)$. Under the conditions of the theorem $0 < C^2_{2|1}(\alpha) < \infty$. Indeed, we have, $C_{2|1}(\alpha) = 0$, if and only if $y_1^{<\alpha/2-1>}y_2 = c|y_1|^{\alpha/2}$ a.e. (Γ) for some constant c, i.e., $y_2 = cy_1$ a.e. (Γ), i.e. $X_2 = cX_1$ a.s., which contradicts the assumption of linear independence. On the other hand $C^2_{2|1}(\alpha) = \infty$ if $\Gamma\{(0,1)\} > 0$; a simple example is when Γ has mass on $\pm(0,1)$ and $\pm(\cos\theta, \sin\theta)$, which implies that the conditional distribution of X_2 given X_1 is α-stable (see Adler et al. [1]), and therefore $Var(X_2|X_1) = \infty$.

As for large x the conditional standard deviation is approximately linear, $S(x;\alpha) \approx$ Const. $|x|$, it follows that $E[Var(X_2|X_1)]^p < \infty$ if and only if $0 < p < \alpha/2$. In particular, since $1 < \alpha < 2$, the mean conditional standard deviation is finite, $E[SD(X_2|X_1)] = C_{2|1}(\alpha)E[S(\sigma_1^{-1}X_1;\alpha)] < \infty$.

Proof of Theorem 1. The regular conditional characteristic function of X_2 given $X_1 = x_1$ is expressed for all $t_2, x_1 \in \mathbf{R}^1$ as follows,

$$\psi(t_2,x_1) = E(e^{it_2X_2}|X_1 = x_1) = \frac{1}{f(x_1)}\int_{-\infty}^{\infty} e^{it_2x_2}f(x_1,x_2)dx_2$$

$$= \frac{1}{2\pi f(x_1)} \int_{-\infty}^{\infty} e^{-it_1x_1}\phi(t_1, t_2)dt_1$$

(cf. Theorem 3.1 in Zabell [6]). Therefore

$$E(X_2^2|X_1 = x_1) < \infty \quad \text{for all} \quad x_1 \in \mathbf{R}^1,$$

if and only if

$$\lim_{t_2 \to 0} \frac{1}{t_2^2}\{2 - \psi(t_2, x_1) - \psi(-t_2, x_1)\} \text{ exists and is finite for all } x_1 \in \mathbf{R}^1.$$

But

$$\begin{aligned} I(t_2, x_1) &:= \frac{1}{t_2^2}\{2 - \psi(t_2, x_1) - \psi(-t_2, x_1)\} \\ &= \frac{1}{2\pi f(x_1)} \frac{1}{t_2^2} \int_{-\infty}^{\infty} e^{-it_1x_1}\{2\phi(t_1, 0) - \phi(t_1, t_2) - \phi(t_1, -t_2)\}dt_1 \\ &= \frac{1}{\pi f(x_1)} \frac{1}{t_2^2} \int_0^{\infty} \cos(t_1x_1)\phi(t_1, 0)\{2 - exp[-\int_{S_2}(|t_1y_1 + t_2y_2|^\alpha - |t_1y_1|^\alpha)d\Gamma] \\ &\quad -exp[-\int_{S_2}(|t_1y_1 - t_2y_2|^\alpha - |t_1y_1|^\alpha)d\Gamma]\}dt_1 \end{aligned}$$

and putting

$$\Delta(t_1, t_2) = \int_{S_2}(|t_1y_1 + t_2y_2|^\alpha - |t_1y_1)^\alpha)d\Gamma(y_1, y_2)$$

we can write

$$\begin{aligned} I(t_2, x_1) &= \frac{1}{\pi f(x_1)} \int_0^{\infty} \cos(t_1x_1)\phi(t_1, 0)\frac{1}{t_2^2}\{1 - \Delta(t_1, t_2) \\ &- e^{-\Delta(t_1,t_2)} + 1 - \Delta(t_1, -t_2) - e^{-\Delta(t_1,-t_2)}\}dt_1 \\ &+ \frac{1}{\pi f(x_1)} \int_0^{\infty} \cos(t_1x_1)\phi(t_1, 0)\frac{1}{t_2^2}\{\Delta(t_1, t_2) + \Delta(t_1, -t_2)\}dt_1 \\ &=: I_1(t_2, x_1) + I_2(t_2, x_1) \end{aligned}$$

where I_1, I_2 denote the first and second term respectively. We first show that the limit of $I_1(t_2, x_1)$ as $t_2 \to 0$ exists and is finite for all $x_1 \in \mathbf{R}^1$. By Taylor's expansion we can write

$$e^{-\Delta(t_1,t_2)} = 1 - \Delta(t_1, t_2) + \frac{1}{2}\Delta^2(t_1, t_2)e^{-\theta(t_1,t_2)}$$

where $|\theta(t_1, t_2)| \le |\Delta(t_1, t_2)|$. Using the inequality

$$||x_1 + x_2|^\alpha - |x_1|^\alpha| < |x_2|^\alpha + \alpha|x_1|^{\alpha-1}|x_2|$$

for all $x_1, x_2 \in \mathbf{R}^1$ and $1 \le \alpha \le 2$ (cf. Theorem 7.1 in Masry and Cambanis [3]) we obtain

$$|\theta(t_1,t_2)| \le |t_2|^\alpha \int_{S_2} |y_2|^2 d\Gamma + \alpha |t_1|^{\alpha-1}|t_2| \int_{S_2} |y_1|^{\alpha-1}|y_2| d\Gamma \le c(|t_2|^\alpha + |t_1|^{\alpha-1}|t_2|)$$

for some finite constant c. It follows that the absolute value of the integrand of the term $I_1(t_2, x_1)$ is upper-bounded by

$$\begin{aligned} &|cos(t_1x_1)\,\phi(t_1,0)\frac{1}{t_2^2}\{\frac{1}{2}\Delta^2(t_1,t_2)e^{-\theta(t_1,t_2)} + \frac{1}{2}\Delta^2(t_1,-t_2)e^{-\theta(t_1,-t_2)}\}| \\ &\le e^{-\sigma_1^\alpha |t_1|^\alpha}\frac{1}{t_2^2}c(|t_2|^\alpha + |t_1|^{\alpha-1}|t_2|)^2 e^{c(|t_2|^\alpha + |t_1|^{\alpha-1}|t_2|)} \\ &\le e^{-\sigma_1^\alpha |t_1|^\alpha} c(1 + 2|t_1|^{\alpha-1} + |t_1|^{2(\alpha-1)})e^{c(1+|t_1|^{\alpha-1})} \quad \text{for } |t_2| < 1, \end{aligned}$$

which is in $L_1(\mathbf{R}^1)$ as a function of t_1 . In order to apply the dominated convergence theorem we need to examine the convergence of the integrand of the term $I_1(t_2, x_1)$ as $t_2 \to 0$. Notice that $\Delta(t_1, 0) = 0$, and by writing

$$\frac{1}{t_2^2}(1 - \Delta - e^{-\Delta}) = (\frac{\Delta}{t_2})^2 \frac{1 - \Delta - e^{-\Delta}}{\Delta^2}$$

we have

$$\lim_{\Delta \to 0} \frac{1 - \Delta - e^{-\Delta}}{\Delta^2} = -\frac{1}{2}$$

and

$$\begin{aligned} \lim_{t_2 \to 0} \frac{\Delta(t_1,t_2)}{t_2} &= \int_{S_2} \lim_{t_2 \to \infty} \frac{1}{t_2}\{|t_1y_1 + t_2y_2|^\alpha - |t_1y_1|^\alpha\} y_2 d\Gamma(y_1,y_2) \\ &= \int_{S_2} \alpha(t_1y_1)^{<\alpha-1>} y_2 d\Gamma(y_1,y_2) \\ &= \alpha(\int_{S_2} y_1^{<\alpha-1>} y_2 d\Gamma) t_1^{<\alpha-1>} \end{aligned}$$

by the dominated convergence theorem, as

$$\begin{aligned} |\frac{1}{t_2}\{|t_1y_1 + t_2y_2|^\alpha - |t_1y_1|^\alpha\}| &\le |t_2|^{\alpha-1}|y_2|^\alpha + \alpha|t_1y_1|^{\alpha-1}|y_2| \\ &\le |y_2|^\alpha + \alpha|t_1y_1|^{\alpha-1}|y_2| \quad \text{for } |t_2| < 1 \end{aligned}$$

and the upper bound is Γ-integrable. Thus

$$\lim_{t_2\to 0}\frac{1}{t_2^2}(1-\Delta-e^{-\Delta}) = -\frac{1}{2}\alpha^2\Big(\int_{S_2} y_1^{<\alpha-1>}y_2 d\Gamma\Big)^2|t_1|^{2(\alpha-1)}.$$

Dominated convergence therefore applies and gives

$$\lim_{t_2\to 0} I_1(t_2,x_1) = -\frac{1}{\pi f(x_1)}\alpha^2\Big(\int_{S_2} y_1^{<\alpha-1>}y_2 d\Gamma\Big)^2\int_0^\infty \cos(t_1x_1)e^{-\sigma_1^\alpha|t_1|^\alpha}|t_1|^{2(\alpha-1)}dt_1.$$

Since this limit exists and is finite for all $x_1\in\mathbf{R}^1$ (as $\alpha>1$), it follows that $E(X_2^2|X_1=x_1)<\infty$ for all $x_1\in\mathbf{R}^1$ if and only if $lim_{t_2\to 0}I_2(t_2,x_1)$ exists and is finite for all $x_1\in\mathbf{R}^1$.

First assume $\int|y_1|^{\alpha-2}y_2^2d\Gamma<\infty$. We can write

$$\begin{aligned} I_2(t_2,x_1) &= \frac{1}{\pi f(x_1)}\int_0^\infty dt_1\cos(t_1x_1)e^{-\sigma_1^\alpha t_1^\alpha} \\ &\quad \int_{S_2} d\Gamma(y_1,y_2)\frac{1}{t_2^2}\{|t_1y_1+t_2y_2|^\alpha+|t_1y_1-t_2y_2|^\alpha-2|t_1y_1|^\alpha\}. \end{aligned}$$

We will use the inequality

$$||x_1+x_2|^\alpha+|x_1-x_2|^\alpha-2|x_1|^\alpha|\le 2\alpha(\alpha-1)|x_1^{\alpha-2}x_2^2$$

for all $x_1,x_2\in\mathbf{R}^1$ and $1<\alpha\le 2$, which follows from Taylor's expansion:

$$|x_1+x_2|^\alpha = |x_1|^\alpha+\alpha x_1^{<\alpha-1>}x_2+\frac{1}{2}\alpha(\alpha-1)|x_1|^{\alpha-2}x_2^2\,\theta(x_1,x_2)$$

where $|\theta(x_1,x_2)|\le 1$ and $x_1\le 0$. The inner integrand in the expression of I_2 is thus upper-bounded in absolute value by

$$2\alpha(\alpha-1)|t_1y_1|^{\alpha-2}y_2^2$$

for all $t_2\ne 0$, and the resulting integral is finite in view of $1<\alpha\le 2$ and the assumption. Thus the dominated convergence theorem applies and leads to

$$\begin{aligned} \lim_{t_2\to 0} I_2(t_1,x_1) &= \frac{1}{\pi f(x_1)}\int_0^\infty dt_1\cos(t_1x_1)e^{-\sigma_1^\alpha t_1^\alpha}\int_{S_2}d\Gamma(y_1,y_2)\alpha(\alpha-1)|t_1y_1|^{\alpha-2}y_2^2 \\ &= \frac{1}{\pi f(x_1)}\alpha(\alpha-1)\Big(\int_{S_2}|y_1|^{\alpha-2}y_2^2d\Gamma\Big)\int_0^\infty\cos(t_1x_1)e^{-\sigma_1^\alpha t_1^\alpha}t_1^{\alpha-2}dt_1. \end{aligned}$$

Hence this limit exists and is finite, so $E(X_2^2|X_1 = x_1) < \infty$ for all $x_1 \in \mathbf{R}^1$. Also in this case

$$\begin{aligned} E(X_2^2|X_1 = x_1) &= \lim_{t_2\to 0} I(t_2, x_1) = \lim_{t_2\to 0}\{I_1(t_2,x_1) + I_2(t_2,x_1)\} \\ &= \frac{1}{\pi f(x_1)}\int_0^\infty \cos(t_1x_1)e^{-\sigma_1^\alpha t_1^\alpha}\{\alpha(\alpha-1)(\int_{S_2}|y_1|^{\alpha-2}y_2^2 d\Gamma)t_1^{\alpha-2} \\ &\quad -\alpha^2(\int_{S_2} y_1^{<\alpha-1>}y_2 d\Gamma)^2 t_1^{2(\alpha-1)}\}dt_1. \end{aligned}$$

Conversely, assume the limit of $I_2(t_2, x_1)$ as $t_2 \to 0$ exists and is finite for all $x_1 \in \mathbf{R}^1$. Then since

$$|x_1 + x_2|^\alpha + |x_1 - x_2|^\alpha - 2|x_1|^\alpha \geq 0$$

($|2x_1|^\alpha = |x_1 + x_2 + x_1 - x_2|^\alpha \leq 2^{\alpha-1}\{|x_1+x_2|^\alpha + |x_1 - x_2|^\alpha\}$), by Fatou's lemma we have

$$\begin{aligned} \infty &> \lim_{t_2\to 0} I_2(t_2, 0) \\ &= \lim_{t_2\to 0}\frac{1}{\pi f(0)}\int_0^\infty dt_1 e^{-\sigma_1^\alpha t_1^\alpha}\int_{S_2} d\Gamma(y_1,y_2)\frac{1}{t_2^2}\{|t_1y_1 + t_2y_2|^\alpha + |t_1y_1 - t_2y_2|^\alpha - 2|t_1y_1|^\alpha\} \\ &= \frac{1}{\pi f(0)}\int_0^\infty dt_1 e^{-\sigma_1^\alpha t_1^\alpha}\int_{S_2} d\Gamma(y_1,y_2)\lim_{t_2\to 0}\frac{1}{t_2^2}\{|t_1y_1 + t_2y_2|^\alpha + |t_1y_1 - t_2y_2|^\alpha - 2|t_1y_1|^\alpha\} \\ &= \frac{1}{\pi f(0)}\int_0^\infty d_1te^{-\sigma_1^\alpha t_1^\alpha}\int_{S_2} d\Gamma(y_1,y_2)\alpha(\alpha-1)|t_1y_1|^{\alpha-2}y_2^2 \\ &= \frac{\alpha(\alpha-1)}{\pi f(0)}\cdot\int_0^\infty e^{-\sigma_1^\alpha t_1^\alpha}t_1^{\alpha-2}dt_1\cdot\int_{S_2}|y_1|^{\alpha-2}y_2^2 d\Gamma(y_1,y_2) \end{aligned}$$

and thus $\int |y_1|^{\alpha-2}y_2^2 d\Gamma < \infty$. □

Proof of Theorem 2. Putting

$$a = \sigma_1^{-\alpha}\int_{S_2} y_1^{<\alpha-1>}y_2 d\Gamma, \;\; b = \int_{S_2}|y_1|^{\alpha-2}y_2^2 d\Gamma,$$

the expression of the regular conditional second moment in the proof of Theorem 1 can be written as follows

$$E(X_2^2|X_1 = x_1) = \frac{1}{\pi f(x_1)}\int_0^\infty \cos(t_1x_1)e^{-\sigma_1^\alpha t_1^\alpha}\{\alpha(\alpha-1)bt_1^{\alpha-2} - \alpha^2a^2\sigma_1^{2\alpha}t_1^{2(\alpha-1)}\}dt_1.$$

Now $\phi_1(t_1) = \phi(t_1, 0) = \exp\{-|t_1|^\alpha\sigma_1^\alpha\}$, and using

$$\phi_1''(t_1) = \phi_1(t_1)[\sigma_1^{2\alpha}\alpha^2|t_1|^{2\alpha-2} - \sigma_1^\alpha\alpha(\alpha-1)|t_1|^{\alpha-2}],$$

we can write

$$\begin{aligned} E(X_2^2|X_1 = x_1) &= -\frac{a^2}{\pi f(x_1)} \int_0^\infty \cos(t_1 x_1)\phi_1''(t_1)dt_1 \\ &+ \frac{\alpha(\alpha-1)(b-a^2\sigma_1^\alpha)}{\pi f(x_1)} \int_0^\infty \cos(t_1 x_1) e^{-\sigma_1^\alpha t_1^\alpha} t_1^{\alpha-2} dt_1. \end{aligned}$$

Integrating by parts twice the term involving ϕ_1'', we obtain

$$\begin{aligned} E(X_2^2|X_1 = x_1) &= a^2 x_1^2 + \alpha(\alpha-1)(b-a^2\sigma_1^\alpha)\frac{1}{\pi f(x_1)} \int_0^\infty \cos(t_1 x_1) t_1^{\alpha-2} e^{-\sigma_1^\alpha t_1^\alpha} dt_1 \\ &= a^2 x_1^2 + C_{2|1}^2(\alpha) h(x_1;\alpha). \end{aligned}$$

where

$$h(x_1;\alpha) = \frac{\sigma_1^{\alpha-2}}{\pi f(x_1)} \int_0^\infty \cos(t_1 x_1) t_1^{\alpha-2} e^{-\sigma_1^\alpha t_1^\alpha} dt_1.$$

Since $E(X_2|X_1) = aX_1$ (see Kanter [2]), it follows that

$$Var(X_2|X_1 = x_1) = E(X_2^2|X_1 = x_1) - [E(X_2|X_1 = x_1)]^2 = C_{2|1}^2(\alpha) h(x_1;\alpha).$$

Using symmetry we can write

$$\begin{aligned} h(x;\alpha) &= \sigma_1^{\alpha-2} \frac{\int_0^\infty \cos(tx) t^{\alpha-2} e^{-\sigma_1^\alpha t^\alpha} dt}{\int_0^\infty \cos(tx) e^{-\sigma_1^\alpha t^\alpha} dt} \\ &= \frac{\int_0^\infty \cos(ux/\sigma_1) u^{\alpha-2} e^{-u^\alpha} du}{\int_0^\infty \cos(ux/\sigma_1) e^{-u^\alpha} du} = S^2(x/\sigma_1;\alpha). \end{aligned}$$

Note that X_1/σ_1 has ch.f. $e^{-|t|^\alpha}$, hence p.d.f. $f(\cdot;\alpha)$. Thus the p.d.f. $f(\cdot)$ of X_1 is given by $f(x_1) = \sigma_1^{-1} f(x_1/\sigma_1;\alpha)$. Replacing this and $h(x;\alpha) = S^2(x/\sigma_1;\alpha)$ in the defining equation of $h(x;\alpha)$, the scaling σ_1 cancels to give

$$S^2(x;\alpha) f(x;\alpha) = \frac{1}{\pi} \int_0^\infty \cos(tx) t^{\alpha-2} e^{-t^\alpha} dt.$$

It follows that

$$\begin{aligned} [S^2(x;\alpha) f(x;\alpha)]' &= -\frac{1}{\pi} \int_0^\infty \sin(tx) t^{\alpha-1} e^{-t^\alpha} dt \\ &= \frac{1}{\alpha\pi} \int_0^\infty \sin(tx)(e^{-t^\alpha})' dt \\ &= -\frac{x}{\alpha\pi} \int_0^\infty \cos(tx) e^{-t^\alpha} dt \quad = \quad -\frac{1}{\alpha} x f(x;\alpha), \end{aligned}$$

and thus

$$S^2(x;\alpha)f(x;\alpha) = \frac{1}{\alpha}\int_{|x|}^{\infty} uf(u;\alpha)du.$$

This last expression will be used to derive the asymptotics of $S(\cdot;\alpha)$. Using the fact that $\lim_{x\to\infty}[x^{\alpha+1}f(x,\alpha)] = \frac{\alpha}{\pi}\Gamma(\alpha)\sin(\frac{\pi\alpha}{2}) = d$, we find

$$\begin{aligned}
\lim_{S\to\infty}\frac{S^2(x,\alpha)}{x^2} &= \lim_{x\to\infty}\frac{1}{\alpha x^2 f(x;\alpha)}\int_x^{\infty} uf(u;\alpha)du = \frac{1}{\alpha d}\lim_{x\to\infty}\frac{1}{x^{1-\alpha}}\int_x^{\infty} uf(u;\alpha)du \\
&= \frac{1}{\alpha d}\frac{1}{\alpha-1}\lim_{x\to\infty}[x^{\alpha+1}f(x,\alpha)] = \frac{1}{\alpha(\alpha-1)}.
\end{aligned}$$

We also have

$$\begin{aligned}
\lim_{x\to 0}\frac{S^2(x,\alpha)-S^2(0,\alpha)}{x^2} &= \lim_{x\to 0}\frac{[S^2(x,\alpha)]'}{2x} \\
&= \frac{1}{2\alpha}\lim_{x\to 0}\frac{1}{x}\{-x - \frac{f'(x;\alpha)}{f^2(x;\alpha)}\int_{|x|}^{\infty} uf(u;\alpha)du\} \\
&= \frac{1}{2\alpha}[-1 - \frac{f''(0;\alpha)}{f^2(0;\alpha)}\int_0^{\infty} uf(u;\alpha)du]
\end{aligned}$$

since $f(x;\alpha)$ is an even function, and thus $f'(0;\alpha) = 0$. But

$$\begin{aligned}
f(0;\alpha) &= \frac{1}{\pi\alpha}\Gamma(\frac{1}{\alpha}), \\
f''(0;\alpha) &= \frac{1}{\pi}\int_0^{\infty} t^2 e^{-t^\alpha}dt &&= -\frac{1}{\pi\alpha}\Gamma(\frac{3}{\alpha}), \\
\int_0^{\infty} uf(u;\alpha)du &= \frac{\alpha}{\pi}\int_0^{\infty} t^{\alpha-2}e^{-t^\alpha}dt &&= \frac{1}{\pi}\Gamma(1-\frac{1}{\alpha}).
\end{aligned}$$

It follows that

$$\lim_{x\to 0}\frac{S^2(x;\alpha)-S^2(0,\alpha)}{x^2} = \frac{1}{2\alpha}[-1 + \frac{\alpha\Gamma(\frac{3}{\alpha})\Gamma(1-\frac{1}{\alpha})}{\Gamma^2(\frac{1}{\alpha})}] = 2S(0;\alpha)A(\alpha),$$

and thus

$$\lim_{x\to 0}\frac{S(x;\alpha)-S(0,\alpha)}{x^2} = A(\alpha).$$

Finally

$$S^2(0;\alpha) = \frac{1}{f(0;\alpha)}\int_0^{\infty} uf(u;\alpha)du = \alpha\Gamma(1-\frac{1}{\alpha})/\Gamma(\frac{1}{\alpha}).$$

□

3. EXAMPLES

For various important classes of symmetric stable random variables and processes we examine here the dependence on the joint distribution of X_1 and X_2 of the normalized conditional standard deviation coefficient

$$r_{2|1}(\alpha) = C_{2|1}(\alpha)\sigma_2^{-1}.$$

3.1. Scale mixtures of Gaussian variables

Let $(X_1, X_2) = A^{\frac{1}{2}}(G_1, G_2)$, where A is positive, $(\alpha/2)$-stable, with Laplace transform $E\exp(-uA) = \exp(-u^{\alpha/2}), u \geq 0$, and independent of the jointly normally distributed G_1, G_2 with mean zero and correlation coefficient ρ. *This scale mixture of Gaussian variables produces jointly symmetric α-stable variables with finite conditional variance and*

$$r_{2|1}(\alpha) = [\alpha(1-\rho^2)]^{\frac{1}{2}} = r_{1|2}(\alpha).$$

In particular, let $X_t = A^{\frac{1}{2}}G_t, t \in T$, where A is positive, $(\alpha/2)$-stable as above, independent of the Gaussian process $\{G_t, t \in T\}$ with mean zero and covariance function $R(t_1, t_2)$. This scale mixture of a Gaussian process produces a symmetric α-stable process $X_t, t \in T$, with

$$r_{t_2|t_1}(\alpha) = [\alpha\{1 - R(t_1, t_2)\}]^{\frac{1}{2}} = r_{t_1|t_2}(\alpha).$$

Notice that for all $t_1, t_2 \in T$,

$$0 \leq r_{t_2|t_1}(\alpha) \leq (2\alpha)^{\frac{1}{2}}$$

with equality if and only if $[R(t_1, t_1)]^{-\frac{1}{2}}X_{t_1} = \pm[R(t_2, t_2)]^{-\frac{1}{2}}X_{t_2}$ (+ for the left equality and – for the right equality). Thus as a function of t_1, t_2 ($t_2 - t_1$ in the stationary case) the normalized conditional standard deviation coefficient $r_{t_2|t_1}(\alpha)$ is uniformly bounded, for every index set T.

Proof. It follows from the Gaussian scale mixture that

$$\phi(t_1, t_2) = \exp\{-[\frac{1}{2}(s_1^2 t_1^2 + 2s_1 s_2 \rho t_1 t_2 + s_2^2 t_2^2)]^{\alpha/2}\}$$

where s_i^2 is the variance of G_i. But also $\phi(t_1, t_2) = \int_{S_2} |t_1 y_1 + t_2 y_2|^\alpha d\Gamma$. Equating the two expressions of $\phi(t_1, 0)$ and $\phi(0, t_2)$ we find

$$\sigma_1 = \frac{s_1}{2^{\frac{1}{2}}}, \qquad \sigma_2 = \frac{s_2}{2^{\frac{1}{2}}}.$$

From

$$-\frac{\partial}{\partial t_2}\ell n\phi(t_1,t_2) = \alpha\int_{S_2}(t_1y_1+t_2y_2)^{<\alpha-1>}y_2d\Gamma$$
$$= \frac{1}{2^{\alpha/2}}\frac{\alpha}{2}(s_1^2t_1^2+2s_1s_2\rho t_1t_2+s_2^2t_2^2)^{\frac{\alpha}{2}-1}(2s_1s_2t_1+2s_2^2t_2)$$

we find, putting $t_2=0$,

$$\int_{S_2} y_1^{<\alpha-1>}y_2d\Gamma = \frac{\rho s_1^{\alpha-1}s_2}{2^{\frac{\alpha}{2}}}$$

Finally from

$$-\frac{\partial^2}{\partial t_2^2}\ell n\phi(t_1,t_2) = \alpha(\alpha-1)\int_{S_2}|t_1y_1+t_2y_2|^{\alpha-2}y_2^2d\Gamma$$
$$= \frac{1}{2^{\alpha/2}}\frac{\alpha}{2}\{(\frac{\alpha}{2}-1)(s_1^2t_1^2+2s_1s_2\rho t_1t_2+s_2^2t_2^2)^{\frac{\alpha}{2}-1}(2s_1s_2\rho t_1+2s_2^2t_2^2)^2$$
$$+ (s_1^2t_1^2+2s_1s_2\rho t_1t_2+s_2^2t_2^2)^{\frac{\alpha}{2}-1}2s_2^2\},$$

with $t_1\neq 0, t_2=0$, we find

$$\int_{S_2}|y_1|^{\alpha-2}y_2^2d\Gamma = [1-(2-\alpha)\rho^2]s_1^{\alpha-2}s_2^2/\{2^{\frac{\alpha}{2}}(\alpha-1)\}.$$

The finiteness of this expression shows that $\int|y_1|^{\alpha-2}y_2^2d\Gamma<\infty$ thus $Var(Y|X)<\infty$ a.s. (as shown also in Corollary 4.1 of Samorodnitsky and Taqqu [5]). Replacing in the expression of $C_{2|1}^2(\alpha)$ we finally obtain $r_{2|1}^2(\alpha)=\alpha(1-\rho^2)$. □

3.2. Moving averages

Consider the moving average process

$$X_t = \int_{-\infty}^{\infty} f(t-s)dM(s), -\infty<t<\infty,$$

where $\{M(s),\infty<s<\infty\}$ is symmetric α-stable motion (i.e., has stationary, independent, symmetric α-stable increments and Lebesgue control measure) and $f\in L_\alpha=L_\alpha(Leb)$. *If*

$$\int_{f(s)\neq 0} f^2(t_2-t_1+s))|f(s)|^{\alpha-2}ds<\infty$$

then $Var(X_{t_2}|X_{t_1}) < \infty$ *a.s. and*

$$r^2_{t_2|t_1}(\alpha) = \alpha(\alpha-1)\|f\|_\alpha^{-2}\{\|f\|^\alpha - \alpha\int_{f(s)\neq 0} f^2(t_2-t_1+s)|f(s)|^{\alpha-2}ds - [\int_{-\infty}^{\infty} f(t_2-t_1+s)f(s)^{<\alpha-1>}ds]^2\}.$$

If in addition f *is an even function then*

$$r_{t_2|t_1}(\alpha) = r_{t_1|t_2}(\alpha).$$

If furthermore f *is even and nonincreasing on the positive real line then the range of values of* $r_{t_2|t_1}(\alpha)$ *is* $[0,\infty)$, *since as is easily seen,*

$$r_{t_2|t_1}(\alpha) \to \infty \qquad as \qquad |t_2-t_1| \to \infty.$$

This is due to the asymptotic independence of X_{t_2}, X_{t_1} *as* $|t_1 - t_1| \to \infty$, *i.e.*

$$E\exp\{i(u_1X_{t_1}+u_2X_{t_2})\} - E\exp\{iu_1X_{t_1}\}E\exp\{iu_2X_{t_2}\} \to 0 \text{ as } |t_2-t_1| \to \infty$$

which is easily established. Thus for a moving average process, the normalized conditional standard deviation coefficient $r_{t_2|t_1}(\alpha)$ is unbounded over the entire real line. An example satisfying all the conditions mentioned above is the two-sided Ornstein-Uhlenbeck stable process $X_t = \int_{-\infty}^{\infty} e^{-\lambda|t-s|}dM(s)$, where $\lambda > 0$, for which we have, with $\tau = t_2 - t_1$,

$$r^2_{t_2|t_1}(\alpha) = \alpha(\alpha-1)(\frac{\lambda\alpha}{2})^{\frac{2}{\alpha}-2}\{\frac{2e^{\lambda\tau(2-\alpha)}}{4-\alpha} - \frac{e^{2\lambda\tau(1-\alpha)}}{(2-\alpha)^2} - \frac{2(\alpha-1)e^{-\lambda\tau\alpha}}{(2-\alpha)^2} + \frac{(4-3\alpha)e^{-2\lambda\tau}}{(4-\alpha)(2-\alpha)^2}\}$$

Proof. Putting $\tau = t_2 - t_1$, we have

$$\int_{S_2} |u_1y_1+u_2y_2|^\alpha d\Gamma(y_1,y_2) = -\ell n E\exp\{i(u_1X_{t_1}+u_2X_{t_2})\}$$

$$= \int_{-\infty}^{\infty} |u_1f(t_1-s)+u_2f(t_2-s)|^\alpha ds = \int_{-\infty}^{\infty} |u_1f(\tau+s)+u_2f(s)|^\alpha ds.$$

It follows that $\Gamma = \mu \circ T^{-1}$, where $d\mu(s) = [f^2(\tau+s)+f^2(s)]^{\alpha/2}ds$ and $T : \{s; f(\tau+s) \neq 0 \neq f(s)\} \to S_2$ is defined by $T(s) = (f(\tau+s), f(s))[f^2(\tau+s)+f^2(s)]^{-\frac{1}{2}}$ (see also Theorem

2.4.2 in Miller [4] and Theorem 4.1 in Samorodnitsky and Taqqu [5]). Then as is checked in the latter reference $\int_{S_2} |y_1|^{\alpha-2} y_2^2 d\Gamma < \infty$ is equivalent to $\int_{f(s)\neq 0} f^2(\tau+s)|f(s)|^{\alpha-2} ds < \infty$. Also

$$\begin{aligned}\sigma_{t_1} &= \sigma_{t_2} = \|f\|_\alpha, \\ \int_{S_2} y_1^{<\alpha-1>} y_2 d\Gamma(y_1, y_2) &= \int_{-\infty}^{\infty} f(\tau+s) f(s)^{<\alpha-1>} ds, \\ \int_{S_2} |y_1|^{\alpha-2} y_2^2 d\Gamma(y_1, y_2) &= \int_{f(s)\neq 0} f^2(\tau+s)|f(s)|^{\alpha-2} ds.\end{aligned}$$

The expression of $r^2_{t_2|t_1}(\alpha)$ then follows.

3.3 Harmonizable variables

Consider the harmonizable process

$$X_t = Re \int_{-\infty}^{\infty} e^{it\lambda} dZ(\lambda), \; -\infty < t < \infty,$$

where the complex process $\{Z(\lambda), -\infty < \lambda < \infty\}$ has independent, radially symmetric α-stable increments and finite control measure m. Then

$$\begin{aligned} Eexp\{i(u_1 X_{t_1} + u_2 X_{t_2})\} &= exp\{-\int_{-\infty}^{\infty} |u_1 e^{it_1\lambda} + u_2 e^{it_2\lambda}|^\alpha dm(\lambda)\} \\ &= exp\{-\int_{-\infty}^{\infty} |u_1^2 + 2\cos[(t_2-t_1)\lambda] u_1 u_2 + u_2^2|^{\alpha/2} dm(\lambda)\}, \end{aligned}$$

$Var(X_{t_2}|X_{t_1}) < \infty$ a.s. and

$$r^2_{t_2|t_1}(\alpha) = \alpha\{\frac{\alpha}{2} + (1-\frac{\alpha}{2}) r[2(t_2-t_1)] - (\alpha-1) r^2(t_2-t_1)\}$$

where $r(\tau) = \int_{-\infty}^{\infty} \cos(\tau\lambda) dm(\lambda)/m(\mathbf{R}^1)$. In this case $r_{t_2|t_1}(\alpha) = r_{t_1|t_2}(\alpha)$ and for all t_1, t_2,

$$0 \le r_{t_2|t_1}(\alpha) \le \alpha.$$

Proof. From the form of the joint ch.f. of X_{t_1}, X_{t_2} and $E\exp\{i(u_1 X_{t_1} + u_2 X_{t_2})\} = \exp\{-\int |u_1 y_1 + u_2 y_2|^\alpha d\Gamma(y_1, y_2)\}$, it follows that $\Gamma(\cdot) = \int_{-\infty}^{\infty} \Gamma_\lambda(\cdot) dm(\lambda)$ where Γ_λ is the spectral measure of scale mixture of Gaussian variables (Example 3.1) with variances 2 and correlation coefficient $\cos(\tau\lambda), \tau = t_2 - t_1$;

$$|u_1^2 + 2\cos(\tau\lambda)u_1u_2 + u_2^2|^{\alpha/2} = \int_{S_2} |u_1y_1 + u_2y_2|^2 d\Gamma_\lambda(y_1, y_2).$$

We thus have $\sigma_{t_2}^2 = m(\mathbf{R}^1) = \sigma_{t_1}^2$,

$$\begin{aligned}
\int_{S_2} y_1^{<\alpha-1>} y_2 d\Gamma(y_1, y_2) &= \int_{-\infty}^{\infty} \{\int_{S_2} y_1^{<\alpha-1>} y_2 d\Gamma_\lambda(y_1, y_2)\} dm(\lambda) \\
&= \int_{-\infty}^{\infty} \cos(\tau\lambda) dm(\lambda) = m(\mathbf{R}^1) r(\tau), \\
\int_{S_2} |y_1|^{\alpha-2} y_2^2 d\Gamma(y_1, y_2) &= \int_{-\infty}^{\infty} \{\int_{S_2} |y_1|^{\alpha-1} y_2^2 d\Gamma_\lambda(y_1, y_2)\} dm(\lambda) \\
&= \int_{-\infty}^{\infty} [1 - (2-\alpha)\cos^2(\tau\lambda)] dm(\lambda) = \frac{m(\mathbf{R}^1)}{2(\alpha-1)} [\alpha - (2-\alpha) r(2\tau)],
\end{aligned}$$

from which the results follows. □

REFERENCES

[1] R.J. Adler, S. Cambanis and G. Samorodnitsky "On stable Markov processes", *Stochastic Proc. Appl.* 34 (1990), 1-17.

[2] M. Kanter "Linear sample spaces and stable processes", *J. Funct. Anal.* 9 (1972), 441-459.

[3] E. Masry and S. Cambanis "Spectral density estimation for stationary stable processes", *Stochastic Proc. Appl.* 18 (1984), 1-31.

[4] G. Miller "Some results on symmetric stable distributions and processes", Institute of Statistics Mimeo Series No. 1121 (1977), Univ. of North Carolina.

[5] G. Samorodnitsky and M. Taqqu "Conditional moments of stable random variables", preprint (1989).

[6] S. Zabell "Continuous versions of regular conditional distributions", *Ann. Probab.* 7 (1979), 159-165.

Research supported by the Air Force Office of Scientific Research Contract No. F49620 85C 0144.

Department of Statistics
University of North Carolina
Chapel Hill, NC 27599-3260

Bounded Stationary Stable Processes and Entropy

JOHN P. NOLAN

We show that a bounded stationary stable process has a finite metric entropy integral. Necessary conditions for sample boundedness and sample path continuity are given.

Introduction

In this paper we will show that any stationary or stationary increment p-stable process, $1 \le p < 2$, that is sample bounded has a finite metric entropy integral. The result is an application of Talagrand's work on majorizing measures for stable processes [7]. We combine this result with earlier results to give necessary conditions for a stationary increment stable process to have a.s. bounded or a.s. continuous sample paths.

To state the theorem, let $X = \{X(t), t \in T\}$ be a real, continuous in probability, symmetric p-stable process, $1 \le p < 2$, where the index set T is a locally compact Abelian group with translation invariant metric τ. Let $d(t,s) = [-\log\{E\exp(i(X(t)-X(s)))\}]^{1/p}$ be the pseudo-metric associated with X. Since X is continuous in probability, d is continuous on $T \times T$. For any compact $K \subseteq T$ and any $\varepsilon > 0$, let $N(K,\varepsilon)$ = minimum number of d-balls with centers in K and radius ε needed to cover K. Let q be the dual index of p, i.e. $1/p + 1/q = 1$, and define the metric entropy:

$$(1) \qquad H_q(K,\varepsilon) = \begin{cases} (\log N(K,\varepsilon))^{1/q} & 2 \le q < \infty \\ \log^+ \log N(K,\varepsilon) & q = \infty \end{cases}$$

Theorem 1. *Let X be a p-stable process, $1 < p < 2$, with stationary increments and let $K \subseteq T$ be a compact set. If $\{X(t), t \in K\}$ has a.s. bounded sample paths, then* $\int_0^\infty H_q(K,\varepsilon)\, d\varepsilon < \infty$.

It should be noted that this necessary condition depends only on the "incremental standard deviation" $d(t,s)$. Another known condition, (C2) below, is stated in terms of a representation for the process: there is a measure space (U,m) and a collection $\{f(t,u),\ t\in T\} \subseteq L^p(U,m)$ such that $E\exp(i\sum_{j=1}^n a_j X(t_j)) = \exp(-||\sum_{j=1}^n a_j f(t_j,\cdot)||_{p,m}^p)$. It is known [3] that the stochastic integral $\int_U f(t,u)\, W_m(du)$ gives a version of X. We will not go into the details of this integral, only remark that the pair $\langle (U,m),\{f(t,\cdot),t\in T\}\rangle$ determines all finite dimensional distributions of X. In particular, $d(t,s) = ||f(t,\cdot)-f(s,\cdot)||_{p,m}$, so we can define the metric entropy in terms of the pair. We shall call such a pair a representation of X, m a control measure and $f(t,u)$ a kernel. We note that there are many possible representations for any X.

A kernel $f_0(t,u)$ is a modification of $f(t,u)$ if for all $t\in T$, $f(t,\cdot) = f_0(t,\cdot)$ m - a.e. Define the following three conditions:

(C1) f has a modification f_0 such that for every $u\in U$, $f_0(\cdot,u)$ is continuous on (T,τ).

(C2) For K a compact set, $f^*(u) = \sup_{t\in K} |f(t,u)|$ is in $L^p(U,m)$. (For measurability, the sup is taken over a countable separant of K).

(C3) For K a compact set, $\int_0^\infty H_q(K,\varepsilon)\, d\varepsilon < \infty$.

Combining Theorem 1 with Theorem 3 of [6], we have the following.

Corollary 2. *Let X be a p-stable process, $1\le p<2$, with stationary increments and let $K\subseteq T$ be a compact set.*
(i) If $\{X(t),t\in K\}$ has a.s. bounded sample paths, then (C2) holds for every representation of X and (C3) holds for the process.
(ii) If $\{X(t),t\in K\}$ has a.s. continuous sample paths, then (C1) and (C2) hold for every representation of X and (C3) holds for the process.

We will give a brief summary of results and conjectures from [6], then discuss some ideas for further study. For $0<p<2$, sample continuity of X is equivalent to (C1) and continuity at each point. When $0<p<1$, sample boundedness is equivalent to (C2) and sample path continuity is equivalent to (C1) and (C2), even when X has nonstationary increments. Conjecture 1 is that (C1), (C2) and (C3) are sufficient for sample path continuity when $p\ge 1$. If this is true, then that combined with the Theorem 1 would prove Conjecture 2: a stationary p-stable process with $1\le p<2$ is sample continuous if and only if (C1), (C2) and (C3) hold. In light of these ideas, it seems natural to put forth another conjecture. If true, it would imply Conjecture 2 because a stationary stable process satisfying (C1) is sample continuous when and only when it is sample bounded [1].

Conjecture 3. *A stationary p-stable process, $1\le p<2$, is sample bounded on compact sets if and only if (C2) and (C3) hold.*

The main idea of [6] was to characterize path regularity of a stable process through its representations. If one is studying a particular class of stable processes, e.g. harmonizable, moving average, etc., then this is a natural approach. In these cases, the kernel and the control measure are the objects that define the process. For example in Marcus and Pisier [4], the continuity problem was solved for harmonizable p-stable processes, $1<p<2$. There the kernel is specified to be a character of T, so (C1) and (C2) hold, and the characterization is in terms of the metric entropy integral (C3).

However, for an abstract stable process, such a characterization seems less useful because we have no idea what possible representations are. In fact there is a rigidity in possible representations of path continuous stable processes, but it is not clear how useful this is for an abstract process. Unlike the Gaussian case, metric entropy and majorizing measures are not sufficient tools for analyzing regularity for the stable case. From Cambanis and Soltani [2], we know that the classes of harmonizable, moving average and subgaussian stable processes are disjoint. At this point, we seem to need to treat these cases separately in studying different properties of X. For example, Remark 1.7 [4] shows that the metric entropy approach to regularity, which works for harmonizable processes, fails for subgaussian processes.

It does seem worthwhile to further study the question of the rigidity of representations. In particular, if X is stationary, it is known [3] that the kernel is given by the orbit of a group of operators on L^p: $f(t,\cdot) = T_t f(0,\cdot)$. If the process is continuous, then the group of operators can be taken to be continuous contractions on L^p, and each of the sections $f(\cdot,u)$ can be taken to be continuous. This kind of structure should be a useful tool in studying the process.

Proof of Theorem 1

It suffices to prove the result when K is a compact symmetric neighborhood of the origin. For if K′ is any compact symmetric neighborhood of the origin, let n=number of translates of K′ needed to cover K and a straightforward covering argument shows $N(K,\varepsilon)\le nN(K',\varepsilon)$. Hence the finiteness of the entropy integral for K′ will show the finiteness of the entropy integral for K. For the rest of the proof, we assume K is a compact symmetric neighborhood of the origin.

Define $D=\sup\{d(t,s):s,t\in K\}$ and the d-balls $B(t,\varepsilon)=\{s\in T:d(t,s)<\varepsilon\}$. Since X has stationary increments, $d(t,s)$ is translation invariant and $B(t,\varepsilon) = t+B(0,\varepsilon)$. Let μ be Haar measure on T, normalized so that $\mu(K) = 1$. For $t>0$, define

$$h_q(t) = \begin{cases} [\log(1/t)]^{1/q}\, 1_{(0,1]}(t) & 2\le q<\infty \\ \log^+ \log(1/t)\, 1_{(0,1]}(t) & q=\infty. \end{cases}$$

Step 1. Let $K+K=\{s+t:s,t\in K\}$, then

$$\int_0^\infty h_q(\mu(B(0,\varepsilon)\cap(K+K))\, d\varepsilon = \int_0^D h_q(\mu(B(0,\varepsilon)\cap(K+K))\, d\varepsilon < \infty. \tag{2}$$

Since the balls $B(t,\varepsilon)$ are translates of $B(0,\varepsilon)$, for $t\in K$ we have

$$t+(B(0,\varepsilon)\cap(K+K)) \supseteq B(t,\varepsilon)\cap K. \tag{3}$$

For a sample bounded process, Theorem A of Talagrand [7] shows there is a discrete majorizing measure m on (K,d) and a finite constant $M=M(p,X)$ such that for all $t\in K$, $\int_0^\infty h_q(\sup\{\ m\{s\} : d(s,t) \le \varepsilon\ \})\, d\varepsilon \le M$. Since h_q is nonincreasing, this implies that for all $t\in K$,

$$\int_0^\infty h_q(m(B(t,\varepsilon)))\, d\varepsilon \le M. \tag{4}$$

Now m is a probability measure and μ is Haar measure, so

$$\begin{aligned}\mu(B(0,\varepsilon)\cap(K+K)) &= \int_T m(t+[B(0,\varepsilon)\cap(K+K)])\ \mu(dt)\\ &\ge \int_K m(t+[B(0,\varepsilon)\cap(K+K)])\ \mu(dt).\end{aligned}$$

Using (3) and the fact that m is supported on K,

$$\mu(B(0,\varepsilon)\cap(K+K)) \ge \int_K m(B(t,\varepsilon)\cap K)\ \mu(dt) = \int_K m(B(t,\varepsilon))\ \mu(dt). \tag{5}$$

h_q is decreasing and convex on $(0,\exp(-1/p))$, so there is a decreasing convex function g on $(0,\infty)$ such that $h_q(u)=g(u)$ for $u\in(0,\alpha]$, where $\alpha := \exp(-1/p)/2$. Let δ be small enough so that $\varepsilon\le\delta$ implies $\mu(B(0,\varepsilon)\cap(K+K))\le\alpha$. Then for any $\varepsilon\le\delta$, (5) and Jensen's inequality gives

$$\begin{aligned}h_q(\mu(B(0,\varepsilon)\cap(K+K))) &= g(\mu(B(0,\varepsilon)\cap(K+K)))\\ &\le g\left(\int_K m(B(t,\varepsilon))\ \mu(dt)\ \right)\\ &\le \int_K g(m(B(t,\varepsilon)))\ \mu(dt).\end{aligned}$$

Split $K = K_1 \cup K_2$, where $K_1 := \{t\in K: m(B(t,\varepsilon))\le\alpha\}$ and $K_2 := K - K_1$. Then

$$\begin{aligned}h_q(\mu(B(0,\varepsilon)\cap(K+K))) &\le \int_{K_1} h_q(m(B(t,\varepsilon)))\ \mu(dt) + \int_{K_2} g(\alpha)\ \mu(dt)\\ &\le \int_K h_q(m(B(t,\varepsilon)))\ \mu(dt) + h_q(\alpha)\ \mu(K_2)\end{aligned}$$

Integrate this for $0<\varepsilon\le D$ and use (4) to get (2).

Step 2. $\int_0^\infty H_q(K,\varepsilon)\, d\varepsilon = \int_0^D h_q(N(K,\varepsilon)^{-1})\, d\varepsilon < \infty.$

Covering arguments, e.g. Lemmas II.1.1 and II.1.3 of [5], show that $\mu(B(0,\varepsilon)\cap(K+K))\le c(K)N(K,4\varepsilon)^{-1}$. Combining this inequality with Step 1, we see that

$$\int_0^D h_q(\ N(K,\varepsilon)^{-1})\ d\varepsilon = 4 \int_0^{D/4} h_q(\ N(K,4\varepsilon)^{-1})\ d\varepsilon$$

$$\leq 4 \int_0^{D/4} h_q(\mu(B(0,\varepsilon)\cap(K+K))/c(K))\, d\varepsilon < \infty.$$

This completes the proof of Theorem 1. Q.E.D.

REFERENCES

[1] Cambanis, S., Nolan, J. P. and Rosinski, J. "On the oscillation of infinitely divisible and some other processes." To appear in Stochastic Processes and Their Applications (1990).

[2] Cambanis, S. and Soltani, A. R. "Prediction of stable processes: spectral and moving average representations." Z. Wahrsch. verw. Gebiete 66, 593-612 (1983).

[3] Hardin, C. D. "On the spectral representation of symmetric stable processes." J. Multivar. Analysis 12,385-401 (1982).

[4] Marcus, M. B. and Pisier, G. "Characterizations of almost surely continuous p-stable random Fourier series and strongly stationary processes." Acta Math. 152, 245-301 (1984).

[5] Marcus, M. B. and Pisier, G. Random Fourier Series with Applications to Harmonic Analysis, Princeton University Press, Princeton, NJ (1981).

[6] Nolan, J. P. "Continuity of symmetric stable processes." J. Multivar. Analysis 29, 84-93 (1989).

[7] Talagrand, M. "Necessary conditions for sample boundedness of p-stable processes." Annals of Probability 16, 1584-1595 (1988).

John P. Nolan
Department of Mathematics and Statistics
American University
Washington, DC

Alternative multivariate stable distributions and their applications to financial modeling*

STEFAN MITTNIK SVETLOZAR T. RACHEV

Introduction

It is commonly accepted that the distribution of returns on many financial assets is nonnormal. Mandelbrot [5] and Fama [2] proposed the α-stable distribution for modeling stock returns. In [9] we find that the geometric summation scheme provides a better model for univariate stock index data than various stable alternatives, including the α-stable model. Here we extend the geometric summation model to multivariate settings which allows us to model portfolios of financial assets.

Univariate Stable Models

Summarizing some of the results in [9], we report in [8] that, among all the stable probabilistic schemes listed in Table 1, the geometric summation scheme provides the best model for describing the stability properties of stock returns computed from the Standard and Poor's (S&P) 500 index.

Letting X_i denote the return on an asset in period i, we assume that $X_1, X_2, \ldots$ are independently and identically distributed (ii) real-valued random variables (r.v.'s). To categorize the probabilistic schemes under consideration, we write

$$X_1 \stackrel{d}{=} a_M(X_1 \circ X_2 \circ \cdots \circ X_M) + b_M, \tag{1}$$

where $\stackrel{d}{=}$ denotes equality in distribution; $\circ$ stands for summation, min, max, or multiplication; and M is a deterministic or random integer. The standard summation scheme produces the α-stable distribution. The maximum and minimum schemes lead to extreme-value distributions. For the minimum scheme, the Weibull distribution is one of these extreme-value distributions. The multiplication scheme yields the multiplication-stable

*We thank S. Resnick and G. Samorodnitsky for helpful discussions.

Scheme	Stability Property
Summation	$a_n(X_1 + \cdots + X_n) + b_n \stackrel{d}{=} X_1$
Maximum	$a_n \max_{1\le i\le n} X_i + b_n \stackrel{d}{=} X_1$
Minimum	$a_n \min_{1\le i\le n} X_i + b_n \stackrel{d}{=} X_1$
Multiplication	$A_n(X_1 \cdots X_n)^{C_n} \stackrel{d}{=} X_1$
Geometric Summation	$a(p)(X_1 + \cdots + X_{T(p)}) + b(p) \stackrel{d}{=} X_1$
Geometric Maximum	$a(p) \max_{1\le i\le T(p)} X_i + b(p) \stackrel{d}{=} X_1$
Geometric Minimum	$a(p) \min_{1\le i\le T(p)} X_i + b(p) \stackrel{d}{=} X_1$
Geometric Multiplication	$A(p)(X_1 \cdots X_{T(p)})^{C(p)} \stackrel{d}{=} X_1$

Table 1: Stable Probabilistic Schemes

distribution of which the log-normal is the basic example. If M is random, we set $M = T(p)$, where $T(p)$ is specified as a geometrically distributed r.v. with parameter $p \in (0,1)$ independent of X_i,

$$Pr(T(p) = k) = (1-p)^{k-1}p, \; k = 1, 2, \ldots . \tag{2}$$

Here $T(p)$ represents the moment at which the probabilistic structure governing the asset returns breaks down. The break down could, for example, be due to new information affecting the fundamentals of the underlying asset. Such news could, for example, affect future profits of a company in a presently unknown manner. Thus, the stability properties of X_i are only preserved up to period $T(p)$, the moment of the break down. The Weibull distribution arises as the limit distribution of the geometric summation scheme; the geometric maximum, minimum and multiplication schemes produce the (geometric,max)-, the (geometric,min)- and the (geometric,multiplication)-stable distributions, respectively.

In [9] we examine the goodness-of-fit of the alternative stable distributions, by fitting them to daily returns on the S&P 500 index (measured in first differences of the natural logarithm of the level of the index). The Kolmogorov distance, $\rho = sup|F_s(x) - \hat{F}(x)|$ (multiplied by 100), and its normalization $N^{1/2}$ was used to measure the goodness of fit. Here, $F_s(x)$ denotes the sample distribution, $\hat{F}(x)$ is the estimated distribution function, and N is the sample size. Table 2 reports the fit of the distributions for 1, 2, 3, 4, and 5 year samples. (All samples end on December 31, 1986.)

The results show that among the distributions considered, the Weibull distribution, generated by the min-stable as well as the geometric summation scheme, provides the best model. In all 5 sample periods the Weibull distribution fits substantially better than any of the other models, including

Distribution	Sample (years)[a]				
	1	2	3	4	5
α-stable	5.00	6.68	8.32	6.09	7.47
	0.80	1.50	2.29	1.94	2.66
max-stable	5.73	5.17	5.27	4.86	4.71
	0.91	1.16	1.45	1.55	1.67
min- and *(geo,sum)*-	2.27	2.32	1.62	1.07	1.07
stable (Weibull)	0.36	0.52	0.45	0.34	0.38
mult.-stable	3.66	3.16	4.52	4.45	4.28
(Log-normal)	0.58	0.71	1.25	1.42	1.52
(geo,sum)-stable	8.27	8.02	7.44	7.61	7.48
(Laplace)	1.34	1.80	2.05	2.42	2.66
(geo,max)- and	3.91	3.38	2.88	2.85	3.07
(geo,min)-stable	0.62	0.76	0.79	0.91	1.09
(geo,mult.)-stable	8.27	7.59	7.91	7.60	8.30
	1.32	1.71	2.18	2.42	2.95

[a]The entries in the first of each case represent the Kolmogorov distance, ρ multiplied by 100. The entries in the second row represent $N^{1/2}\rho$, where N denotes the sample size.

Table 2: Comparison of Fit

the α-stable distribution. When comparing Weibull and α-stable distributions over sample periods of up to 20 years, the Weibull distribution fits consistently better (see [9]). In addition to its superior fit, we find that for the Weibull distribution the normalized Kolmogorov distances remain fairly constant as N varies.

As discussed in [9], in addition to its superior fit, the Weibull distribution possesses a number of interesting properties, which we only mention here:

1. While permitting fatter than normal tails, the Weibull distribution has exponential tails. Consequently, and in contrast to α-stable distribution, all moments exist. This is an attractive implication of our findings, since the existence of variances is crucial for a variety of economic/financial theories as well as econometric techniques.

2. In contrast to the α-stable distribution, estimating the parameters of the Weibull distribution via maximum likelihood methods or regression-type estimators is straightforward. Also, given the existence of closed-form expressions, the construction of estimated Weibull distribution functions presents no problem.

3. When fitting daily and monthly data over the same sampling periods,

the Weibull distribution exhibits invariance of the shape parameter with respect to the sampling interval, while the α-stable distribution violates this stability property.

4. Simulation indicate that when fitting the α-stable distribution to data drawn from a Weibull distribution, the estimated characteristic exponents of the α-stable distribution increase when the sampling interval is lengthened. The increases are similar to the ones observed empirically when fitting the S&P 500 data.

5. In reliability theory the Weibull distribution with shape parameter α is known to be a distribution with *increasing failure rate* if $\alpha \geq 1$ and a distribution with *decreasing failure rate* if $\alpha \leq 1$. In the context of modeling asset returns, one could characterize a Weibull distribution as a distribution with *reversion tendency* when $\alpha \geq 1$ and as a distribution with *diversion tendency* when $\alpha \leq 1$. Fitting the Weibull distribution to various samples of the S&P 500 data, we find that, except for one case, all estimated shape parameters are greater than unity. The exception is the sample covering 1986, where the shape parameter for (the absolute values of) the negative returns is 0.919, indicating that increasing negative stock price movements are likely. Whether this may be interpreted as an omen for the stock market crash in the following year remains to be investigated.

Multivariate Geometric $(\alpha, +)$-stable Distributions

As in the univariate case it is assumed that the probabilistic structure of the asset returns holds up to the geometrically distributed random moment $T(p)$ specified in Equation 2. Let the vector of returns during period i of d dependent assets be denoted by $X^{(i)} = (X_1^{(i)}, \ldots, X_d^{(i)})$ and assume $\{X^{(i)}\}_{i \geq 1}$ to be a sequence of iid random vectors. Then, the geometric sum

$$G_p = \sum_{i=1}^{T(p)} X^{(i)} \tag{3}$$

describes the vector $(G_{p,1}, \ldots, G_{p,d})$ of total returns up to $T(p)$

$$G_{p,k} = \sum_{i=1}^{T(p)} X_k^{(i)}, \; k = 1, 2, \ldots, d \tag{4}$$

for the d assets. Note that even in the case of independent assets, i.e., when $\{X_j^{(i)}\}_{j=1,\ldots,d}$ are independent, the total returns up $T(p)$ will be dependent variables. Typically, the assets in vector X are from the same "sector," implying that they are subjected to the same market forces (common factors), such as macroeconomic conditions or speculative bubbles. It is, therefore, appropriate to impose the same p for each asset. Any other choice is difficult to justify.

We assume that the joint distribution function H of $X^{(i)}$ is *geometric (α)-stable*; or, more general, H belongs to the domain of attraction of geometric (α)-stable distributions.

Definition 1 *The m-dimensional random vector Y with d.f. G is said to be multivariate geometric stable with respect to the summation scheme (in short, (m-geo,+)-stable), if there exists a sequence of iid random vectors $X^{(1)}$, $X^{(2)}$, ..., a geometric r.v. $T(p)$, and constants $a = a(p) > 0$ and $b = b(p) \in \mathbf{R}^d$ such that*

$$a(p) \sum_{i=1}^{T(p)} (X^{(i)} + b(p)) \xrightarrow{d} Y, \text{ as } p \to 0. \tag{5}$$

Proposition 1 (Explicit representation for the characteristic function of (m-geo,+)-stable distributions) *A nondegenerate multivariate d.f. is (m-geo,+)-stable if and only if its ch.f. f_g has the form*

$$f_g(\theta) = \frac{1}{1 - ln\phi(\theta)}, \; \theta \in \mathbf{R}^d, \tag{6}$$

where $\phi(\theta)$ is a ch.f. of some α-stable multivariate distribution, i.e., there exist $\alpha \in (0,2]$, $\mu \in \mathbf{R}^d$ and a spectral measure Γ on the sphere S^d such that

$$f_g(\theta) = \left(1 + \int_{S^d} |(\theta, s)|^\alpha (1 - i\,sign((\theta, s))tan\frac{\pi\alpha}{2})\Gamma(ds) + i(\theta, \mu)\right)^{-1}, \tag{7}$$

if $\alpha \neq 1$, or

$$f_g(\theta) = \left(1 + \int_{S^d} |(\theta, s)|(1 + i\frac{2}{\pi}\,sign((\theta, s))ln|(\theta, s)|\Gamma(ds) + i(\theta, \mu)\right)^{-1}, \tag{8}$$

if $\alpha = 1$.

Proof. Suppose g is the ch.f. of an (m-geo,+)-stable distribution; then (5) holds if and only if

$$\frac{p\, f(a\theta;b)}{1-q\, f(a\theta;b)} \xrightarrow{p\to 0} g(\theta), \quad \text{for all } \theta \in \mathbf{R}^d, \tag{9}$$

where $q := 1-p$ and $f(a\theta;b) = f(a\theta)e^{i(\theta,ab)}$ is the ch.f. of the $X^{(i)}$'s in (5). Relationship (9) is equivalent to

$$\frac{1-q\, f(a\theta;b)}{p\, f(a\theta;b)} \xrightarrow{p\to 0} \frac{1}{g(\theta)} \tag{10}$$

or

$$-\frac{1-f(a\theta;b)}{pf(a\theta;b)} \xrightarrow{p\to 0} \left(1-\frac{1}{g(\theta)}\right), \tag{11}$$

$$\left|\frac{1-f(a\theta;b)}{f(a\theta;b)}\right| \xrightarrow{p\to 0} 0, \tag{12}$$

and thus $f(a(p)\theta;b(p)) \to 1$ as $p \to 0$. Combining this with (11), we obtain

$$-\frac{1}{p}(1-f(a\theta;b)) \xrightarrow{p\to 0} 1-\frac{1}{g(\theta)} \tag{13}$$

and

$$\frac{1}{p} log\, f(a\theta;b) \xrightarrow{p\to 0} 1-\frac{1}{g(\theta)} \tag{14}$$

or

$$f^{1/p}(a(p)\theta;b(p)) \xrightarrow{p\to 0} exp\{1-\frac{1}{g(\theta)}\}. \tag{15}$$

Thus, $\phi(\theta) = exp\{1-1/g(\theta)\}$ is a ch.f. of the (α,+)-stable distribution and, therefore, (6) holds. Using (6), we get (7) and (8). Conversely, if (6) holds, then $\phi(\theta) = exp\{1-1/g(\theta)\}$, where $g(\theta) = f_G(\theta)$ is a ch.f. of a multivariate α-stable distribution. Going backward from (15) to (9) we conclude that (5) holds. Q.E.D.

Example 1 (Bivariate Marshall-Olkin distribution) In the bivariate case ($d = 2$) the exponential Marshall-Olkin distribution has the form

$$G(x_1,x_2) = Pr(X_1 \ge x_1, X_2 \ge x_2) = exp\{-\lambda_1 x_1 - \lambda_2 x_2 - \lambda_{12} max(x_1,x_2\},$$

for $x_1, x_2 \ge 0$ (see [1], [4], [6], and [7]). The "survival" function $G(x_1,x_2)$ can also be written as

$$G(x_1,x_2) = \begin{cases} exp\{-\delta x_2 - \theta_1(x_1-x_2)\}, & \text{for } x_1 \ge x_2 \\ exp\{-\delta x_1 - \theta_2(x_2-x_1)\}, & \text{for } x_1 \le x_2, \end{cases}$$

where $\lambda_1 = \delta - \theta_2$, $\lambda_2 = \delta - \theta_1$ and $\lambda_{12} = \theta_1 + \theta_2 - \delta$. The Marshall-Olkin distribution is completely determined by its values on the axes $x_1 = 0$, $x_2 = 0$ and the diagonal $x_1 = x_2$. The singular part of the distribution is concentrated on the diagonal. Hence, assuming for the sake of convenience that $\lambda_1 = \lambda_2 = 0$, the ch.f. of G has the form

$$\begin{aligned} f_G(t_1, t_2) &= \int_{R^2_+} exp\{i(t_1x_1 + t_2x_2)\}\, d\, exp\{-\lambda_{12} max(x_1, x_2)\} \\ &= \int_0^\infty exp\{i(t_1 + t_2)x\}\, \lambda_{12}\, exp\{-\lambda_{12}x\}\, dx \\ &= \frac{\lambda_{12}}{\lambda_{12} - i(t_1 + t_2)}, \end{aligned}$$

which, by Proposition 1, shows that G is a bivariate (geo,+)-stable distribution that corresponds to the discrete distribution in the set of bivariate α-stable distributions. The corresponding "normal" distribution will be the *bivariate symmetric* Marshall-Olkin distribution concentrated on the line $X_2 = kX_1$, where it has an exponential density, i.e., $X_2 = kX_1$ *a.s.*, where X_1 has a Laplace distribution.

The Marshall-Olkin distribution has a slight generalization (see [7] and [10]), namely the *bivariate Weibull-Marshall-Olkin distribution*

$$\begin{aligned} G_{(\overline{\lambda}, \overline{\alpha}, k)}(x_1, x_2) &= Pr(X_1 \geq x_1, X_2 \geq x_2) \\ &= exp\{-\lambda_1 x_1^{\alpha_1} - \lambda_2 x_2^{\alpha_2} - \lambda_{12} max(x_1^{\alpha_1}, k x_2^{\alpha_2})\}, \end{aligned}$$

for all $x_1, x_2 \geq 0$, where $\alpha = (\alpha_1, \alpha_2) \in \mathbf{R}^2_+$, $\lambda = (\lambda_1, \lambda_2, \lambda_{12}) \in \mathbf{R}^3_+$ and k is a nonnegative constant.

Example 2 (Multivariate Marshall-Olkin distribution) In the general d-dimensional case any vector $\mathbf{X} = (k_1X, k_2X, \ldots, k_dX)$, with X being exponentially distributed, has an (m-geo,+)-distributions which corresponds to the discrete random variable for the usual stable case; and, if X is Laplace distributed, $\mathbf{X}$ has an (m-geo,+)-distribution which is related to the multivariate normal distribution for the stable scheme of iid random vectors. The multivariate Marshal-Olkin distribution is of the form

$$G(X_1, \ldots, X_d) = exp\{ -\sum_{i=1}^{d} \lambda_i x_i$$

$$
\begin{aligned}
&-\sum_{i<j} \lambda_{ij} max(x_i, x_j) \\
&- \sum_{i<j<k} \lambda_{ijk} max(x_i, x_j, x_k) \\
&\vdots \\
&-\lambda_{12\ldots d} max(x_1, \ldots, x_d)\}
\end{aligned}
$$

for all $X_i \geq 0$, with the λ's being nonnegative constants. Clearly,

$$\overline{G}(X_1, \ldots, X_d) = exp\{-\lambda_{12\ldots d} max(x_1, k_2 x_2 \ldots, k_d x_d)\}$$

is an (m-geo,+)-stable distribution. The d-dimensional Weibull-Marshall-Olkin distribution $Pr_{\overline{X}}\left(\overline{X} = (\overline{X}_1, \ldots, \overline{X}_d)\right)$ is defined by

$$\overline{X}_i = X_i^{1/\alpha_i}, \quad \alpha_i > 0, \quad i = 1, \ldots, d,$$

where $X = (X_1, \ldots, X_d)$ is a multivariate Marshall-Olkin distributed random vector.

Proposition 2 (Domain of attraction of (m-geo,+)-stable distributions) *A multivariate d.f. H belongs to the domain of attraction of the (m-geo,+)-stable distribution G, that is, (5) holds with $H(x) = Pr(Y^{(i)} \leq x), x \in \mathbf{R}^d$, if and only if H belongs to the domain of attraction of certain α-stable distributions.*

The proof follows directly from the proof of Proposition 1.

If in (5) $a(p) = p^{1/\alpha}$ and $b(p) = 0$, we obtain the following version of the univariate Rényi Theorem.

Proposition 3 *Let $\{X^{(i)}\}_{i\geq 1}$ denote the iid random vectors $X^{(i)} = (X_1^{(i)}, \ldots, X_d^{(i)})$ with $\mathbf{E}|X_j^{(i)}|^\alpha < \infty$, for some $\alpha > 0$ and all $j = 1, \ldots, d$. Then,*

$$Z_{p,\alpha} = p^{1/\alpha} \sum_{j=1}^{T(p)} X^{(i)} \xrightarrow{d} W, \tag{16}$$

where W is a symmetric (m-geo,+)-stable distributed random vector, i.e.,

$$W \stackrel{d}{=} p^{1/\alpha} \sum_{j=1}^{T(p)} W^{(i)}, \tag{17}$$

with the $W^{(i)}$'s being iid copies of W.

In general, the derivation of the d.f. of W from (6), (7) or the ch.f. of W,

$$f_w(\theta) = \frac{p f_w(p^{1/\alpha}\theta)}{1-(1-p)f_w(p^{1/\alpha}\theta)}, \quad \theta \in \mathbf{R}^d, \tag{18}$$

is a nontrivial task. For this reason, we introduce an approximation for the distribution of W based on the limit relation (16).

Proposition 4 *Let π be the Prokhorov distance in the space of probability distributions:*

$$\begin{aligned}\pi(X,Y) &= \pi(Pr_X, Pr_Y)\\ &= inf\{\epsilon>0: Pr(X\in A) \le Pr(Y\in A^\epsilon)+\epsilon \ \ \forall\, A\in\mathcal{B}(\mathbf{R}^d)\},\end{aligned} \tag{19}$$

where A^ϵ is the ϵ-neighborhood of A and $A^\epsilon := \{x: \| x - A \| \le \epsilon\}$, with $\|\cdot\|$ denoting the Euclidean norm in $\mathbf{R}^d$. Let $0 < \alpha < 2$. Then,

(a) if $0 < \alpha < 1$, for any $r \in (\alpha, 1)$,

$$\pi(Z_{p,\alpha}, W) \le \left(p^{1-r/\alpha} C_{r,\alpha} \frac{\Gamma(1+p)}{\Gamma(1+r)} \left(\mathbf{E} \| X^{(1)} \|^r + \mathbf{E} \| W \|^r\right)\right)^{1/(1+r)}, \tag{20}$$

with $C_{r,\alpha}$ being an absolute constant;

(b) if, $1 \le \alpha < 2$, relationship (20) holds for any $r \in (\alpha, 2)$, provided that $\mathbf{E}X_j^{(1)} = \mathbf{E}W_j$ for all $j = 1, \ldots, d$.

Proof. First we establish some facts pertaining to the theory of probability metrics. Let $\mathcal{X}^d = \mathcal{X}(\mathbf{R}^d)$ be the space of all random vectors $\mathbf{X} = (X_1, \ldots, X_d)$ given on the probability space $(\Omega, \mathcal{A}, Pr)$ and $\mathcal{L}^d$ be the space of their probability distributions $Pr_X(A) = Pr(X \in A)$, $A \in \mathcal{B}(\mathbf{R}^d)$, the Borel σ-algebra on $\mathbf{R}^d$. A metric $\mu(X,Y) = \mu(Pr_X, Pr_Y)$ in $\mathcal{L}^d$ is said to be of *ideal order* r, if, for any random vectors X' and X'' independent of the random vector Z and any non-zero constant c, the following two properties are satisfied (see [11]):

(a) *regularity:* $\mu(X'+Z, X''+Z) \le \mu(X', X'')$,

(b) *homogeneity of order r:* $\mu(cX', cX'') = |c|^r \mu(X', X'')$.

Zolotarev [11] showed the existence of an ideal metric of a given order $r \ge 0$ and defined the ideal metric

$$\zeta_r(X',X'') := \sup\left\{|\mathbf{E}(f(X')-f(X''))| : |f^{(m)}(x)-f^{(m)}(y)| \le \| x-y\|^{\beta}\right\}, \tag{21}$$

where $m = 1,2,3$ and $\beta \in (0,1]$ satisfy $m+\beta = r$, and $f^{(m)}$ denotes the mth Fréchet derivative of f for $m \ge 0$ and $f^{(0)}(x) = f(x)$. An upper bound for ζ_r in terms of the rth absolute moment is also obtained in [11]. If

$$\mathbf{E} \parallel X' \parallel^r + \mathbf{E} \parallel X'' \parallel^r < \infty,$$

$$\mathbf{E}\left((X_1')^{\alpha_1}\dots(X_d')^{\alpha_d}\right) = \mathbf{E}\left((X_1'')^{\alpha_1}\dots(X_d'')^{\alpha_d}\right), \tag{22}$$

for all $\alpha_i = 0,1,\dots,$ and $\alpha_1+\dots+\alpha_d = m$, then

$$\zeta_r(X',X'') \le \frac{\Gamma(1+\beta)}{\Gamma(1+r)}\left(\mathbf{E} \parallel X' \parallel^r + \mathbf{E} \parallel X'' \parallel^r\right) \le \infty. \tag{23}$$

The ζ_r-metric admits the following bound from below:

$$\pi^{1+r}(X',X'') \le C_{r,d}\zeta_r(X',X''), \tag{24}$$

where $C_{r,d} = C_r d^{r/2}$ and C_r is an absolute constant depending only on r (see [12]).

Since ζ_r is an ideal metric, we have (see (21))

$$\begin{aligned}
\zeta_r\left(p^{1/\alpha}\sum_{i=1}^{T(p)} X^{(i)}, Y\right) &= \zeta_r\left(p^{1/\alpha}\sum_{i=1}^{T(p)} X^{(i)}, p^{1/\alpha}\sum_{i=1}^{T(p)} Y^{(i)}\right) \\
&\le p^{r/\alpha}\zeta_r\left(\sum_{i=1}^{T(p)} X^{(i)}, \sum_{i=1}^{T(p)} Y^{(i)}\right) \\
&\le p^{r/\alpha}\sum_{j=1}^{\infty} P(T(p)=j)\zeta_r\left(\sum_{i=1}^{j} X^{(i)}, \sum_{i=1}^{j} Y^{(i)}\right) \\
&\le p^{r/\alpha}\sum_{j=1}^{\infty} P(T(p)=j)\sum_{i=1}^{j}\zeta_r(X^{(i)},Y^{(i)}) \\
&= p^{r/\alpha-1}\zeta_r(X^{(1)},Y),
\end{aligned} \tag{25}$$

where $\{Y^{(i)}\}_{i\ge 1}$ are iid copies of Y. Using the upper and lower bounds (see (23) and (24)) for ζ_r, it follows from (25) that

$$\pi\left(p^{1/\alpha}\sum_{i=1}^{T(p)} X^{(i)}, Y\right) \leq \left(p^{r/\alpha-1}C_{r,d}\zeta_r(X^{(1)}, Y)\right)^{\frac{1}{r+1}}$$

$$\leq\left(C_{r,d}\frac{\Gamma(1+\beta)}{\Gamma(1+r)}(\mathbf{E}\|X^{(1)}\|^2+\mathbf{E}\|Y\|^2)p^{\frac{r}{\alpha}-1}\right)^{\frac{1}{r+1}} \quad (26)$$

provided that (22) holds. Q.E.D.

An Application to Stock Return Data

For illustrative purposes, we fitted the bivariate normal (BN), the bivariate exponential Marshall-Olkin (BEMO) and the bivariate Weibull-Marshall-Olkin (BWMO) distributions to daily stock return data of Chrysler and Ford covering the two-year period 1987-1988. The returns are measured by $r_t = lnp_t^* - lnp_{t-1}^*$, where p_t^* denotes the dividend-corrected price of a stock. The sample consists of 506 pairs of observations. Since both stocks are from the automobile sector, they are expected to be highly dependent. The chosen sample period is, from a statistical viewpoint, a difficult one, since it includes the stock market crash in October 1987 where both stocks experienced substantial price drops and a subsequent period of high volatility. As in the univariate case, we used the Kolmogorov distance (multiplied by 100) to measure the goodness of fit.

The means and standard deviations of the returns over the sample period are 5.693×10^{-4} and 2.599×10^{-2} for Chrysler and 1.556×10^{-3} and 2.112×10^{-2} for Ford; the correlation coefficient is 0.696. We measure the fit by examining the joint distribution $Pr(-x < X < x, -y < Y < y)$, which is of interest in a number of financial applications. The Kolmogorov distance for the BN distribution is $100\rho_{BN} = 8.86$.

When fitting the BEMO and BWMO distributions, one could estimate separate distributions for each of the four quadrants in $\mathbf{R}^2$ which would give rise to a total of 16 and 24 parameters, respectively. To reduce the parameter space to some extent, we assume symmetry, i.e., we impose the restriction that the parameters are equal for all four quadrants, and, in addition, restrict λ_{12} to unity. This leaves us with three free parameters to estimate for the BEMO distribution and with five free parameters for the BWMO distribution. Since the specification of the BN requires five parameters, we are comparing the BN distribution with distributions that have less or the same number of free parameters.

Despite the restrictions imposed, both the BEMO and the BWMO distributions provide a better fit, namely $100\rho_{BEMO} = 7.40$ and $100\rho_{BWMO} =$

Estimates	BEMO	BWMO	BN
α_1	n.a.	1.003	n.a.
α_2	n.a.	1.147	n.a.
λ_1	27.06	39.15	n.a.
λ_2	33.69	59.55	n.a.
k	1.807	27.42	n.a.
100ρ	7.396	4.398	8.858

Table 3: Estimates and Fit of BEMO and BWMO Distributions

4.40, than the BN distribution. Table 3 reports the parameter estimates of the two distributions. Being specified by the same number of parameters the BWMO distribution yields a substantially better fit than the BN distribution. Even the three-parameter BEMO fits slightly better than the BN distribution.

These findings have consequences for financial asset pricing theories and portfolio analyses based on mean-variance concepts. The particular implication on the capital asset pricing model (CAPM) (see, for example, [3]) will be subject of future research.

REFERENCES

[1] V. Akgiray, G.G. Booth, "The stable-law model of stock returns", J. Bus. Econ. Statist. **6** (1988), 51-57.

[2] E. Fama, "The behavior of stock market prices", J. Bus. **38** (1965), 34-105.

[3] E. Fama, "Risk, return and equilibrium: some clarifying comments", J. Finance **23** (1968), 29-40.

[4] J. Galambos, S. Kotz, *Characterization of Probability Distributions*, Springer-Verlag, 1978.

[5] B. Mandelbrot, "The variation of certain speculative prices", J. Bus. **26** (1963), 394-419.

[6] A.W. Marshall, I. Olkin, "A multivariate exponential distribution", J. Amer. Statist. Assoc. **62** (1967), 30-44.

[7] A.W. Marshall, I. Olkin, "A generalized bivariate distribution", J. Appl. Prob. **4** (1967), 291-302.

[8] S. Mittnik, S.T. Rachev, "Stable distributions for asset returns", Appl. Math. Lett. **3** (1989), 301-304.

[9] S. Mittnik, S.T. Rachev, *Modeling Speculative Prices With Alternative Stable Distributions*, in preparation, 1990.

[10] M.L. Moeschberger, "Life test under competing causes of failure", Technometrics **16** (1974), 39-47.

[11] V.M. Zolotarev, "Approximations of distributions of sums of independent random variables with values in infinite dimensional spaces", Theor. Probab. Appl. **21** (1976), 721-736.

[12] V.M. Zolotarev, "Probability metrics", Theor. Probab. Appl. **28** (1983), 278-302.

Stefan Mittnik
Department of Economics
State University of New York-Stony Brook
Stony Brook, NY 11794-4384

Svetlozar T. Rachev
Department of Statistics and Applied Probability
University of California-Santa Barbara
Santa Barbara, CA 93106

Construction of Multiple Stable Measures and Integrals Using Lepage Representation *†‡

Gennady Samorodnitsky
Boston University

Murad S. Taqqu
Boston University

Abstract

Consider a symmetric α-stable, $0 < \alpha < 2$, random measure M with a Radon control measure, defined on $(\mathbf{R}, \mathcal{B})$. Let $\mathbf{f}$ be a Banach-valued deterministic function defined on $\mathbf{R}^n$, symmetric in its arguments and vanishing on the diagonals. We provide a construction of the product measure

$$M^{(n)}(dx_1, \ldots, dx_n) = n! M(dx_1) \ldots M(dx_n)$$

and of the multiple stochastic integral

$$\mathbf{I}_n(\mathbf{f}) = \int_{\mathbf{R}^n} \mathbf{f}(x_1, \ldots, x_n) M(dx_1) \ldots M(dx_n)$$

using a multiple Lepage representation.

*The authors are supported by the National Science Foundation grant DMS-8805627 at Boston University

†ADDRESS: Department of Mathematics, 111 Cummington Street, Boston University, Boston, MA 02215

‡KEY WORDS: Lepage representation, random measure, vector measure, product measure, multiple stable integral, symmetric α-stable, Banach-valued random variable

1 Introduction

Krakowiak and Szulga ([KS88]) have defined a multiple stochastic integral with respect to a real-valued strictly α-stable random measure when the integrand is a Banach-valued deterministic function. In this paper, we provide a *construction* of this multiple stochastic integral using a multiple Lepage representation, in the special case when the random measure is symmetric. The idea is to represent the multiple Banach-valued integral as a multiple Banach-valued series.

An important step in the definition of the multiple integral is to establish the crucial control inequality (3.3) ((5.1) in [KS88]), and then use that inequality to prove that the product measure is σ-additive. The argument in [KS88] is based on decoupling inequalities. The present paper does not rely on decoupling inequalities. It presents a more direct and constructive way of introducing the product of stable measures by defining them as products of their Lepage series representation, as in [SS89]. A similar method can be applied to multiple integrals with respect to infinitely divisible random measures as shown recently by [Szu89]. Because here we want to integrate Banach-valued functions, we have to obtain special bounds on the moments. These bounds, which are essential to the proof are established in Lemma 2.2 below.

Our series construction of the multiple stochastic integral is used in [ST90] to derive conditions for existence of multiple integrals taking values in Banach spaces of Rademacher-type p, $\alpha < p \leq 2$, and to determine the behavior of their probability tails. These results are then applied in [RST] to study the sample path properties of processes represented by multiple stochastic integrals.

Let S be a real separable Banach space with norm $\| \ \|$ and suppose that $\mathbf{f} : \mathbf{R}^n \to S$ is symmetric and vanishes on the diagonals (i.e., $\mathbf{f}(x_1, \ldots, x_n) = \mathbf{f}(x_{\pi(1)}, \ldots, x_{\pi(n)})$ for any permutation π of $(1, \ldots, n)$

and $\mathbf{f}(x_1, x_2, \ldots, x_n) = 0$ if $x_i = x_j$ for some $i \neq j$.) Let M denote an independently scattered real-valued symmetric α-stable ($S\alpha S$) random measure on $(\mathbf{R}, \mathcal{B})$ with Radon control measure m, that is, $m(A) < \infty$ for each compact set $A \in \mathcal{B}$. A standard example of such an m is the Lebesgue measure.

We prove that if the multiple integral

$$\mathbf{I}_n(\mathbf{f}) = \int_{-\infty}^{+\infty} \cdots \int_{-\infty}^{+\infty} \mathbf{f}(x_1, \ldots, x_n) M(dx_1) \ldots M(dx_n)$$

exists (as an integral with respect to a vector-valued measure) then the multiple sum

$$\mathbf{S}_n(\mathbf{f}) = n! c_\alpha^{n/\alpha} \sum_{\mathbf{j} \in \mathcal{D}_n} [\epsilon_{\mathbf{j}}][\Gamma_{\mathbf{j}}]^{-1/\alpha} [\psi(Y_{\mathbf{j}})]^{-1} \mathbf{f}(\mathbf{Y}_{\mathbf{j}})$$

converges a.s. and

$$\mathbf{I}_n(\mathbf{f}) \stackrel{d}{=} \mathbf{S}_n(\mathbf{f}),$$

where $\stackrel{d}{=}$ denotes equality in distribution.

Here, and throughout the paper

$$\mathcal{D}_n = \{\mathbf{j} = (j_1, \ldots, j_n) \in \mathbf{N}^n : 1 \leq j_1 < j_2 < \ldots < j_n < \infty\},$$

and we write

$$[a_{\mathbf{j}}] = a_{j_1} a_{j_2} \ldots a_{j_n}, \ \mathbf{j} = (j_1, \ldots, j_n),$$

for a sequence $\mathbf{a} = (a_1, a_2, \ldots)$ of real numbers. Moreover, the ϵ_j's are i.i.d. Rademacher random variables, i.e. $P\{\epsilon_j = 1\} = P\{\epsilon_j = -1\} = 1/2$, the Γ_j's, $j = 1, 2, \ldots$ are the arrival times of a Poisson process with unit rate, ψ is a measurable function such that $\psi(x) > 0$ m-almost everywhere and

$$\int_{\mathbf{R}} \psi(x)^\alpha m(dx) = 1;$$

the Y_j's, are i.i.d. random variables with distribution

$$m_\psi(A) = \int_A \psi(x)^\alpha m(dx) \ , \quad A \in \mathcal{B};$$

the sequences $\{\epsilon_j, j = 1,2,\ldots\}$, $\{\Gamma_j, j = 1,2,\ldots\}$ and $\{Y_j, j = 1,2,\ldots\}$ are independent. Finally, we set $\mathbf{Y_j} = (Y_{j_1},\ldots,Y_{j_n})$ when $\mathbf{j} = (j_1,\ldots,j_n)$,

$$c_\alpha = (\int_0^\infty x^{-\alpha}\sin x dx)^{-1} = \begin{cases} \frac{1-\alpha}{\Gamma(2-\alpha)\cos(\pi\alpha/2)} & \text{if } \alpha \neq 1, \\ 2/\pi & \text{if } \alpha = 1, \end{cases}$$

and $\sum_{j_1=1}^{\infty}\cdots\sum_{j_n=1}^{\infty} := \lim_{N\to\infty}\sum_{j_1=1}^{N}\cdots\sum_{j_n=1}^{N}$.

$\mathbf{S}_n(\mathbf{f})$ is called the *multiple Lepage representation of* $\mathbf{I}_n(\mathbf{f})$. The representation of $S_n(\mathbf{f})$ depends on ψ, but, as we will see, its distribution does not depend on ψ.

The next section contains some auxiliary lemmas. The product measure $M(dx_1)\ldots M(dx_n)$ is constructed in Section 3 and the multiple integral $\mathbf{I}_n(\mathbf{f})$ in Section 4.

2 Auxiliary results

We state here some lemmas that will be used in the sequel. Let $\{\Gamma_j, j = 1,2,\ldots\}$ be the arrival times of a Poisson process with unit rate.

LEMMA 2.1 *For every real β there is a finite constant $K = K(\beta) > 0$ such that for every $\mathbf{j} \in \mathcal{D}_n$ with $j_1 > n\beta$ we have*

$$E[\Gamma_\mathbf{j}]^{-\beta} \leq K[\mathbf{j}]^{-\beta}. \tag{2.1}$$

PROOF: By the Hölder inequality, $E[\Gamma_\mathbf{j}]^{-\beta} \leq \prod_{i=1}^n (E\Gamma_{j_i}^{-n\beta})^{1/n}$. Now suppose $j > n\beta$. Then $E\Gamma_j^{-n\beta} < \infty$, and by the Stirling's formula, $E\Gamma_j^{-n\beta} \leq Kj^{-n\beta}$. ∎

The following lemma is the main ingredient in the proof of the control inequality (3.3).

LEMMA 2.2 *Suppose that $Y_1, Y_2, \ldots$ are i.i.d. with probability distribution m. Then for every $0 < \gamma < \alpha$ and every $r > \alpha$, there is a constant c such that for each $A \in \mathcal{B}_n^{(s)}$,*

$$c^{-1}(m^{(n)}(A))^{\gamma/\alpha} \le E\left(\sum_{\mathbf{j}\in\mathcal{D}_n} [\Gamma_{\mathbf{j}}]^{-2/\alpha} 1(\mathbf{Y_j} \in A)\right)^{\gamma/2} \le c(m^{(n)}(A))^{\gamma/r}. \tag{2.2}$$

PROOF: It is sufficient to establish the right-hand inequality in (2.2) for $r \in (\alpha, 2)$, since if it holds for $r_1 = r$, it holds for $r_2 > r_1$, by increasing the constant c. Fix then $\alpha < r < 2$, and for $k = 1, 2, \ldots, n, n+1$, define

$$\nu_k = \sum_{i=1}^{k} [(n-i+1)r/\alpha]$$

where [ ] denotes here the integer part. Let

$$U := E\left(\sum_{\mathbf{j}\in\mathcal{D}_n} [\Gamma_{\mathbf{j}}]^{-2/\alpha} 1(\mathbf{Y_j} \in A)\right)^{\gamma/2}. \tag{2.3}$$

We can express U as

$$U = E\Big(\sum_{k=0}^{n} \sum_{i_1=1}^{\nu_1} \sum_{i_2=i_1+1}^{\nu_2} \cdots \sum_{i_{n-k}=i_{n-k-1}+1}^{\nu_{n-k}} \Gamma_{i_1}^{-2/\alpha} \Gamma_{i_2}^{-2/\alpha} \cdots \Gamma_{i_{n-k}}^{-2/\alpha}$$

$$\cdot \sum_{\mathbf{j}\in\mathcal{D}_k^{(\nu_{n-k+1}+1)}} [\Gamma_{\mathbf{j}}]^{-2/\alpha} 1((Y_{i_1}, \ldots, Y_{i_{n-k}}, Y_{j_1}, \ldots, Y_{j_k}) \in A)\Big)\Big)^{\gamma/2},$$

where

$$\mathcal{D}_k^{(l)} = \{\mathbf{j} = (j_1, \ldots, j_k) \in \mathcal{D}_k : j_1 \ge l\}.$$

To verify this decomposition note that the case $k = n$ corresponds to $\mathbf{j} \in \mathcal{D}_n^{(\nu_1+1)}$, i.e. $j_1 \ge \nu_1 + 1$; the case $k = n-1$ forces the previous j_1 (now denoted i_1) to take values $\le \nu_1$, and since the new $\mathbf{j} \in \mathcal{D}_{n-1}^{(\nu_2+1)}$, we observe that the previous j_2 (now called j_1) has to be $\ge \nu_2 + 1$, etc.. Inserting the exponent $\gamma/2 < 1$ inside the summation signs, yields

$$U \le \sum_{k=0}^{n} \sum_{i_1=1}^{\nu_1} \sum_{i_2=i_1+1}^{\nu_2} \cdots \sum_{i_{n-k}=i_{n-k-1}+1}^{\nu_{n-k}} V_{i_1,\ldots,i_{n-k}}^{(n,k)} \tag{2.4}$$

where

$$V^{(n,k)}_{i_1,\dots,i_{n-k}} = E\left\{\Gamma_{i_1}^{-\gamma/\alpha}\Gamma_{i_2}^{-\gamma/\alpha}\dots\Gamma_{i_{n-k}}^{-\gamma/\alpha}\left[\sum_{\mathbf{j}\in\mathcal{D}_k^{(\nu_{n-k+1}+1)}} R^{(\mathbf{j},k)}_{i_1,\dots,i_{n-k}}\right]^{\gamma/2}\right\},$$

$$R^{(\mathbf{j},k)}_{i_1,\dots,i_{n-k}} = [\Gamma_{\mathbf{j}}]^{-2/\alpha}1((Y_{i_1},\dots,Y_{i_{n-k}},\tilde{Y}_{j_1},\dots,\tilde{Y}_{j_k})\in A)$$

and where $\{\tilde{Y}_j,\ j=1,2,\dots\}$ is an independent copy of $\{Y_j,\ j=1,2,\dots\}$. Now,

$$V^{(n,k)}_{i_1,\dots,i_{n-k}} \le (E\Gamma_1^{-\gamma/\alpha})^{n-k} E\left[\sum_{\mathbf{j}\in\mathcal{D}_k^{[kr/\alpha]+1}} R^{(\mathbf{j},k)}_{i_1,\dots,i_{n-k}}\right]^{\gamma/2}. \tag{2.5}$$

To verify the last inequality, note that $\Gamma_{i_1}\ge\Gamma_1$ and

$$\Gamma_{i_2} = (\Gamma_{i_2}-\Gamma_{i_1})+\Gamma_{i_1} \ge \Gamma_{i_2}-\Gamma_{i_1} \stackrel{d}{=} \tilde{\Gamma}_{i_1} \ge \tilde{\Gamma}_1 \text{ etc...},$$

where $\{\tilde{\Gamma}_1,\tilde{\Gamma}_2,\dots\}$ is an independent copy of $\{\Gamma_1,\Gamma_2,\dots\}$. Moreover,

$$\Gamma_j = (\Gamma_j-\Gamma_{i_{n-k}})+\Gamma_{i_{n-k}} \ge \Gamma_j-\Gamma_{i_{n-k}} \ge \Gamma_j-\Gamma_{\nu_{n-k}} \stackrel{d}{=} \tilde{\Gamma}_{j-\nu_{n-k}}.$$

Finally, $\nu_{n-k+1}+1-\nu_{n-k} = [kr/\alpha]+1$, so that $\mathbf{j}' := \mathbf{j}-\nu_{n-k}\in\mathcal{D}_k^{[kr/\alpha]+1}$ if $\mathbf{j}\in\mathcal{D}_k^{(\nu_{n-k+1}+1)}$, and $E\Gamma_1^{-\gamma/\alpha}<\infty$ since $\gamma<\alpha$.

The Hölder inequality gives

$$V^{(n,k)}_{i_1,\dots,i_{n-k}} \le c\left[E\left(\sum_{\mathbf{j}\in\mathcal{D}_k^{[kr/\alpha]+1}} R^{(\mathbf{j},k)}_{i_1,\dots,i_{n-k}}\right)^{r/2}\right]^{\gamma/r}$$

$$\le c\left[\sum_{j\in\mathcal{D}_k^{[kr/\alpha]+1}} E(R^{(\mathbf{j},k)}_{i_1,\dots,i_{n-k}})^{r/2}\right]^{\gamma/r}$$

since $0 < r/2 < 1$. Using the expression for $R^{(\mathbf{j},k)}_{i_1,\ldots,i_{n-k}}$ and Lemma 2.1, we get

$$V^{(n,k)}_{i_1,\ldots,i_{n-k}} \le c\left[\sum_{\mathbf{j}\in\mathcal{D}_k^{[kr/\alpha]+1}} E[\Gamma_{\mathbf{j}}]^{-r/\alpha} m^{(n)}(A)\right]^{\gamma/r}$$

$$\le c\left[\left(\sum_{j=[kr/\alpha]+1}^{\infty} j^{-r/\alpha}\right)^k m^{(n)}(A)\right]^{\gamma/r}$$

$$\le c(m^{(n)}(A))^{\gamma/r}. \tag{2.6}$$

The constant c may be different from line to line. The right-hand inequality in (2.2) follows from (2.4) and (2.6).

We now turn to the left-hand inequality in (2.2). By keeping only the first term in the first $n-1$ summations in (2.3), we get

$$U \ge E\left\{\Gamma_1^{-\gamma/\alpha}\ldots\Gamma_{n-1}^{-\gamma/\alpha}\left[\sum_{j=n}^{\infty}\Gamma_j^{-2/\alpha}1((Y_1,\ldots,Y_{n-1},Y_j)\in A)\right]^{\gamma/2}\right\}$$

$$\ge E\left\{\Gamma_1^{-\gamma/\alpha}\ldots\Gamma_{n-1}^{-\gamma/\alpha}1(\Gamma_{n-1}\le 1)\left[\sum_{j=n}^{\infty}\Gamma_j^{-2/\alpha}1((Y_1,\ldots,Y_{n-1},Y_j)\in A)\right]^{\gamma/2}\right\}$$

$$\ge E\left\{1(\Gamma_{n-1}\le 1)\left[\sum_{j=n}^{\infty}\Gamma_j^{-2/\alpha}1((Y_1,\ldots,Y_{n-1},Y_j)\in A)\right]^{\gamma/2}\right\}$$

since $\Gamma_{n-1}\le 1$ implies $\Gamma_j^{-1}\ge\Gamma_{n-1}^{-1}\ge 1$, $j=1,\ldots,n-1$. Therefore

$$U \ge c_1 E\Delta^{\gamma/2} \tag{2.7}$$

where $c_1 = P(\Gamma_{n-1}\le 1)$,

$$\Delta = \sum_{j=1}^{\infty}(\Gamma_j+1)^{-2/\alpha}1((Y_1,\ldots,Y_{n-1},\tilde{Y}_j)\in A) \tag{2.8}$$

and $\{\tilde{Y}_j, j\ge 1\}$ is an independent copy of $\{Y_j, j\ge 1\}$. We used the fact that $\Gamma_{n-1}\le 1$ implies

$$\Gamma_j = (\Gamma_j-\Gamma_{n-1})+\Gamma_{n-1}\le(\Gamma_j-\Gamma_{n-1})+1 \stackrel{d}{=} \tilde{\Gamma}_{j-n+1}+1.$$

There is only one sum left in (2.7) which appears in the expression (2.8) defining Δ. The idea now is to choose a term *at random* in that sum, and express the fact that this term is less or equal to the whole sum. To formalize this idea, we introduce the random variable

$$F(Y_1, \ldots, Y_{n-1}) = P((Y_1, \ldots, Y_{n-1}, \tilde{Y}_j) \in A | Y_1, \ldots, Y_{n-1}).$$

Observe that

(i) $EF(Y_1, \ldots, Y_{n-1}) = P((Y_1, \ldots, Y_{n-1}, \tilde{Y}_j) \in A) = m^{(n)}(A)$,

(ii) since any random variable X with $EX = 1$ satisfies $P(X \geq \frac{1}{2}) > 0$, it follows that there is a Borel set $A_{n-1} \subset \mathbf{R}^{n-1}$ with $P((Y_1, \ldots, Y_{n-1}) \in A_{n-1}) > 0$ such that for any $(Y_1, \ldots, Y_{n-1}) \in A_{n-1}$ we have

$$F(Y_1, \ldots, Y_{n-1}) \geq \frac{m^{(n)}(A)}{2}. \tag{2.9}$$

Then, from (2.7), we get

$$U \geq c_1 E\{\Delta^{\gamma/2} 1((Y_1, \ldots, Y_{n-1}) \in A_{n-1})\}$$

$$= c_1 \int_{1((Y_1, \ldots, Y_{n-1}) \in A_{n-1})} E(\Delta^{\gamma/2} | Y_1, \ldots, Y_{n-1}) dP.$$

Fix $(Y_1, \ldots, Y_{n-1}) \in A_{n-1}$ and let

$$T_A = \inf\{j : (Y_1, \ldots, Y_{n-1}, \tilde{Y}_j) \in A\},$$

i.e., T_A is the first j such that $1((Y_1, \ldots, Y_{n-1}, \tilde{Y}_j) \in A) = 1$. Then, keeping only the term with index T_A in (2.8), we get

$$W := E(\Delta^{\gamma/2} | Y_1, \ldots, Y_{n-1})$$

$$\geq E\left((\Gamma_{T_A} + 1)^{-\gamma/\alpha} | Y_1, \ldots, Y_{n-1}\right). \tag{2.10}$$

Since the random variables $\tilde{Y}_j$, $j \geq 1$ are independent, T_A has a geometric distribution with parameter (probability of success) $\rho :=$

$F(Y_1, \ldots, Y_{n-1})$. Therefore, by (2.10),

$$\begin{aligned}
W &\geq \sum_{i=1}^{\infty} \rho(1-\rho)^{i-1} E(\Gamma_i + 1)^{-\gamma/\alpha} \\
&= \sum_{i=1}^{\infty} \rho(1-\rho)^{i-1} \int_0^{\infty} (x+1)^{-\gamma/\alpha} e^{-x} \frac{x^{i-1}}{(i-1)!} dx \\
&= \int_0^{\infty} (x+1)^{-\gamma/\alpha} e^{-x} \rho \sum_{i=1}^{\infty} \frac{(x(1-\rho))^{i-1}}{(i-1)!} dx \\
&= \int_0^{\infty} (x+1)^{-\gamma/\alpha} e^{-x} \rho e^{x(1-\rho)} dx \\
&= \left(\int_0^{\infty} (y+\rho)^{-\gamma/\alpha} e^{-y} dy \right) \rho^{\gamma/\alpha} \\
&\geq \left(\int_0^{\infty} (y+1)^{-\gamma/\alpha} e^{-y} dy \right) \rho^{\gamma/\alpha} \\
&\geq c(m^{(n)}(A))^{\gamma/\alpha}
\end{aligned}$$

by (2.9). This concludes the proof of the lemma. ∎

In the same way one obtains

COROLLARY 2.1 *For every $0 < \gamma < \alpha$ and $r > \alpha$, there is a constant c such that for each measurable function $g : \mathbf{R}^n \to \mathbf{R}_+$,*

$$E \left(\sum_{\mathbf{j} \in \mathcal{D}_n} [\Gamma_{\mathbf{j}}]^{-2/\alpha} g(\mathbf{Y}_{\mathbf{j}}) \right)^{\gamma/2} \leq c \left(E(g(Y_1, Y_2, \ldots, Y_n))^{r/2} \right)^{\gamma/r}.$$

The next lemma gives a generalized Khinchine inequality (see [KS86a], Relation (2.4)).

LEMMA 2.3 *For every $\gamma > 0$, there is a constant $c > 0$ such that, for any real-valued array $x_{\mathbf{j}}$, $\mathbf{j} \in \mathcal{D}_n$,*

$$c^{-1} (\sum_{\mathbf{j} \in \mathcal{D}_n} x_{\mathbf{j}}^2)^{\gamma/2} \leq E \left| \sum_{\mathbf{j} \in \mathcal{D}_n} [\epsilon_{\mathbf{j}}] x_{\mathbf{j}} \right|^{\gamma} \leq c (\sum_{\mathbf{j} \in \mathcal{D}_n} x_{\mathbf{j}}^2)^{\gamma/2}.$$

The following lemma is due to [KS86b], (Theorem 2.4).

LEMMA 2.4 *Let* $\{\mathbf{a}_{\mathbf{j}}^{(k)}, \mathbf{j} \in \mathcal{D}_n\}$, $k = 1, 2, \ldots$ *be* S*-valued arrays such that the series* $\boldsymbol{\xi}^{(k)} = \sum_{\mathbf{j} \in \mathcal{D}_n} [\epsilon_{\mathbf{j}}]\mathbf{a}_{\mathbf{j}}^{(k)}$ *converges for each* $k \geq 1$*. If* $\boldsymbol{\xi}^{(k)} \to \boldsymbol{\xi}$ *in probability, then there is an* S*-valued array* $\mathbf{a}_{\mathbf{j}}$, $\mathbf{j} \in \mathcal{D}_n$ *such that*

(i) $\mathbf{a}_{\mathbf{j}}^{(k)} \to \mathbf{a}_j$ *as* $k \to \infty$*, for each* $\mathbf{j} \in \mathcal{D}_n$*,*

(ii) $\sum_{\mathbf{j} \in \mathcal{D}_n} [\epsilon_{\mathbf{j}}]\mathbf{a}_{\mathbf{j}}$ *converges a.s. to* $\boldsymbol{\xi}$*.*

3 Construction of the product stable measure

Let M be an independently scattered $S\alpha S$ random measure on $(\mathbf{R}, \mathcal{B})$ with a control measure m. Assume at this point that m is *finite* (this assumption will be relaxed later). We may suppose without loss of generality that the random measure M is given by its Lepage representation

$$M(A) = c_\alpha^{1/\alpha} \sum_{j=1}^{\infty} \epsilon_j \Gamma_j^{-1/\alpha} \psi(Y_j)^{-1} 1(Y_j \in A)$$

where c_α, ψ, the ϵ_j's, Γ_j's and Y_j's are defined in Section 1 ([Ros89].) Recall in particular that the Y_j's are i.i.d. with probability distribution $m_\psi(A) = \int_A \psi(x)^\alpha m(dx)$, $A \in \mathcal{B}$.

Let $n \geq 1$ and denote by $\mathcal{B}_n^{(s)}$ the σ-algebra of sets in $\mathbf{R}^n$ generated by the "symmetric rectangles" of the type

$$A = \bigcup_{(\ell_1, \ell_2, \ldots, \ell_n)} (A_{\ell_1} \times A_{\ell_2} \times \cdots \times A_{\ell_n}),$$

where $A_i \in \mathcal{B}, A_i \cap A_j = \emptyset$ for any $i, j = 1, \ldots, n, i \neq j$ and where the union is taken over all permutations of the indices $1, 2, \ldots, n$. One defines then a random (product) measure $M^{(n)}$ on "symmetric rectangles" by putting

$$M^{(n)} \left(\bigcup_{(\ell_1, \ell_2, \ldots, \ell_n)} (A_{\ell_1} \times A_{\ell_2} \times \cdots \times A_{\ell_n}) \right) \stackrel{def}{=} n! M(A_1) M(A_2) \cdots M(A_n)$$

$$= n! c_\alpha^{n/\alpha} \sum_{\mathbf{j} \in \mathbf{N}_n} \epsilon_{j_1} \ldots \epsilon_{j_n} \Gamma_{j_1}^{-1/\alpha} \ldots \Gamma_{j_n}^{-1/\alpha}$$

$$\cdot \psi(Y_{j_1})^{-1} \cdots \psi(Y_{j_n})^{-1} 1(Y_{j_1} \in A_1) \ldots 1(Y_{j_n} \in A_n)$$

$$= c_\alpha^{n/\alpha} \sum_{\mathbf{j} \in \mathbf{N}_n} [\epsilon_\mathbf{j}][\Gamma_\mathbf{j}]^{-1/\alpha}[\psi(Y_\mathbf{j})]^{-1} \sum_{(\ell_1, \ldots, \ell_n)} 1(Y_{j_1} \in A_{\ell_1}) \ldots 1(Y_{j_n} \in A_{\ell_n})$$

$$= n! c_\alpha^{n/\alpha} \sum_{\mathbf{j} \in \mathcal{D}_n} [\epsilon_\mathbf{j}][\Gamma_\mathbf{j}]^{-1/\alpha}[\psi(Y_\mathbf{j})]^{-1} 1(\mathbf{Y}_\mathbf{j} \in A). \tag{3.1}$$

We want to extend this representation to all sets in $\mathcal{B}_n^{(s)}$ and show that the distribution of the extension does not depend on ψ. We first focus on sets in $\mathcal{F}_n^{(s)}$, the family of *finite unions* of "symmetric rectangles". Clearly each $A \in \mathcal{F}_n^{(s)}$ can be represented as a finite union of *disjoint* symmetric rectangles and hence, by additivity, we can extend (3.1) to all $A \in \mathcal{F}_n^{(s)}$.

Viewing $\{M^{(n)}(A), A \in \mathcal{F}_n^{(s)}\}$ as a *stochastic process*, we observe that different ψ's yield different versions of M. However all the versions have the same finite-dimensional distributions because the characteristic function of $M(A), A \in \mathcal{B}$,

$$\begin{aligned} Ee^{-i\theta M(A)} &= \exp\{-|\theta|^\alpha \int_{-\infty}^{+\infty} |\psi^{-1}(x) 1(x \in A)|^\alpha m_\psi(dx)\} \\ &= \exp\{-|\theta|^\alpha m(A)\} \end{aligned}$$

does not depend on ψ.

We are now going to view $M^{(n)}$ as a *vector-valued measure* taking values in $L^\gamma(\Omega, \mathcal{F}, P)$, $\gamma > 0$ and we write

$$|||M^{(n)}(A)|||_\gamma = [E|M^{(n)}(A)|^\gamma]^{1/\gamma},$$

letting $||| \ |||_\gamma$ denote the L^γ quasi-norm. The semi-variation $|M^{(n)}|_\gamma$ of $M^{(n)}$ is the set function

$$|M^{(n)}(A)|_\gamma = \sup |||\sum_{i=1}^m \theta_i M^{(n)}(C_i)|||_\gamma$$

where the sup is over all integers $m \geq 1$, real numbers θ_i with $|\theta_i| \leq 1, i = 1, \ldots, m$, and disjoint subsets C_i, $i = 1, \ldots, m$ of the set A ([DS58] or [DJ77]).

The measure $M^{(n)}$ has been so far defined only for sets in $\mathcal{F}_n^{(s)}$. In the following theorem, we show that it can be extended to the whole of $\mathcal{B}_n^{(s)}$, that the extension does not depend on the chosen version and that the extension admits a Lepage representation. The control inequality (3.3) below is due to Krakowiak and Szulga ([KS88], Theorem 5.4). We provide here a direct proof of that inequality using Lepage representation.

THEOREM 3.1 *The measure $M^{(n)}$ can be extended to a L^γ-valued σ-additive vector measure on $\mathcal{B}_n^{(s)}$ for any given $0 < \gamma < \alpha$. The extension can be represented as*

$$M^{(n)}(A) = n! c_\alpha^{n/\alpha} \sum_{\mathbf{j} \in \mathcal{D}_n} [\epsilon_{\mathbf{j}}][\Gamma_{\mathbf{j}}]^{-1/\alpha}[\psi(Y_{\mathbf{j}})]^{-1} 1(\mathbf{Y}_{\mathbf{j}} \in A) \tag{3.2}$$

for every $A \in \mathcal{B}_n^{(s)}$. The finite-dimensional distributions of $\{M^{(n)}(A), A \in \mathcal{B}_n^{(s)}\}$ do not depend on the function ψ.

Moreover, the semi-variation $|M^{(n)}|_\gamma$ of $M^{(n)}$ satisfies the following relation: for every $r > \alpha$, there is a finite positive constant c such that for every $A \in \mathcal{B}_n^{(s)}$,

$$c^{-1}(m^{(n)}(A))^{1/\alpha} \leq |M^{(n)}|_\gamma(A) \leq c(m^{(n)}(A))^{1/r}. \tag{3.3}$$

PROOF: Note first that if $0 < \gamma < \alpha/n$, then

$$S := \sum_{\mathbf{j} \in \mathcal{D}_n} [\Gamma_{\mathbf{j}}]^{-2/\alpha}[\psi(Y_{\mathbf{j}})]^{-2} 1(\mathbf{Y}_{\mathbf{j}} \in A) \ , \ A \in \mathcal{B}_n^{(s)}$$

satisfies, by Khinchine inequality,

$$\begin{aligned} ES^{\gamma/2} &\leq E(\sum_{j=1}^{\infty} \Gamma_j^{-2/\alpha} \psi(Y_j)^{-2})^{\gamma n/2} \\ &\leq cE(|\sum_{j=1}^{\infty} \epsilon_j \Gamma_j^{-1/\alpha} \psi(Y_j)^{-1}|^{\gamma n}) \\ &< \infty \end{aligned} \tag{3.4}$$

since $\sum_{j=1}^{\infty} \epsilon_j \Gamma_j^{-1/\alpha} \psi(Y_j)^{-1}$ is a $S\alpha S$ random variable and $\gamma n < \alpha$. In particular $S < \infty$ a.s.

The proof of the theorem will be divided in four parts.

(A) Suppose $0 < \gamma < \alpha/n$. To show that $M^{(n)}$ as defined in (3.2) is a σ-additive L^γ-valued random measure, note that it is clearly finitely additive, and if $C_1, C_2, \ldots$ are disjoint sets in $\mathcal{B}_n^{(s)}$, we have

$$|||M^{(n)}(\bigcup_{i=1}^{k} C_i) - M^{(n)}(\bigcup_{i=1}^{\infty} C_i)|||_\gamma = |||M^{(n)}(\bigcup_{i=k+1}^{\infty} C_i)|||_\gamma$$

$$= \left(E|M^{(n)}(\bigcup_{i=k+1}^{\infty} C_i)|^\gamma \right)^{1/\gamma}$$

$$= \left(E\left[n! c_\alpha^{n/\alpha} \left| \sum_{\mathbf{j}\in\mathcal{D}_n} [\epsilon_\mathbf{j}][\Gamma_\mathbf{j}]^{-1/\alpha}[\psi(Y_\mathbf{j})]^{-1} 1(\mathbf{Y_j} \in \bigcup_{i=k+1}^{\infty} C_i) \right| \right]^\gamma \right)^{1/\gamma}$$

$$\le c_{\alpha,n} \left(E\left[\sum_{\mathbf{j}\in\mathcal{D}_n} [\Gamma_j]^{-2/\alpha}[\psi(Y_\mathbf{j})]^{-2} 1(\mathbf{Y_j} \in \bigcup_{i=k+1}^{\infty} C_i) \right]^{\gamma/2} \right)^{1/\gamma} \tag{3.5}$$

by the generalized Khinchine inequality (Lemma 2.3). By (3.4) and the monotone convergence theorem, the right hand side of (3.5) converges to 0 as $k \to \infty$. This shows that $M^{(n)}$, as defined in (3.2) is a σ-additive L^γ-valued random measure when $0 < \gamma < \alpha/n$. In part (C) below we show that this is true, as well, for all $0 < \gamma < \alpha$.

(B) The finite dimensional distributions of $\{M^{(n)}(A), A \in \mathcal{B}_n^{(s)}\}$ do not depend on ψ because the finite-dimensional distributions of $\{M^{(n)}(A), A \in \mathcal{F}_n^{(s)}\}$ do not depend on ψ, and because of the following claim:

CLAIM. *For any $A \in \mathcal{B}_n^{(s)}$, there is a sequence of sets $A_k \in \mathcal{F}_n^{(s)}, k = 1, 2, \ldots$ such that $M^{(n)}(A_k) \to M^{(n)}(A)$ in probability.*

PROOF OF THE CLAIM. Because of the geometric properties of $\mathbf{R}$, there is a sequence of sets $\mathrm{B}_k \in \mathcal{F}_n^{(s)}$, $k = 1, 2, \ldots$, not necessarily monotone, such that $m^{(n)}(A\triangle B_k) \to 0$ as $k \to \infty$. Since

$$\begin{aligned} p_k : &= P((Y_1, \dots, Y_n) \in A\triangle B_k) \\ &= \int_{A\triangle B_k} \psi(x_1)^\alpha \cdots \psi(x_n)^\alpha m(dx_1)\dots m(dx_n) \end{aligned}$$

tends to zero as $k \to \infty$, there is, for each $i = 1, 2, \dots$, an index $k(i)$ such that $p_{k(i)} \leq 2^{-i}$. Assume, without loss of generality that $k(1) < k(2) < \dots$ and let $A_i = B_{k(i)}, i = 1, 2, \dots$. Then, as in (3.5), for $0 < \gamma < \alpha/n$

$$E|M^{(n)}(A) - M^{(n)}(A_i)|^\gamma$$

$$= E\left| n! c_\alpha^{n/\alpha} \sum_{\mathbf{j}\in\mathcal{D}_n} [\epsilon_{\mathbf{j}}][\Gamma_{\mathbf{j}}]^{-1/\alpha}[\psi(Y_{\mathbf{j}})]^{-1}(1(\mathbf{Y_j} \in A) - 1(\mathbf{Y_j} \in A_i))\right|^\gamma$$

$$\leq cES_i^{\gamma/2}$$

where

$$S_i = \sum_{\mathbf{j}\in\mathcal{D}_n} [\Gamma_j]^{-2/\alpha}[\psi(Y_{\mathbf{j}})]^{-2} 1(\mathbf{Y_j} \in A\triangle A_i)$$

and where c is a constant which may change from line to line.

To show that $ES_i^{\gamma/2} \to 0$ as $i \to \infty$, denote by Ω_2 the probability space of the Y's and Γ's and, for each $\mathbf{j} \in \mathcal{D}_n$, let

$$\Omega_0^{(\mathbf{j})} = \{\omega \in \Omega_2 : \mathbf{Y_j} \in A\triangle A_i \text{ for only finitely many } i's\}.$$

Since $P(\mathbf{Y_j} \in A\triangle A_i) = p_{k(i)} \leq 2^{-i}$, the Borel-Cantelli lemma yields $P(\Omega_0^{(\mathbf{j})}) = 1$ and hence $P(\Omega_0) = 1$ where $\Omega_0 = \bigcap_{\mathbf{j}\in\mathcal{D}_n} \Omega_0^{(\mathbf{j})}$. Now $S_i < \infty$ a.s. by (3.4) and hence by the dominated convergence theorem, for a.e. $\omega \in \Omega_0$, we have $S_i \to 0$ as $i \to \infty$. Since, by (3.4), $S_i^{\gamma/2}$ is bounded by a random variable with finite expectation, we can apply the dominated convergence theorem once again, to conclude that $ES_i^{\gamma/2} \to 0$ as $i \to \infty$. Therefore $M^{(n)}(A_i) \to M^{(n)}(A)$ in L^γ and hence in probability, proving the claim.

(C) We now establish (3.3) for any $0 < \gamma < \alpha$. In view of Part (B), we may set $\psi(x) = 1/m^{1/\alpha}(\mathbf{R})$ without loss of generality, i.e. , suppose that the $Y_j, j = 1, 2, \dots$ are i.i.d. with distribution $\hat{m} = m/m(\mathbf{R})$.

For any $\mathcal{B}_n^{(s)}$ sets $C_1, \ldots, C_m$ in A disjoint, and any real $\theta_1, \ldots, \theta_m$ such that $|\theta_i| \le 1$, $i = 1, 2, \ldots, m$, we have by the generalized Khinchine inequality (Lemma 2.3),

$$\||\sum_{i=1}^{m} \theta_i M^{(n)}(C_i)\||_\gamma$$

$$= \left[E|n! c_\alpha^{n/\alpha} m(\mathbf{R})^{n/\alpha} \sum_{\mathbf{j}\in\mathcal{D}_n} [\epsilon_{\mathbf{j}}][\Gamma_{\mathbf{j}}]^{-1/\alpha} \sum_{i=1}^{M} \theta_i 1(\mathbf{Y_j} \in C_i)|^\gamma \right]^{1/\gamma}$$

$$\le c \left[E \left(\sum_{\mathbf{j}\in\mathcal{D}_n} [\Gamma_{\mathbf{j}}]^{-2/\alpha} |\sum_{i=1}^{m} \theta_i 1(\mathbf{Y_j} \in C_i)|^2 \right)^{\gamma/2} \right]^{1/\gamma}$$

$$\le c \left[E \left(\sum_{\mathbf{j}\in\mathcal{D}_n} [\Gamma_{\mathbf{j}}]^{-2/\alpha} 1(\mathbf{Y_j} \in A) \right)^{\gamma/2} \right]^{1/\alpha}$$

since $|\theta_i| \le 1$ and the C_i's are disjoint subsets of A. Therefore,

$$|M^{(n)}|_\gamma(A) \le c \left[E \left(\sum_{\mathbf{j}\in\mathcal{D}_n} [\Gamma_{\mathbf{j}}]^{-2/\alpha} 1(\mathbf{Y_j} \in A) \right)^{\gamma/2} \right]^{1/\gamma}$$

$$\le c(m^{(n)}(A))^{1/r}$$

by Lemma 2.2, establishing the right-hand inequality in (3.3). The left-hand inequality in (3.3) is a consequence of the generalized Khinchine inequality and Lemma 2.2:

$$|M^{(n)}|_\gamma(A) \ge \||M^{(n)}(A)\||_\gamma = (E|M^{(n)}(A)|^\gamma)^{1/\gamma}$$

$$\ge c \left(E \left[\sum_{\mathbf{j}\in\mathcal{D}_n} [\Gamma_{\mathbf{j}}]^{-2/\alpha} 1(\mathbf{Y_j} \in A) \right]^{\gamma/2} \right)^{1/\gamma}$$

$$\ge c(m^{(n)}(A))^{1/\alpha}.$$

(D) Fix $0 < \gamma < \alpha$. For any disjoint sets $A_1, A_2, \ldots$ in $\mathcal{B}_n^{(s)}$, we have, by (3.3)

$$|||M^{(n)}(\bigcup_{i=1}^{k} A_i) - M^{(n)}(\bigcup_{i=1}^{\infty} A_i)|||_\gamma \leq |M^{(n)}|_\gamma(\bigcup_{i=k+1}^{\infty} A_i)$$

$$\leq cm^{(n)}(\bigcup_{i=k+1}^{\infty} A_i),$$

which tends to zero as $k \to \infty$, proving that $M^{(n)}$, as defined in (3.2) is a σ-additive L^γ-valued random measure. This completes the proof of the theorem. ∎

We have supposed throughout this section that the control measure m is finite. We now relax this assumption, and suppose that m is *Radon*, that is $m(A) < \infty$ for all compact sets $A \in \mathcal{B}$. We can no longer extend the product measure $M^{(n)}$ to all sets A in $\mathcal{B}_n^{(s)}$, even with $m^{(n)}(A) < \infty$.

COUNTEREXAMPLE: Let $n = 2$, and m be the Lebesgue measure on $\mathbf{R}^1$. Choose a positive sequence $1 \geq a_n \downarrow 0$ such that

$$\sum_{n=1}^{\infty} a_n^2 < \infty, \tag{3.6}$$

$$\sum_{n=1}^{\infty} a_n^2 |\log a_n| = \infty. \tag{3.7}$$

Construct in $\mathbf{R}_2^+$ a sequence of disjoint squares $\{A_n\}$ of dimension $a_n \times a_n$ whose sides are parallel to the axes, with the Northwest corner of the square A_n located at the coordinate point (n, n). Let A be the union of these squares. Clearly, $m^{(2)}(A) = \sum_{n=1}^{\infty} a_n^2 < \infty$ by (3.6). On the other hand, if $M^{(2)}$ were extendable to A, we would have $M^{(2)}(A) = \sum_{n=1}^{\infty} M^{(2)}(A_n)$. Now, $M^{(2)}(A_n)$, $n = 1, 2, \ldots$ are independent random variables and $M^{(2)}(A_n) \stackrel{d}{=} a_n^{2/\alpha} X_n^2$, where $X_1, X_2, \ldots$ are i.i.d. $S\alpha S$.

Thus, if $M^{(2)}(A)$ were well defined, we would have

$$\sum a_n^{2/\alpha} X_n^2 < \infty \ a.s.. \tag{3.8}$$

The Three Series Theorem shows that (3.8) implies $\sum_{n=1}^{\infty} a_n^2 |\log a_n| < \infty$, contradicting (3.7).

Although $M^{(n)}$ cannot be extended to all $A \in \mathcal{B}_n^{(s)}$ with $m(A) < \infty$, it can be extended to the "domain of $M^{(n)}$":

DEFINITION 3.1 We say that $A \in \mathcal{B}_n^{(s)}$ is in the domain of $M^{(n)}$ if $M^{(n)}(A \cap [-K, K]^n)$ converges in probability as $K \to \infty$. In this case, we write $A \in \mathcal{D}(M^{(n)})$ and define

$$M^{(n)}(A) = \lim_{K \to \infty} M^{(n)}(A \cap [-K, K]^n). \tag{3.9}$$

THEOREM 3.2 *If m is Radon, then the representation* (3.2) *holds for $A \in \mathcal{D}(M^{(n)})$. The finite-dimensional distributions of $\{M^{(n)}(A),\ A \in \mathcal{D}(M^{(n)})\}$ do not depend on the function ψ.*

PROOF: We first show that (3.2) holds for $A \in [-K, K]^n$, $A \in \mathcal{B}^{(n)}$ when m is Radon. (The distribution m_ψ of the Y_j's should not depend on K.) Because the claim stated in the proof of Theorem 3.1 holds, there is a sequence $B_k \in \mathcal{F}_n^{(s)}$, $B_k \in [-K, K]^n$, $k = 1, 2, \ldots$, such that $m^{(n)}(A \Delta B_k) \to 0$ and $M^{(n)}(B_k) \to M^{(n)}(A)$ as $k \to \infty$. We may assume without loss of generality that the convergence is a.s.. Then by Fubini theorem, there is a.s. convergence for every $\{\Gamma_j\}$ and $\{Y_j\}$. By Lemma 2.4, for every $\{\Gamma_{\mathbf{j}}\}, \{Y_{\mathbf{j}}\}[\Gamma_{\mathbf{j}}]^{-1/\alpha}[\psi(Y_{\mathbf{j}})]^{-1} 1(\mathbf{Y_j} \in B_k)$ converges a.s. as $k \to \infty$ to some $a_{\mathbf{j}}$ which depends on $\{\Gamma_{\mathbf{j}}\}$ and $\{Y_{\mathbf{j}}\}$. To identify this $a_{\mathbf{j}}$, focus on $1(\mathbf{Y_j} \in B_k)$. Since $m^{(n)}(A \Delta B_k) \to 0$, we have $m_\psi^{(n)}(A \Delta B_k)$ and thus $1(\mathbf{Y_j} \in B_k) \to 1(\mathbf{Y_j} \in A)$ in probability. This implies $a_{\mathbf{j}} = [\Gamma_{\mathbf{j}}]^{-1/\alpha}[\psi(Y_{\mathbf{j}})]^{-1} 1(\mathbf{Y_j} \in A)$ a.s., and by Lemma 2.4, relation (3.2) holds for $M^{(n)}(A)$.

Now suppose $A \in \mathcal{D}(M^{(n)})$. We have just shown that (3.2) holds for $M^{(n)}(A \cap [-K, K]^n)$. Applying Lemma 2.4, we conclude that (3.2) also holds for $M^{(n)}(A)$.

Finally, as shown in Theorem 3.1 the finite-dimensional distributions of $\{M^{(n)}(A), A \in \mathcal{D}(M^{(n)})\}$ do not depend on ψ. ∎

4 Construction of the multiple integral

Let $\mathbf{f}$ be a symmetric vanishing on the diagonals strongly measurable function from $\mathbf{R}^n$ into a separable Banach space S. (Strong measurability in a Banach space setting corresponds to our usual notion of measurability.) Clearly, $\mathbf{f}$ is $\mathcal{B}_n^{(s)}$-measurable. If $\mathbf{f}$ is a simple function of the type $\mathbf{f}(x_1, \ldots, x_n) = \sum_{i=1}^{N} \mathbf{a}_i \mathbf{1}((x_1, \ldots, x_n) \in A_i)$ with $\mathbf{a}_1, \ldots, \mathbf{a}_n$ belonging to S and $A_1, \ldots, A_n$ disjoint and belonging to $\mathcal{B}_n^{(s)}$, then one defines the multiple stable integral of $\mathbf{f}$ with respect to M by

$$
\begin{aligned}
\mathbf{I}_n(\mathbf{f}) &= \int_{-\infty}^{\infty} \cdots \int_{-\infty}^{\infty} \mathbf{f}(x_1, \ldots, x_n) M(dx_1) \ldots M(dx_n) \\
&\stackrel{def}{=} \sum_{i=1}^{N} \mathbf{a}_i M^{(n)}(A_i).
\end{aligned}
$$

In general, $\mathbf{f}$ is said to be n *times integrable* with respect to M if

(i) there is a sequence of simple functions $\{\mathbf{f}^{(k)}, k = 1, 2, \ldots\}$ as above converging to $\mathbf{f}$ in semivariation, (i.e., if for some $0 < \gamma < \alpha$ and every $\epsilon > 0$

$$\lim_{k \to \infty} |M^{(n)}|_\gamma((x_1, \ldots, x_n) : \|\mathbf{f}_k(x_1, \ldots, x_n) - \mathbf{f}(x_1, \ldots, x_n)\| > \epsilon) = 0 \ ),$$

or, equivalently, by Theorem 3.1, in the measure $m^{(n)} = m \times m \times \cdots \times m$.

(ii) if for *every* $C \in \mathcal{B}_n^{(s)}$ the sequence $\{\mathbf{I}_n(\mathbf{f}^{(k)} 1_C), k = 1, 2, \ldots\}$ converges in $||| \ |||_\gamma$, $0 < \gamma < \alpha$, or, equivalently, in probability by [KS88], Proposition 2.1(ii). In that case we define

$$\mathbf{I}_n(\mathbf{f}) = \text{plim}_{k \to \infty} \mathbf{I}(\mathbf{f}^{(k)}).$$

By this we mean

$$\mathbf{I}_n(\mathbf{f}1_C) = \text{plim}_{k\to\infty}\mathbf{I}_n(\mathbf{f}^{(k)}1_C)$$

for every $C \in \mathcal{B}_n^{(s)}$, including, obviously, $C = \mathbf{R}^n$ (see [Bar56]).

Let $\mathbf{S}_n(\mathbf{f})$ denote the (formal) series

$$n!c_\alpha^{n/\alpha} \sum_{\mathbf{j}\in\mathcal{D}_n} [\epsilon_\mathbf{j}][\Gamma_\mathbf{j}]^{-1/\alpha}[\psi(Y_\mathbf{j})]^{-1}\mathbf{f}(\mathbf{Y}_\mathbf{j}),$$

where $Y_1, Y_2, \ldots$ are i.i.d. random variables having common distribution $m_\psi(dx) = \psi(x)^\alpha m(dx)$, and where $\mathbf{f}(\mathbf{Y}_\mathbf{j}) = \mathbf{f}(Y_{j_1}, Y_{j_2}, \ldots, Y_{j_n})$ if $\mathbf{j} = (j_1, j_2, \ldots, j_n)$.

THEOREM 4.1 *Suppose that* $\mathbf{f}$ *is* n *times integrable with respect to* M*. Then* $\mathbf{S}_n(\mathbf{f})$ *converges a.s. and*

$$\mathbf{I}_n(\mathbf{f}) \stackrel{d}{=} \mathbf{S}_n(\mathbf{f}) \tag{4.1}$$

PROOF: Let $\mathbf{f}$ be a simple function of the type $\mathbf{f}(x_1, \ldots, x_n) = \sum_{i=1}^{N} \mathbf{a}_i \mathbf{1}((x_1, \ldots, x_n) \in A_i)$, $A_i \in \mathcal{B}_n^{(s)}, i = 1, \ldots, N$. With $M^{(n)}$ given by (3.2) we have

$$\begin{aligned}
\mathbf{I}_n(\mathbf{f}) &= \sum_{i=1}^{N} \mathbf{a}_i M^{(n)}(A_i) \\
&= n!c_\alpha^{n/\alpha} \sum_{\mathbf{j}\in\mathcal{D}_n} \sum_{i=1}^{N} \mathbf{a}_i [\epsilon_\mathbf{j}][\Gamma_\mathbf{j}]^{-1/\alpha}[\psi(Y_\mathbf{j})]^{-1}1(\mathbf{Y}_\mathbf{j} \in A_i) \\
&= n!c_\alpha^{n/\alpha} \sum_{\mathbf{j}\in\mathcal{D}_n} [\epsilon_\mathbf{j}][\Gamma_\mathbf{j}]^{-1/\alpha}[\psi(Y_\mathbf{j})]^{-1}\mathbf{f}(\mathbf{Y}_\mathbf{j}) = \mathbf{S}_n(\mathbf{f}),
\end{aligned}$$

so that (4.1) always holds for simple functions.

Suppose now that a general symmetric vanishing on diagonals function $\mathbf{f}$ is n times integrable with respect to M. Then there is a sequence of simple functions $\mathbf{f}^{(k)}$ as above, converging to $\mathbf{f}$ as $k \to \infty$ in the measure $m^{(n)}$, such that $\mathbf{I}_n(\mathbf{f}^{(k)})$ converges as $k \to \infty$ to $\mathbf{I}_n(\mathbf{f})$. Of course we may assume (choose, if necessary, a subsequence) that $\mathbf{I}_n(\mathbf{f}^{(k)})$ converges a.s. to $\mathbf{I}_n(\mathbf{f})$. Then $\mathbf{S}_n(\mathbf{f}^{(k)})$ also converges a.s. to $\mathbf{I}_n(\mathbf{f})$, as $k \to \infty$. By Fubini's theorem we conclude that $\mathbf{S}_n(\mathbf{f}^{(k)})$ converges a.s. to $\mathbf{I}_n(\mathbf{f})$ for almost any realization of the sequences $\Gamma_1, \Gamma_2, \ldots$, and $Y_1, Y_2, \ldots$.

We use now Lemma 2.4 to conclude that there is an array $\mathbf{g}_{\mathbf{j}}(Y_1, Y_2, \ldots, \Gamma_1, \Gamma_2, \ldots)$, $\mathbf{j} \in \mathcal{D}_n$ such that

(i) $\mathbf{f}^{(k)}(\mathbf{Y}_{\mathbf{j}})[\Gamma_{\mathbf{j}}]^{-1/\alpha} \to \mathbf{g}_{\mathbf{j}}(Y_1, Y_2, \ldots, \Gamma_1, \Gamma_2, \ldots)$ as $k \to \infty$ for any $\mathbf{j} \in \mathcal{D}_n$,

(ii) the multiple series

$$\sum_{\mathbf{j} \in \mathcal{D}_n} \mathbf{g}_{\mathbf{j}}(Y_1, Y_2, \ldots, \Gamma_1, \Gamma_2, \ldots)[\epsilon_{\mathbf{j}}]$$

converges a.s. to $\mathbf{I}_n(\mathbf{f})$.

Since $\mathbf{f}^{(k)}$ converges to $\mathbf{f}$ in the measure $m^{(n)}$, we conclude from (i) that

$$\mathbf{g}_{\mathbf{j}}(Y_1, Y_2, \ldots, \Gamma_1, \Gamma_2, \ldots) = \mathbf{f}(\mathbf{Y}_{\mathbf{j}})[\Gamma_{\mathbf{j}}]^{-1/\alpha}, \ \mathbf{j} \in \mathcal{D}_n.$$

Therefore, by (ii), $\mathbf{S}_n(\mathbf{f})$ converges a.s. to $\mathbf{I}_n(\mathbf{f})$ for almost every choice of $\Gamma_1 \Gamma_2, \ldots, Y_1, Y_2, \ldots$, so by Fubini's Theorem, $\mathbf{S}_n(\mathbf{f})$ converges a.s. to $\mathbf{I}_n(\mathbf{f})$. ∎

References

[Bar56] R.G. Bartle. A general bilinear vector integral. *Studia Math.*, 15:337–352, 1956.

[DJ77] J. Diestel and J.J. Uhl Jr. Vector measures. *Math. Surveys*, 15, 1977.

[DS58] N. Dunford and J.T. Schwartz. *Linear Operators, Part 1: General Theory*. Wiley, New York, 1958.

[KS86a] W. Krakowiak and J. Szulga. Random multilinear forms. *Ann. Probab.*, 14:955–973, 1986.

[KS86b] W. Krakowiak and J. Szulga. Summability and contractivity of random multilinear forms. Preprint 86-61, Case Western Reserve University, 1986.

[KS88] W. Krakowiak and J. Szulga. A multiple stochastic integral with respect to a strictly p-stable measure. *Ann. Probab.*, 16:764–777, 1988.

[Ros89] J. Rosiński. On path properties of certain infinitely divisible processes. *Stoch. Proc. Appl.*, 33:73–87, 1989.

[RST] J. Rosiński, G. Samorodnitsky, and M.S. Taqqu. Sample path properties of stochastic processes represented as multiple stable integrals. Preprint, 1989. To appear in the *Journal of Multivariate Analysis*, 1991.

[SS89] G. Samorodnitsky and J. Szulga. An asymptotic evaluation of the tail of a multiple symmetric α-stable integral. *Ann. Probab.*, 17:1503–1520, 1989.

[ST90] G. Samorodnitsky and M.S. Taqqu. Multiple stable integrals of Banach-valued function. *J. Theor. Prob.*, 3:267–287, 1990.

[Szu89] J. Szulga. Multiple stochastic integrals with respect to symmetric infinitely divisible random measures. Preprint, 1989.

Numerical computation of non-linear stable regression functions *†‡

Clyde D. Hardin Jr. Gennady Samorodnitsky
Murad S. Taqqu

Abstract

A previous paper by the authors gives explicit formulas for the regression function of one stable random variable upon another. Although the regression may sometimes be linear, it is in general not a linear function. It involves the quotient of two integrals which cannot be computed analytically and must therefore be approximated numerically. Although the general problem of computing the integrals is straightforward in principle, the specific task is fraught with difficulties. In order to allow the practioner to apply the formulas, this paper presents a self-contained exposition of the regression problem and a software package, written in the C language, which overcomes the numerical difficulties and allows the user control over the accuracy of the approximation. The package also allows the user to compute numerically the probability density function of a stable random variable.

1 Introduction

One of the first steps towards understanding the conditional structure of a random vector (X_1, X_2) is the determination of the conditional expectation, or regression, $E(X_2|X_1 = x)$. For jointly α-stable (X_1, X_2), this problem is now completely solved ([JST91]). Although the regression is linear and has a simple characterization when (X_1, X_2) is symmetric, the regression, in the non-symmetric case, is

*The last two authors were supported by the AFOSR grant 89-0115 and the ONR grant 90-J-1287 at Boston University.

†**AMS 1980 subject classifications.** Primary 60E07, 62J02, 65-04. Secondary 65D30.

‡**Keywords and phrases** Stable random vectors, linear regression, numerical integration, software package.

usually non-linear and has a much more complicated form. It involves two integrals which cannot be computed analytically. To make matters worse, these integrals have features which make them not amenable to standard numerical integration procedures. In the past, the lack of *usable* formulas has been an impediment to the application of stable distribution to real-life phenomena. In this paper, we develop efficient algorithms for computing the regression formulas and we present them in a form useful to practioners. The procedures described here can also be used to evaluate other integrals which appear in the context of stable distributions.

We present a self-contained exposition of the problem of computing stable regression functions. To this end, we include all necessary formulas (Section 2), and describe some of the numerical difficulties and our numerical solution (Section 3). To make the exposition clear and of greatest use to the practitioner, and to show the details of the techniques, we include a complete software package which computes regression functions for any choice of the defining parameters. This package allows the user to select parameters to control the accuracy of the approximation, and includes a means for experimenting with these parameters in order to determine their effect in particular situations. As a byproduct of the regression computation, the software also computes stable density functions. Section 4 describes the functioning of the software in more detail, Section 5 gives a sample output, and Section 6 describes the usage and contains the source code, written in the C language.

We have investigated several numerical integration techniques, such as midpoint trapezoidal and Simpson's rule. A straightforward application of these techniques gives a very poor approximation. We developed variants with variable stepsize and cutoffs. A variety of representations of the integrals were also examined. While we make no claim that the procedures described here represent the best possible way to compute these regressions, they represent the most appropriate among those examined. They were used to compute the regression functions graphed in [JST91].

The techniques described here and the programs included in the paper should be useful not only to practioners but also to mathematicians working on stable distributions who want to link theory with application.

2 The regression formulas

The expression of the regression formulas depends on how the random vector (X_1, X_2) is specified. The most convenient way to specify α-stable random vectors $(X_1, X_2, \ldots, X_d)$ or stochastic processes $\{X_t,\ t \in T\}$ with $0 < \alpha < 2$, is through an integral representation. In this section, we present succinctly the integral representation, list some examples, give conditions for the existence of the regression and state the regression formulas. Proofs and additional details can be found in [Har84], [ST89] and [JST91].

2.1 Integral representation

Let $(E, \mathcal{E}, m)$ be a measure space and let β be a function from E to $[-1, 1]$. (In many examples, E is the real line, $\mathcal{E}$ is the Borel σ-field $\mathcal{B}$ and m is Lebesgue measure.) Let $\{f_t,\ t \in T\}$ be a family of measurable functions on E satisfying for each $t \in T$,

$$f_t \in L^\alpha(m)$$

and also $f_t \beta \ln |f_t| \in L^1(m)$ if $\alpha = 1$.

An α-stable stochastic process $\{X(t),\ t \in T\}$ is said to have an integral representation if it can be expressed as

$$X_t = \int_E f_t(x) M(dx),\ t \in T, \tag{2.1}$$

where the equality is in the sense of the finite-dimensional distributions. The integrator M is an α-stable random measure with control measure m and skewness intensity β. This means that for disjoint sets $A_1, \ldots, A_d$ in $\mathcal{E}$, with finite m-measure, the random variables $M(A_1), \ldots, M(A_d)$ are independent and each $M(A) = \int_E 1_A(x) M(dx)$ is an α-stable random variable, with scale parameter

$$(m(A))^{1/\alpha},$$

and skewness parameter

$$\frac{\int_A \beta(x) m(dx)}{m(A)}.$$

Physically, $X_t = \int_E f_t(x) M(dx)$ is obtained by weighing independent α-stable random variables $M(dx)$ by $f_t(x)$ and summing. Essentially any stable process admits an integral representation.

2.2 Examples

The following are typical examples of α-stable processes. The *moving average process*

$$X_t = \int_{-\infty}^{+\infty} f(t-x)M(dx)$$

(where m is Lebesgue), the *real harmonizable process*

$$X_t = \mathrm{Re}\int_{-\infty}^{+\infty} e^{itx}M(dx)$$

(here M is complex valued and m is a finite measure), and the *sub-Gaussian process*

$$X_t = \sqrt{A}G_t$$

where G_t is a mean zero Gaussian process and A is an $\frac{\alpha}{2}$-stable random variable, totally skewed to the right and independent of the process G_t.

Here are some more specific examples. The *Lévy stable motion*

$$X_t = \int_0^t M(dx)$$

(m is Lebesgue), is a process with stationary independent increments which is the α-stable counterpart to Brownian motion. The *linear fractional stable motion*

$$X_t = \int_{-\infty}^{+\infty} [a((t-x)_-^{H-1/\alpha} - (-x)_-^{H-1/\alpha}) + b((t-x)_+^{H-1/\alpha} - (-x)_+^{H-1/\alpha})]M(dx),$$

where $0 < H < 1$, $H \neq 1/\alpha$, m Lebesgue, has stationary dependent increments. So does the *log-fractional Lévy motion*

$$X_t = \int_{-\infty}^{+\infty} (\ln|t-x| - \ln|x|)M(dx),\ 1 < \alpha < 2,$$

m Lebesgue. The *Ornstein-Uhlenbeck process*

$$X_t = \int_{-\infty}^{t} e^{-\lambda(t-x)}M(dx),\ (\lambda > 0,\ m \text{ Lebesgue}),$$

is stationary and Markov,and so is the reverse *Ornstein-Uhlenbeck process*

$$X_t = \int_t^{\infty} e^{-\lambda(x-t)}M(dx),\ (\lambda > 0,\ m \text{ Lebesgue}).$$

2.3 Existence of conditional moments

Regression problems involve only a bivariate vector (X_{t_1}, X_{t_2}) which we denote simply

$$(X_1, X_2) = (\int_E f_1(x)M(dx),\ \int_E f_2(x)M(dx)). \tag{2.2}$$

When $\alpha > 1$, the regression $E(X_2|X_1 = x)$ is always defined because $E|X_2| < \infty$. This is not always the case when $0 < \alpha \leq 1$. For example the regression $E(X_{t_2}|X_{t_1} = x)$ is never defined for $0 < \alpha \leq 1$, when $t_2 > t_1$ and when X_t has the non-anticipating representation $X_t = \int_{-\infty}^{t} f_t(x)M(dx)$ with $\int_{t_1}^{t_2} |f_{t_2}(x)|^\alpha m(dx) \neq 0$. For example, when $\alpha \leq 1$ the regression is not defined for Lévy stable motion or the Ornstein-Uhlenbeck process. This makes sense physically because in these cases there is a constant $a = a(t_1, t_2)$ such that $X_{t_2} - aX_{t_1}$ is independent of X_{t_1} and $E|X_{t_2} - aX_{t_1}| = \infty$ when $\alpha \leq 1$.

Nevertheless, the regression is sometimes defined when $\alpha \leq 1$. A sufficient condition for the regression $E(X_2|X_1 = x)$ to be defined when $0 < \alpha \leq 1$ is

$$\int_{E_+} \frac{|f_2(x)|^{\alpha+\nu}}{|f_1(x)|^\alpha} m(dx) < \infty,$$

for some $\nu > 1 - \alpha$, where $E_+ = \{x \in E : f_1^2(x) + f_2^2(x) \neq 0\}$ (see [JST91].) For example, the regression is defined for all $0 < \alpha < 2$, when the process is the reverse Ornstein-Uhlenbeck or when it is sub-Gaussian or harmonizable.

2.4 Analytic expression of the regression function

The regression function $E(X_2|X_1 = x)$ is always linear in x in the symmetric α-stable case ($\beta(\cdot) \equiv 0$) but is usually not linear when $\beta(\cdot) \not\equiv 0$. The following quantities enter in the expression of the regression:

$$\begin{aligned}
\sigma_1^\alpha &= \int_E |f_1(x)|^\alpha m(dx), \\
\beta_1 &= \frac{1}{\sigma_1^\alpha} \int_E f_1(x)^{<\alpha>} \beta(x) m(dx), \\
\kappa &= \frac{[X_2, X_1]_\alpha}{\sigma_1^\alpha} = \frac{1}{\sigma_1^\alpha} \int_{E_+} f_2(x) f_1(x)^{<\alpha-1>} m(dx), \\
\lambda &= \frac{1}{\sigma_1^\alpha} \int_{E_+} f_2(x) |f_1(x)|^{\alpha-1} \beta(x) m(dx), \\
\mu_1 &= -\frac{2}{\pi} \int_{E_+} f_1(x) (\ln \frac{|f_1(x)|}{\sqrt{f_1^2(x) + f_2^2(x)}}) \beta(x) m(dx), \qquad (\text{case } \alpha = 1), \\
k_0 &= \frac{1}{\sigma_1} \int_{E_+} f_2(x) (\ln \frac{|f_1(x)|}{\sqrt{f_1^2(x) + f_2^2(x)}}) \beta(x) m(dx), \qquad (\text{case } \alpha = 1).
\end{aligned}$$

σ_1 is the scale parameter of X_1, β_1 is the skewness parameter of X_1, κ is the normalized covariation of X_2 on X_1, λ is the normalized skewed covariation of X_2 on X_1, μ_1 and k_0 are shift parameters that appear in the case $\alpha = 1$. The notation $a^{<b>} = |a|^b \operatorname{sign} a$ denotes a signed power.

When $\beta(\cdot) \equiv 0$, the regression is linear

$$E(X_2|X_1 = x) = \kappa x,$$

with slope equal to κ, the normalized covariation. Observe that $\kappa = \frac{EX_2X_1}{EX_1^2}$ when $\alpha = 2$.

When $\beta(\cdot) \not\equiv 0$, the formulas for the regression are different in the cases $\alpha \neq 1$ and $\alpha = 1$.

- In the case $\alpha \neq 1$,

$$E(X_2|X_1 = x) = \kappa x + \alpha(\tan\frac{\pi\alpha}{2})(\lambda - \beta_1\kappa)\frac{r(x/\sigma_1)}{s(x/\sigma_1)}\sigma_1, \tag{2.3}$$

where

$$r(x) = \int_0^\infty e^{-t^\alpha} t^{\alpha-1}\cos(xt - (\tan\frac{\pi\alpha}{2})\beta_1 t^\alpha)dt, \tag{2.4}$$

$$s(x) = \int_0^\infty e^{-t^\alpha}\cos(xt - (\tan\frac{\pi\alpha}{2})\beta_1 t^\alpha)dt. \tag{2.5}$$

The density function of X_1 equals $\frac{1}{\pi\sigma_1}s(\frac{x}{\sigma_1})$.

- When $\alpha = 1$ and $\beta_1 \neq 0$,

$$E(X_2|X_1 = x) = -\frac{2\sigma_1}{\pi}k_0 + \kappa(x - \mu_1) + \frac{\lambda - \beta_1\kappa}{\beta_1}[(x - \mu_1) - \sigma_1\frac{U(x)}{W(x)}], \tag{2.6}$$

where

$$U(x) = \int_0^\infty e^{-\sigma_1 t}\sin(t(x - \mu_1) + \frac{2}{\pi}\beta_1\sigma_1 t\ln t)dt, \tag{2.7}$$

$$W(x) = \int_0^\infty e^{-\sigma_1 t}\cos(t(x - \mu_1) + \frac{2}{\pi}\beta_1\sigma_1 t\ln t)dt. \tag{2.8}$$

- When $\alpha = 1$ and $\beta_1 = 0$,

$$E(X_2|X_1 = x) = -\frac{2\sigma_1}{\pi}k_0 + \kappa(x - \mu_1) - \frac{2\sigma_1}{\pi}\lambda\frac{V(x)}{W(x)}, \tag{2.9}$$

where

$$V(x) = \int_0^\infty e^{-\sigma_1 t}(1 + \ln t)(\cos t(x - \mu_1))dt \tag{2.10}$$

and $W(x)$ is as above. When $\alpha = 1$, the density of X_1 equals $\frac{1}{\pi}W(x)$.

2.5 Linearity of the regression

Suppose $0 < \alpha < 2$. The regression is linear when $\beta_1 = \pm 1$, and more generally, when $\lambda = \beta_1\kappa$. It is then equal to

$$\kappa x$$

if $\alpha \neq 1$ and to

$$\kappa x - (\frac{2\sigma_1}{\pi} k_0 + \kappa\mu_1)$$

if $\alpha = 1$.

The regression is always asymptotically linear. When $\beta_1 \neq \pm 1$,

$$E(X_2|X_1 = x) \sim \frac{\kappa + \lambda}{1 + \beta_1} x, \ x \to \infty,$$

and

$$E(X_2|X_1 = x) \sim \frac{\kappa - \lambda}{1 - \beta_1} x, \ x \to -\infty.$$

3 Numerical Computation of the Integrals

Computation of the regression formulas given in (2.3), (2.6), and (2.9) is straightforward once the functions $r(x), s(x), U(x), V(x)$ and $W(x)$, defined through the integrals in (2.4), (2.5),(2.7),(2.8) and (2.10), have been computed. This section describes the computation of these integral functions.

Integrand Characteristics

In the form given, the integrals defining $r(x), s(x), U(x), V(x)$, and $W(x)$ are improper integrals, and all have integrands which decay exponentially while oscillating, in many cases increasingly rapidly, about zero. The integrand of V, and the integrand of r in the case $\alpha < 1$, have singularities at zero; otherwise the integrands are reasonably well-behaved.

It might be expected that with the exponential decay of the integrands, the improper integrals could be truncated at some adequately high cutoff and subjected to any one of the standard integration techniques, with acceptable results. One problem with this is, of course, the singularities. Another problem is that as the parameter x gets large, the sinusoidal oscillations become more rapid, and the integrals each decay to zero, imposing more stringent requirements on the accuracies of the individual integrals to ensure accuracy of the quotient. For example, the integral $r(x)$ decays as $x^{-\alpha}$ and $s(x)$ decays as $x^{-(\alpha+1)}$ ([JST91]). For larger x, then, both more iterations and higher truncation levels may be required. The parameter α also affects the requirements for truncation due to the slower decay of the integrand for lower α. To illustrate some of the behavior mentioned above, Figure 1 shows a graph of the integrand for $r(x)$ for $\alpha = 0.5$ and $\alpha = 1.5$, when the value of x is 8.

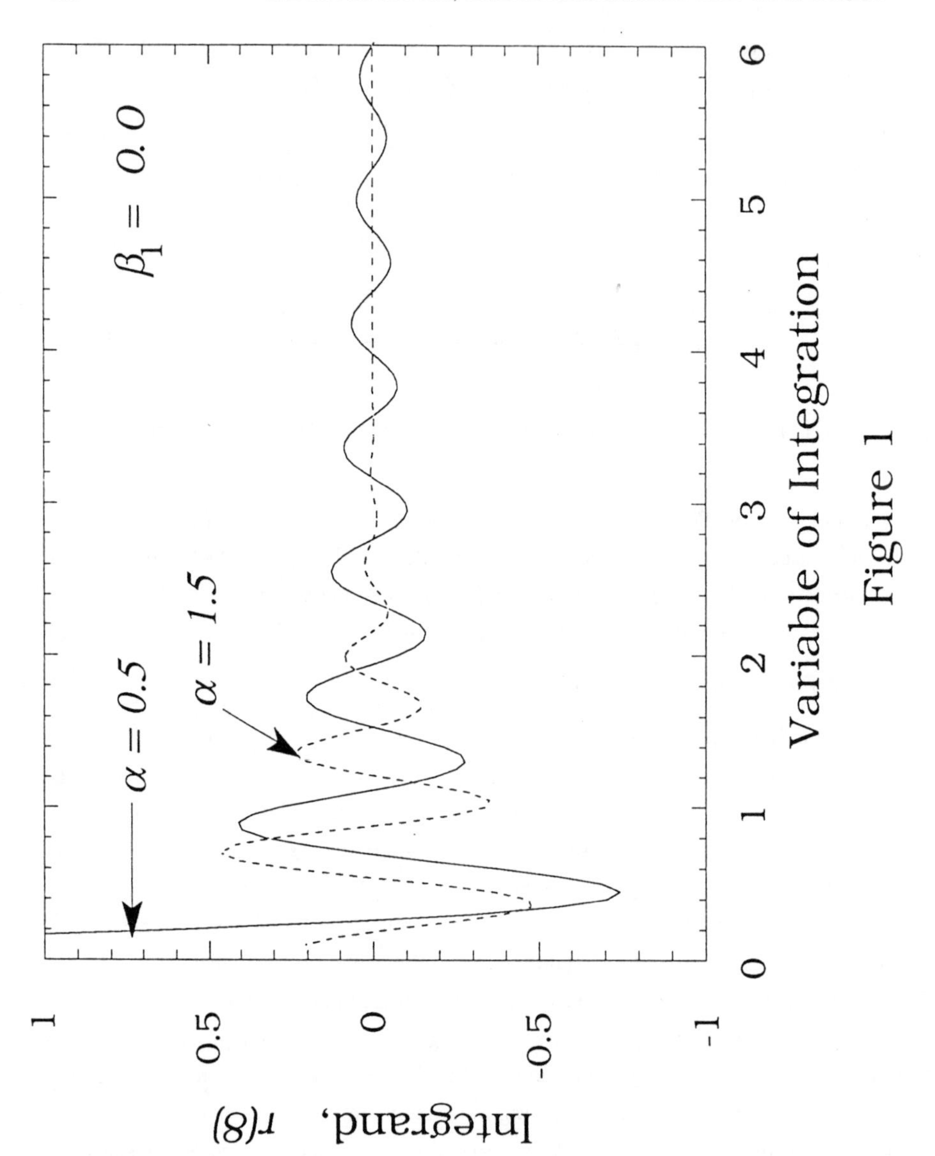

β1 = 0.0
α = 0.5
α = 1.5
Integrand, r(8)
Variable of Integration
1
0.5
0
-0.5
-1
0
1
2
3
4
5
6

Figure 1

One can make the integrals more amenable to numerical techniques by making changes of variables in the integrands. For example, the change of variables $s = t^\alpha$ converts $r(x)$ to

$$r(x) = \frac{1}{\alpha}\int_0^\infty e^{-s}\cos(xs^{\frac{1}{\alpha}} - (\tan\frac{\pi\alpha}{2})\beta_1 s)ds. \tag{3.1}$$

This eliminates the singularity present in the integrand in (2.4) when $\alpha < 1$. The same change of variables in the denominator integral $s(x)$, however, introduces a singularity at zero for values of α greater than one. To eliminate the need for truncation, the change of variables $s = e^{-t^\alpha}$ represents $r(x)$, after some fortuitous cancellations, as

$$r(x) = \frac{1}{\alpha}\int_0^1 \cos(x(-\ln s)^{\frac{1}{\alpha}} + (\tan\frac{\pi\alpha}{2})\beta_1 \ln s)ds. \tag{3.2}$$

The same change of variables in the integrand for $s(x)$ introduces a singularity at zero for $\alpha < 1$, and at one for $\alpha > 1$. Although the representation (3.2) results in a bounded integrand, the integrand oscillates increasingly rapidly about zero. Even though the contribution of the integrand from zero to ϵ is bounded by ϵ, the oscillations for s slightly greater than ϵ may be rapid enough to require an unacceptably small stepsize in the numerical procedure in order to obtain the necessary accuracy.

In the case $\alpha = 1$, the logarithms in the integrands do not cause singularities or discontinuities except in $V(x)$. That singularity may be removed at the expense of introducing an additional improper integral with the change of variables $t = e^{-s}$ for $0 < t \leq 1$:

$$\int_0^1 e^{-\sigma_1 t}(1+\ln t)\cos(t(x-\mu_1))dt = \int_0^\infty e^{-\sigma_1 e^{-s}-s}(1-s)\cos(e^{-s}(x-\mu_1))ds. \tag{3.3}$$

Numerical Procedures

Because of the variation of integrand characteristics over the range of parameters considered, no matter what the representation of the integral, it is important for the integration scheme to allow truncation (if used) and stepsize to vary as the parameters vary. Bounding arguments can be applied to whatever form of the integrand one chooses in order to find a cutoff point for which the truncation error is small. Stepsize is somewhat more difficult to pick, as the usual error formulas involve derivatives which may be difficult to evaluate, or may have extremely high values not indicative of the actual errors.

A standard way of solving the stepsize selection process is to use one of the (by now standard) versions of the classical midpoint and trapezoidal integration formulas which allow the stepsize to be iteratively decreased. The integration sample point mesh is successively refined, and the corresponding numerical approximation is successively refined by updating the previous computations. When successive approximations differ by some small pre-specified amount, the iterations are terminated. Approaches such as these seem suited to the problem at hand since different parameter choices have different requirements. Unfortunately, these methods are slow to converge for our problem, and tend to require more computation than other methods. This problem is not alleviated by any of the changes of variables mentioned. In fact, for some representations, the approximations did not converge.

Another promising possibility is to use a variable stepsize formula, since for all of the representations considered, the integrand varies much more rapidly in one part of the domain than in other parts. The bulk of the computation could take place in the regions of greatest change, enabling faster convergence and more control over the accuracy. Unfortunately, as for the iterative methods, this seemingly good idea fails the test of a practical implementation.

One consideration in the choice of integrand representation and integration technique is computational expense, and tradeoffs involving number of iterations, number and complexity of function evaluations, and numerical accuracy must be considered. For example, in the case $\alpha \neq 1$, the integrands $r(x)$ and $s(x)$ in (2.4) and (2.5) are identical except for the factor $t^{\alpha-1}$. It is tempting to attempt an integration scheme which reuses the function evaluation for s in computing r in order to decrease the computational overhead. The savings realized in doing this, however, do not outweigh the costs incurred, compared with other methods, due to a larger number of iterations.

Many combinations of integration technique and integrand representation were tried. The combination that met with the most success was the version of Simpson's rule known as the "one-third" rule, coupled with the representation (3.1) for the function $r(x)$, the representation (2.5) for the function $s(x)$, the representations (2.7) and (2.8) for $U(x)$ and $W(x)$, and the representation for $V(x)$, obtained by writing the integral in (2.10) as the sum of two integrals, one from 0 to 1 and the other from 1 to ∞, and using the representation (3.3) for the first of these. The remainder of this section describes this approach in more detail.

To choose the stepsizes, the following approach is taken. Require a certain minimum stepsize, but allow it to be chosen smaller, based on the frequency of the sinusoidal term. This frequency is, of course, not well-defined since no unique frequency exists. An attempt is made, however, to approximate the frequency as effected by the regressor value, x, and to force the number of sample points per "period" to be bounded below by an input parameter. A parameter declaring the maximum number of samples is also provided which prevents the algorithm from taking the stepsize guidance to extremes.

Since all of the integrals are represented as improper integrals, they must be truncated. An input parameter, ϵ, specifies the greatest absolute truncation error to be suffered in either integral being computed. Analytically determined cutoff points, M, are then determined for each integral. More specifically, the truncation error, ϵ, must be less than or equal to the integral from the cutoff, M, to infinity. By bounding the sinusoidal terms with unity, M is determined as follows. Solving

$$\epsilon = \int_M^\infty ke^{-ct} = \frac{k}{c}e^{-cM} \tag{3.4}$$

for M results in

$$M = -\frac{1}{c}\ln\left[\frac{c\epsilon}{k}\right] \tag{3.5}$$

for positive c and k. This implies that the truncation error is no greater than ϵ when the integral is truncated at M, where

- $M = -\ln(\alpha\epsilon)$ for $r(x)$
- $M = -\frac{1}{\sigma_1}\ln(\sigma_1\epsilon)$ for $U(x)$ and $W(x)$.

The cutoffs for $s(x)$ and $V(x)$ require the following lemma.

LEMMA 3.1 *For $p > 0$ and $x \geq \max\{0, 4p(\ln(4p) - 1)\}$, $e^{-x}x^p \leq e^{-x/2}$.*

PROOF: The result follows if it is shown that for such $x, \ln x \leq \frac{x}{2p}$. Since $\ln x \leq x - 1$ for all $x > 0$, we have $\ln(cx) \leq cx - 1$ for all $x, c > 0$, in turn implying that $\ln x \leq cx - 1 - \ln c$ for such x and c. Now, for $0 < c < \frac{1}{2p}$, we have $cx - 1 - \ln c \leq \frac{x}{2p}$ if and only if $x \geq (-1 - \ln c)/(\frac{1}{2p} - c)$. Choosing $c = \frac{1}{4p}$ gives the result. ∎

We focus first on the cutoff for $s(x)$. Make the change of variables $u = t^\alpha$ and apply the Lemma to compute

$$\begin{aligned}\int_M^\infty e^{-t^\alpha}dt &= \frac{1}{\alpha}\int_{M^\alpha}^\infty e^{-u}u^{\frac{1}{\alpha}-1}du \\ &\leq \frac{1}{\alpha}\begin{cases} \int_{M^\alpha}^\infty e^{-u}du & \alpha > 1,\ M > 1, \\ \int_{M^\alpha}^\infty e^{-u/2}du & \alpha < 1,\ M^\alpha \geq B, \end{cases}\end{aligned}$$

where $B = 4(\frac{1}{\alpha} - 1)\ln[4(\frac{1}{\alpha} - 1) - 1]$. Consequently,

$$\int_M^\infty e^{-t^\alpha} dt \le \begin{cases} \frac{1}{\alpha} e^{-M^\alpha} & \alpha > 1,\ M > 1, \\ \frac{2}{\alpha} e^{-M^{\alpha/2}} & \alpha < 1,\ M \ge B^{1/\alpha}. \end{cases}$$

This shows that truncating the integral defining $s(x)$ at M, where

$$M = \begin{cases} \{-\ln(\alpha\epsilon)\}^{1/\alpha} \vee 1 & \alpha > 1, \\ \{-2\ln(\frac{\alpha\epsilon}{2})\}^{1/\alpha} \vee B^{1/\alpha} & \alpha < 1, \end{cases}$$

results in truncation error not greater than ϵ.

We now turn to the cutoff for $V(x)$. Use the Lemma to write for $M \ge \frac{4}{\sigma_1}(\ln 4 - 1) \approx \frac{1.545}{\sigma_1}$,

$$\begin{aligned} \int_M^\infty e^{-\sigma_1 t}(1 + \ln t) dt &\le \int_M^\infty e^{-\sigma_1 t} t dt \\ &\le \frac{1}{\sigma_1} \int_M^\infty e^{-\frac{\sigma_1 t}{2}} dt \\ &= \frac{2}{\sigma_1^2} e^{-\frac{\sigma_1 M}{2}}. \end{aligned}$$

Thus, defining $M = \max\{\frac{4}{\sigma_1}(\ln 4 - 1), -\frac{2}{\sigma_1}\ln(\frac{\sigma_1^2\epsilon}{2})\}$, the truncation error incurred by truncating the integral defining $V(x)$ at M is no greater than ϵ.

4 What the Software Package does

The software package included in Section 6 contains the programs **quint**, **reg**, **reg1**, **testints**, **testints1**, **eval**, **eval1**, **simrule**, and **makefile**. It implements in the C programming language the integration techniques outlined in Section 3. Given the parameters $\alpha, \beta_1, \sigma_1, \kappa, \lambda$, and if necessary, μ_1 and k_0, the programs calculate the regression $E(X_2|X_1 = x)$ for an input range of the regressor values, x. The programs allow the user some influence over the accuracy of the numerical integrations involved by providing selectable parameters which control the truncation error, the stepsize, and the number of sample points.

The computations are dissimilar enough in the cases $\alpha \ne 1$ and $\alpha = 1$ that separate routines are provided for those two cases. In the case $\alpha \ne 1$, the integral functions $r(x)$ and $s(x)$ depend only on α and β_1. Since nearly all the computational overhead in computing the regressions is in the computation of these integrals, it is more efficient to compute the integrals and their quotients separately from the regression. The results are then stored in files which may be read for computing regressions for different σ_1, κ, and λ, without having to recompute the integrals. The programs **quint** (quotient of integrals) and **reg** (regression), contained in the ".c" files of the same name carry out this pair of operations. (Files with the extension ".c" are source files written in C. Once compiled, they become object files and the extension is then denoted ".o".) For an input α and β_1, the program **quint** computes the quotient $r(x)/s(x)$ for an input range of regressors, writing the integrals and their quotients to disk. (If the input file containing the values of α and β_1 is called, say, *data*, then the output file written by **quint** will be called *data.quint*.) In addition to the values α and β_1, the input file to **quint** must contain the least regressor value, the greatest regressor value and the increment between regressors, all in units of σ_1. For example, if $\alpha = 1.5$, $\beta_1 = -0.3$, $\sigma_1 = 2$ and we want $E[X_2|X_1 = x]$ for $-10 \leq x \leq 10$ with $\Delta x = 0.02$, then the input file to **quint** should read:

1.5 -0.3 -5 5 0.01

Given values for $\sigma_1, \kappa, \lambda$, and the name of a **quint** output file, **reg** computes the regression for the chosen parameters. If no output file name is specified, the output goes to the standard output (e.g. the monitor), and can thus be piped to other applications, including the UNIX plotter. Since the function $s(x)$ computed by **quint** is, up to a scaling factor and a change of variables, the probability density function for X_1, an option exists in **reg** to compute this density function.

In the case $\alpha = 1$, the integrals depend in an essential way on σ_1 and μ_1, as well as on β_1. Consequently, the computation of the integrals and the regressions are bundled together in the program **reg1** (regression for $\alpha = 1$). This program takes an input parameter file on disk specifying $\sigma_1, \beta_1, \mu_1, k_0$, and the range and spacing of the regressors, and computes the integrals $W(x)$ and $U(x)$ or $V(x)$ and the regressions $E(X_2|X_1 = x)$. The integrals and regressions are written to a disk file. (If the input file is called, say, *data*, the output file will be called *data.reg1*.) The program **reg1** also contains an option for computing the probability density function for $\alpha = 1$.

Both integral computation programs, **quint** and **reg1**, control the integration through several parameters. The parameter ϵ of Section 3 appears as *trunceps* in the code and is the truncation error bound for each integral. This parameter governs selection of the upper limit of integration M (*T1num* and *T1den* in the code) through the functional dependencies derived in Section 3. The parameter *minptsper* (minimum number of points per period) sets the lower bounds for the number of integration sample points in one "period" of the sinusoid in the integrand. Other parameters set the minimum and maximum number of sample points, the minimum stepsize, and the maximum upper limit of integration.

Two auxiliary routines, **testints** (test integrals) and **testints1** (test integrals for $\alpha = 1$) are included for experimentation purposes. They compute the integrals ($r(x)$ and $s(x)$ in **testints**; and $W(x)$ and either $U(x)$ or $V(x)$ in **testints1**) for just a single x value, given the upper limit of integration and the number of sample points to use. To keep the input simple, the other parameters defining the integrals are redefinable but must be changed within the program. These routines allow experimentation to determine the effect of changes in parameter values so that prudent choices can be made in the more computationally intensive programs. The codes to evaluate the integrand in the numerator (i.e. the integrand in $r(x)$ if $\alpha \neq 1$; in $U(x)$ or $V(x)$ if $\alpha = 1$) are called **eval.c** and **eval1.c**, respectively. Since they are included in all programs computing the integrals, **testints** and **testints1** can be used to experiment with different integrand representations as well.

Also included in the package is a function, **simrule**, to perform the Simpson's $\frac{1}{3}$-rule integration. The function is completely general, with the inputs consisting of the integrand, the limits of integration, and the number of sample points; the output gives the numerical approximation of the integral.

Finally, a **makefile** program is included. It provides a mechanism for compiling source programs on a UNIX operating system. It also indicates the programs that are linked together. For example, **quint** uses **simrule** and **eval**.

In summary, to perform the regressions, use the programs **simrule, eval, quint** and **reg** when $\alpha \neq 1$, and use the programs **simrule, eval1** and **reg1** when $\alpha = 1$.

5 How to use the Software Package

Put in one directory all the .c programs as well as the **makefile** program. Then create object files. On a UNIX system, this is done by entering successively

make quint make reg make reg1 make testints make testints1.

We shall now illustrate the use of the software package with an example. Let (X_1, X_2) be defined as in (2.2), where $E = [0, 1]$, M is α-stable with $\alpha = 1.5$,

Lebesgue control measure and constant skewness intensity $\beta(\cdot) = 0.3$, and where

$$f_1(x) = \begin{cases} 1 & 0 \le x \le 0.9, \\ -1 & 0.9 < x \le 1, \end{cases} \quad \text{and} \quad f_2(x) = \begin{cases} 0 & 0 \le x \le 0.9, \\ 1 & 0.9 < x \le 1. \end{cases}$$

The goal is to obtain $E(X_2|X_1 = x)$ for $-10\sigma_1 \le x \le 10\sigma_1$ with increments $\Delta x = 0.1\sigma_1$. The formulas in Section 3 give

$$\sigma_1 = 1, \quad \beta_1 = 0.24, \quad \kappa = -10^{-1/3} = -0.464159, \quad \lambda = 0.139248.$$

Now make an input file for **quint**, called, say *example*, containing[1]

1.5 0.24 -10 10 0.1

These numbers equal respectively α, β_1, the least regressor value, the greatest regressor value and the increments between regressors. (The last three numbers are in units of σ_1). To run **quint**, type

quint example

at the prompt. The output is automatically written to a file called *example.quint*. That file lists parameter values whose meaning is defined in the code, and gives, in tabular form, the values of x, the numerator $r(x)$, the number of iterations and stepsize used in the computation of $r(x)$, the denominator $s(x)$, the number of iterations and stepsize used in the computation of $s(x)$, and finally the ratio $r(x)/s(x)$. This is how the beginning of the *example.quint* file will look like:

```
    alpha: 1.500000
    beta1: 0.240000
    lowbd: -10.000000
     upbd: 10.000000
      inc: 0.100000
  maxiter: 10000
  miniter: 1000
 trunceps: 0.000500
 maxvalue: 500.000000
 mindelta: 0.050000
minptsper: 40.000000
       TO: 1.000000e-20

****
-10.00     -0.02031767       1000   0.00720               0.00257451        1000   0.0076
 -7.89184302
```

[1] An easy way to do this in UNIX is to enter first *cat* > *example*, then type the numbers, press enter and quit with Control-D.

```
-9.90    -0.02058080      1000  0.00720          0.00264550      1000  0.0076
-7.77954249
-9.80    -0.02084428      1000  0.00720          0.00271929      1000  0.0076
-7.66532596
```

The graphs of the numerator $r(x)$, the denominator $s(x)$ and the quotient $r(x)/s(x)$ are displayed in Figure 2.

To compute the regression and to store its values in a file, called, say *result*, type

reg 1 -0.464159 0.139248 example.quint result

The three numbers are the values of σ_1, κ and λ respectively. The file *result* gives the values of x and $E(X_2|X_1 = x)$ in tabular form and begins as follows:

```
-10.00000    7.60868
 -9.90000    7.52004
 -9.80000    7.43068
 -9.70000    7.34173
 -9.60000    7.25419
 -9.50000    7.16885
 -9.40000    7.08617
 -9.30000    7.00620
 -9.20000    6.92862
 -9.10000    6.85277
 -9.00000    6.77778
 -8.90000    6.70269
```

The file *result* is used to obtain the graphical representation of the regression function displayed in Figure 3.

Note that the file *example.quint* can also be used to obtain the probability density function of an α-stable random variable with $\alpha = 1.5$ and skewness parameter $\beta_1 = 0.24$. Entering

reg -d example.quint density

writes the output to a file called *density*.

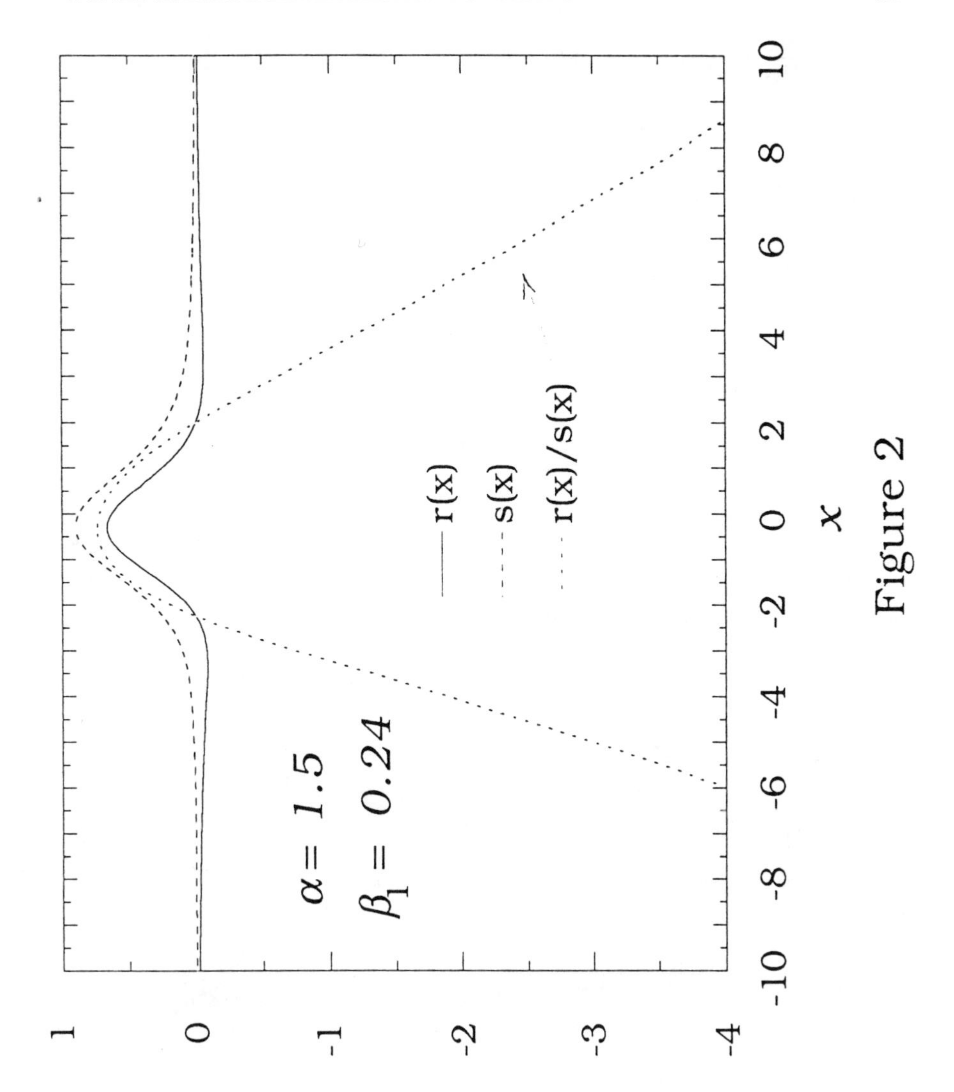

Figure 2

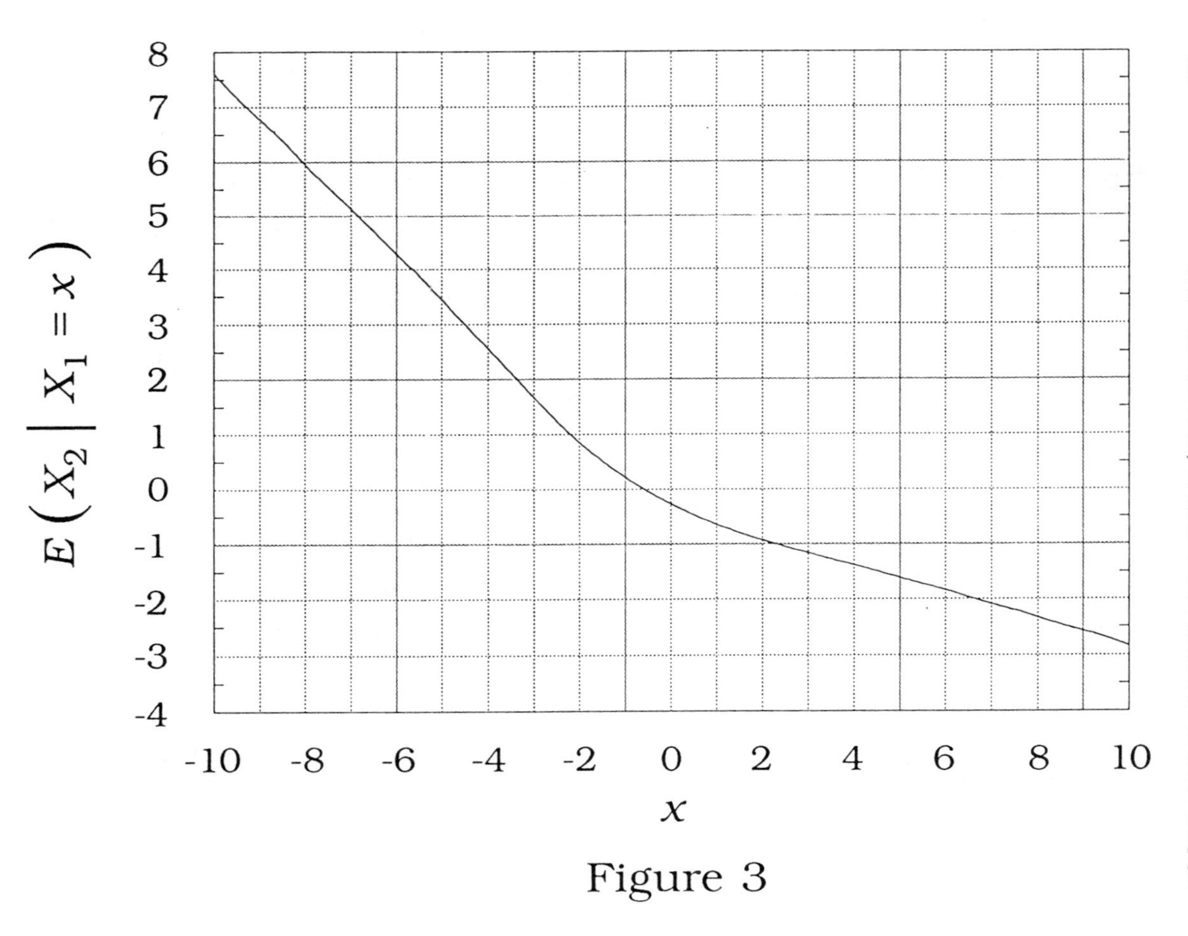

Figure 3

6 Source Code

Following is a listing of the source code implementing the procedures described in the preceding sections. The code has been, we hope, adequately commented. Particularly important are the header comments appearing at the definition of each program. They state the program name and describe the purpose of the program, the usage (command line syntax), the input and output. The code is intentionally elaborate in its I/O since it is important that the user have full view of, and maintain control of, the many parameters involved. Messages are issued at each stage of the processing, and as output files are written the contents are echoed to the terminal. When possible, output files automatically include the parameters used to create them, both for use with other programs, and for bookkeeping purposes.

In **quint** and **reg1**, some of the parameters controlling the integrations have been included as optional command line arguments. The default values for these parameters, and the values of those controlling parameters not definable on the command line, are initialized in the parameter declaration and may be reset at will.

In the code, the parameters σ_1 and β_1 are called *sigma1* and *beta1* respectively. The truncation error ϵ in (3.4) is called *trunceps* and the cutoff M in (3.5) of the upper limit of integration is called *T1num* for the integral in the numerator and *T1den* for the integral in the denominator.

6.1 quint.c

```
/*
 * Program name:
 *    quint
 *
 * Purpose:
 *    This routine computes the two regression integrals and their quotient
 *    for input values of alpha and beta1, with alpha not equal to one.  The
 *    integrals are computed for a range of regressor values, using
 *    Simpson's rule separately on each integral.  Various parameters
 *    controlling the numerical integration may be set on the command line or
 *    recompiled in the code (see Usage, and comments in the code).
 *
 * Usage:
 *    quint  infile  [miniter]  [minptsper]  [trunceps]
 *
 *        infile    - Disk file consisting of a list of the five input
```

```
 *                     parameters: alpha, beta1, least regressor value, great-
 *                     est regressor value, and increment between regressors.
 *        miniter   - Minimum total number of sample points in each numerical
 *                     integration (default=1000).
 *        minptsper - Minimum number of sample points per period to use in
 *                     each numerical integration (default=40).
 *        trunceps  - Maximum truncation error allowed.  This parameter deter-
 *                     mines the upper limit of integration (default=0.0005).
 *
 * Output:
 *    Output is written to a file whose name is the name of the input parameter
 *    file with ".quint" appended.  This file contains a listing of the
 *    parameters used to create the file, followed by tab-delimited data.  One
 *    row of data is produced for each regressor value, with the following
 *    entries in the columns: the regressor value, the numerator integral, the
 *    number of steps used in the numerator integration, the stepsize in the
 *    numerator integration, the denominator integral, the number of steps for
 *    that integration, the corresponding stepsize, and finally the value of
 *    the quotient of the two integrals.
 */

#include <stdio.h>
#include <math.h>
#define MIN(A,B)  (((A)<(B))?(A):(B))
#define MAX(A,B)  (((A)>(B))?(A):(B))

double alpha, beta1, abeta1, x, ooa;


main(argc,argv)
     int argc;
     char *argv[];
{
  int    i, numpts, numiters, deniters;
  double lowbd, upbd, inc, t, beta1, a, oopi, twopi, num, den, T1num, T1den,
         quint, expbound, truncbound, fac, absx, del, minperiod, dendel,numdel;

  char   outfilename[50];
  FILE   *fopen(), *infile, *outfile;
  double simrule(), numeval(), deneval();

  /* Parameters controlling the numerical integration.  Alter if desired     */
  int    maxiter=10000,    /* The maximum number of num'l int'n steps allowed */
         miniter= 1000;    /* The minimum number of num'l int'n steps allowed */
  double trunceps=0.0005, /* The upper bound on the truncation error allowed */
         maxvalue=500,     /* The maximum upper limit of integration          */
         mindelta=0.05,    /* The minimum num'l int'n step-size allowed       */
         minptsper=40,     /* The minimum number of points per period allowed */
         T0=1.0e-20;       /* The initial point for the num'l int'ns (>0)     */

/* Check command line for valid usage */
  if (argc<2) {
    fprintf(stderr,
    "Usage: %s infile [miniter] [minptsper] [trunceps]\n",argv[0]);
```

```
    exit(-1);
  }

/* Open input parameter file */
  if (!(infile = fopen(argv[1],"r"))) {
    fprintf(stderr,"Unable to open input file\n");
    exit(-1);
  }

/* Open output file */
  strcpy(outfilename,argv[1]);
  strcat(outfilename,".quint");
  if (!(outfile = fopen(outfilename,"w"))) {
    fprintf(stderr,"Unable to open output file\n");
    exit(-1);
  }

/* Check for optional command line arguments */
  if (argc>2) miniter   = atof(argv[2]);
  if (argc>3) minptsper = atof(argv[3]);
  if (argc>4) trunceps  = atof(argv[4]);

/* Read input file and echo parameters to standard output and output file */
  fscanf(infile,"%lf %lf %lf %lf %lf",
 &alpha, &beta1, &lowbd, &upbd, &inc);
  fprintf(stdout,"\nalpha = %5.3f,  beta1 = %6.3f\n",alpha,beta1);
  fprintf(stdout,"x ranges from %6.3f to %6.3f by %5.3f.\n\n",lowbd,upbd,inc);
  fprintf(outfile,"    alpha: %lf\n",alpha);
  fprintf(outfile,"    beta1: %lf\n",beta1);
  fprintf(outfile,"    lowbd: %f\n",lowbd);
  fprintf(outfile,"     upbd: %f\n",upbd);
  fprintf(outfile,"      inc: %f\n",inc);
  fprintf(outfile,"  maxiter: %d\n",maxiter);
  fprintf(outfile,"  miniter: %d\n",miniter);
  fprintf(outfile," trunceps: %f\n",trunceps);
  fprintf(outfile," maxvalue: %f\n",maxvalue);
  fprintf(outfile," mindelta: %f\n",mindelta);
  fprintf(outfile,"minptsper: %f\n",minptsper);
  fprintf(outfile,"       T0: %e\n\n",T0);
  fprintf(outfile,"****\n");  /* Start-of-data flag */

 /* Set fixed parameters */
  a = tan(1.570796327*alpha);
  abeta1 = a*beta1;
  oopi = 0.3183098862;  /* 1/pi  */
  twopi = 6.283185307;  /* 2*pi  */
  ooa = 1.0/alpha;
  fac = fabs(abeta1*pow((MAX((fabs(upbd)),(fabs(lowbd)))),(alpha - 1.0)));

 /* Determine number of regressions---abort if none or too many */
  numpts = (upbd - lowbd)/inc;
  if ((numpts<1) || (numpts>1000)) {
    fprintf(stderr,"Number of regressor values, %d, out of bounds\n",numpts);
    exit(-1);
```

```
  }

/* Determine upper and lower limits of integration to be used,
   computed as a function of trunceps */
  if (alpha < 1.0) {
    expbound = 4.0*(ooa-1.0)*(log(4.0*(ooa-1.0))-1.0);
    truncbound = pow((-2.0*log(alpha*trunceps*0.5)),ooa);
    T1den = MAX(truncbound,expbound);
  }
  else  T1den = MAX(-log(trunceps),1.0);

  T1num = -log(alpha*trunceps);

  fprintf(stdout,"Computed limits of integration:\n");
  fprintf(stdout,"       %le to %le for numerator\n",T0,T1num);
  fprintf(stdout,"       %le to %le for denominator\n\n",T0,T1den);
  if ((maxvalue == (T1num = MIN(T1num,maxvalue))) ||
      (maxvalue == (T1den = MIN(T1den,maxvalue)))) {
    fprintf(stdout,"Using: %le to %le for numerator\n",T0,T1num);
    fprintf(stdout,"       %le to %le for denominator\n\n",T0,T1den);
  }


/* Main computation loop */

  for(i=0; i<=numpts; ++i) {

    x = lowbd + ((double) i)*inc;
    absx = fabs(x) + ((x==0)*0.1);

    /* Print column headings every 20th iteration */
    if (i == (i/20)*20) {
      fprintf(stdout,"\n   x-value");
      fprintf(stdout," numerator");
      fprintf(stdout," num_iters");
      fprintf(stdout," num_delta");
      fprintf(stdout," denmnator");
      fprintf(stdout," den_iters");
      fprintf(stdout," den_delta");
      fprintf(stdout,"  quotient\n\n");
    }

    /* Determine final stepsize to use in denominator */
    if (alpha<1.0)     minperiod = twopi/absx;
    else               minperiod = twopi/(MAX(absx,fac));
    del = MIN((minperiod/minptsper),mindelta);
    del = MIN((T1den-T0)/miniter,del);
    deniters = 0.5 + (T1den-T0)/del;
    if (deniters>maxiter) {
      deniters = maxiter;
      fprintf(stdout,"Desired number of denominator iterations too large,");
      fprintf(stdout," using %d iterations\n",deniters);
    }
    dendel = (T1den-T0)/((double)deniters);
```

```
    /* Denominator numerical integration */
    den = simrule(deneval,T0,T1den,deniters);

    /* Determine final stepsize to use in numerator */
    if (alpha<1.0)     minperiod = twopi/(absx*pow(T1num,(ooa-1.0)));
    else               minperiod = twopi/fabs(abeta1);
    del = MIN((minperiod/minptsper),mindelta);
    del = MIN((T1num-T0)/miniter,del);
    numiters = 0.5 + (T1num-T0)/del;
    if (numiters>maxiter) {
      numiters = maxiter;
      fprintf(stdout,"Desired number of numerator iterations too large,");
      fprintf(stdout," using %d iterations\n",numiters);
    }
    numdel = (T1num-T0)/((double)numiters);

    /* Numerator integral */
    num = ooa*simrule(numeval,T0,T1num,numiters);

    /* Determine quotient and write to standard output and to output file */
    quint = num/den;
    fprintf(stdout,"%10.2f %9.5f %9d %9.5f %9.5f %9d %9.5f %9.5f\n",
    x,num,numiters,numdel,den,deniters,dendel,quint);
    fprintf(outfile,
    "%6.2f\t%13.8lf\t%7d\t%8.5f\t%13.8lf\t%7d\t%8.5f\t%13.8f\n",
    x,num,numiters,numdel,den,deniters,dendel,quint);
  }

  fclose(infile);
  fclose(outfile);
}


/* Insert the numerator and denominator integrand evaluation functions */

#include "eval.c"
```

6.2 reg.c

```
/*
 * Program name:
 *    reg
 *
 * Purpose:
 *    This routine computes regressions, E(X2|X1=x), for alpha not equal one,
 *    using parameters read from the command line and from the disk file
 *    output of the program "quint."  In addition, this routine has an option
 *    for computing the stable densities essentially computed in "quint," in
 *    lieu of computing regression functions.
 *
 * Usage:
 *    reg   sigma1  kappa  lambda  quintfile  [outfile]
```

```
 *
 *       sigma1     - The scale parameter of X1
 *       kappa      - The normalized covariation of X2 on X1
 *       lambda     - The normalized skew-covariation of X2 on X1
 *       quintfile - The disk file containing the integral quotients, produced
 *                   by the "quint" program
 *       outfile    - The output file to write the regression results to.  If
 *                   no output file is given, the output goes to the standard
 *                   output file.
 *
 *    Alternate usage:
 *       reg -d quintfile [outfile]
 *
 *       The -d option forces reg to compute the density function only,
 *       and outputs the density function in place of the regression function.
 *
 * Output:
 *    One output row is produced for each regressor processed in the "quint"
 *    program, with each row consisting of the regressor value followed by the
 *    regression value, tab-delimited and suitable for plotting with a variety
 *    of plotting packages.  If no output file name is specified on the
 *    command line, the output goes to the standard output, and can thus
 *    be piped to other applications, including the UNIX plotter.  If the
 *    density option is chosen, the output format is identical, save that
 *    density values appear in place of regressions.
 */

#include <stdio.h>
#include <math.h>


main(argc,argv)
     int argc;
     char *argv[];
{
  int      i, j, k, numiters, deniters, scanret, densflag=0;
  double   a, alpha, sigma1, beta1, kappa, lambda, x, num, numdel, den, dendel,
           quint, regressx, regconst;
  FILE     *quintfile, *outfile, *fopen();
  char     readin[100];

/* Check the command line for proper usage */
  if (argc == 1) {
    fprintf(stderr,
    "Usage:  %s  sigma1 kappa lambda quintfile [outfile],",argv[0]);
    fprintf(stderr," for regression\n");
    fprintf(stderr,
    "        %s  -d  quintfile [outfile], for density\n",argv[0]);
    exit (-1);
  }
  if (!(strcmp(argv[1],"-d"))) {   /* Computing density */
    if (argc < 3) {
      fprintf(stderr,"Usage:  %s  -d  quintfile [outfile]\n",argv[0]);
      exit(-1);
```

```
    }
    else densflag = 2;
  }
  else if (argc < 5) {                /* Computing regressions */
    fprintf(stderr,
    "Usage:  %s  sigma1 kappa lambda quintfile [outfile]\n",argv[0]);
    exit(-1);
  }

/* Open output file if specified */
  if (argc > 5-densflag) {
    if (!(outfile = fopen(argv[5-densflag],"w"))) {
      fprintf(stderr,"Unable to open output file %s\n",argv[5-densflag]);
      exit(-1);
    }
  }
  else outfile = stdout;

/* Get command line arguments and open the quotient file */
  if (!densflag) {
    sigma1 = atof(argv[1]);
    kappa  = atof(argv[2]);
    lambda = atof(argv[3]);
  }
  if (!(quintfile = fopen(argv[4-densflag],"r"))) {
    fprintf(stderr,"Unable to open file containing integral quotients, %s\n",
    argv[4-densflag]);
    exit(-1);
  }

/* Read integral quotient file to determine alpha and beta1 and start of data*/
  fscanf(quintfile,"%s",readin);
  if (strcmp(readin,"alpha:")) {
    fprintf(stderr,"Unable to properly read file of integral quotients\n");
    exit(-1);
  }
  fscanf(quintfile,"%lf",&alpha);
  fscanf(quintfile,"%s",readin);
  if (strcmp(readin,"beta1:")) {
    fprintf(stderr,"Unable to properly read file of integral quotients\n");
    exit(-1);
  }
  fscanf(quintfile,"%lf",&beta1);
  while (strcmp(readin,"****"))  scanret = fscanf(quintfile,"%s",readin);
  if (scanret == EOF) {
    fprintf(stderr,
    "Unable to find start of data in integral quotients file\n");
    exit(-1);
  }

/* Echo parameters used to terminal */
  fprintf(stderr,
  "\nFile containing integral quotients: %s\n",argv[4-densflag]);
  fprintf(stderr,"Parameters used:\n");
```

```
  fprintf(stderr,"  alpha  = %lf\n",alpha);
  fprintf(stderr,"  beta1  = %lf\n",beta1);
  if (!densflag) {
    fprintf(stderr," sigma1  = %lf\n",sigma1);
    fprintf(stderr,"  kappa  = %lf\n",kappa);
    fprintf(stderr,"  lambda = %lf\n",lambda);
    fprintf(stderr,"\nComputing regressions:\n\n");
  }
  else fprintf(stderr,"\nComputing densities:\n\n");

  a = tan(1.570796327*alpha);
  regconst = alpha * a * sigma1 * (lambda - beta1*kappa);

/* Loop to compute regressions */
  while (EOF != fscanf(quintfile,"%lf %lf %d %lf %lf %d %lf %lf",
 &x,&num,&numiters,&numdel,&den,&deniters,&dendel,&quint)) {
    if (densflag) fprintf(outfile,"%10.5f\t%10.5f\n",x,0.318309886*den);
    else {
      x *= sigma1;
      regressx = kappa*x + regconst*quint;
      fprintf(outfile,"%10.5f\t%10.5f\n",x,regressx);
    }
  }
}
```

6.3 reg1.c

```
/*
 * Program name:
 *    reg1
 *
 * Purpose:
 *    This routine computes regressions, E(X2|X1=x), in the case alpha = 1,
 *    using parameters read from a disk file.  Numerical integrals must be
 *    performed for each choice of the parameters, and are computed for a
 *    range of regressor values, using Simpson's rule.   Various parameters
 *    controlling the numerical integration may be set on the command line or
 *    recompiled in the code (see Usage, and comments in the code).  At user
 *    option, this routine computes stable densities in lieu of regressions.
 *
 * Usage:
 *    reg1  parmfile  [miniter]  [minptsper]  [trunceps]
 *
 *        parmfile  - Disk file consisting of a list of the nine input
 *                    parameters: least regressor value, greatest regressor
 *                    value, regressor increment, sigma1, beta1, mu1, k0,
 *                    kappa, and lambda.
 *        miniter   - Minimum total number of sample points in each numerical
 *                    integration (default=1000).
 *        minptsper - Minimum number of sample points per period to use in
 *                    each numerical integration (default=40).
 *        trunceps  - Maximum truncation error allowed.  This parameter deter-
 *                    mines the upper limit of integration (default=0.0005).
```

```
 *
 *    Alternate usage:
 *       reg1 -d parmfile [miniter]  [minptsper]  [trunceps]
 *
 *       The -d option forces reg to compute the density function only,
 *       and outputs the density function in place of the regression function.
 *
 * Output:
 *    Output is written to a file whose name is the name of the input
 *    parameter file with ".reg1" appended.  This file contains a
 *    listing of the parameters used to create the file, followed
 *    by tab-delimited data.  One row of data is produced for each regressor
 *    value, with the following entries in the columns:  the regressor x,
 *    the computed regression E(X2|X1=x), the numerator numerical integral,
 *    its number of steps and stepsize, the denominator numerical integral,
 *    and its number of steps and stepsize.  In the case that a density is
 *    computed, the output file is "parmfile.dens1," and contains in each
 *    column the x value, the density at x, the number of steps used to compute
 *    the integral, and  the stepsize.
 */

#include <stdio.h>
#include <math.h>
#define MIN(A,B)  (((A)<(B))?(A):(B))
#define MAX(A,B)  (((A)>(B))?(A):(B))

double sigma1, beta1, mu1, x, topi;


main(argc,argv)
     int argc;
     char *argv[];
{
  int    i, j, k, numpts, numiters, deniters, densflag=0;
  double k0, kappa, lambda, num, numdel, den, dendel, quint, regressx, density,
         regconst, lowbd, upbd, inc, t, a, twopi, T1num, T1den, T1, T0tr, oopi,
         truncerror, expbound, truncbound, fac, absx, del, minperiod;

  char   outfilename[50];
  FILE   *parmfile, *outfile, *fopen();
  double simrule(), numeval1(), numeval10a(), numeval10b(), deneval1();

  /* Parameters controlling the numerical integration.  Alter if desired     */
  int    maxiter=10000,   /* The maximum number of num'l int'n steps allowed */
         miniter= 1000,   /* The minimum number of num'l int'n steps allowed */
         numiter01=1000;  /* Number of iters. of T0 - 1 int'l when beta1 is 0*/
  double trunceps=0.0005, /* The upper bound on the truncation error allowed */
         maxvalue=500,    /* The maximum upper limit of integration          */
         mindelta=0.05,   /* The minimum num'l int'n step-size allowed       */
         minptsper=40,    /* The minimum number of points per period allowed */
         T0=1.0e-20;      /* The initial point for the num'l int'ns (>0)     */

 /* Check command line for proper usage */
  if (argc<2) {
```

```
    fprintf(stderr,
       "Usage:  %s [-d]  parameter_file  [miniter]  [minptsper]  [trunceps]\n",
argv[0]);
    exit(-1);
  }

/* Check for density computation option */
  if (0==strcmp(argv[1],"-d")) densflag = 1;

/* Open input parameter file */
  if (!(parmfile = fopen(argv[1+densflag],"r"))) {
    fprintf(stderr,"Unable to open input parameter file\n");
    exit(-1);
  }

/* Open output file */
  strcpy(outfilename,argv[1+densflag]);
  if (densflag) strcat(outfilename,".dens1");
  else          strcat(outfilename,".reg1");
  if (!(outfile = fopen(outfilename,"w"))) {
    fprintf(stderr,"Unable to open output file\n");
    exit(-1);
  }

/* Check for optional command line arguments */
  if (argc>2+densflag) miniter   = atof(argv[2+densflag]);
  if (argc>3+densflag) minptsper = atof(argv[3+densflag]);
  if (argc>4+densflag) trunceps  = atof(argv[4+densflag]);

/* Read input file and echo parameters to standard output and output file */
  fscanf(parmfile,"%lf %lf %lf %lf %lf %lf",
 &lowbd,&upbd,&inc,&sigma1,&beta1,&mu1);
  fprintf(stdout,
       "\nalpha =  1.000, sigma1 = %6.3f,  beta1 = %6.3f,      mu1 = %6.3f\n"
 ,sigma1,beta1,mu1);
  fprintf(outfile,"    alpha: 1.000000\n");
  fprintf(outfile,"   sigma1: %lf\n",sigma1);
  fprintf(outfile,"    beta1: %lf\n",beta1);
  fprintf(outfile,"      mu1: %lf\n",mu1);
  if (!densflag) {
    fscanf(parmfile,"%lf %lf %lf",&k0,&kappa,&lambda);
    fprintf(stdout,
"   k0 = %6.3f,  kappa = %6.3f, lambda = %6.3f\n",k0,kappa,lambda);
    fprintf(outfile,"       k0: %lf\n",k0);
    fprintf(outfile,"    kappa: %lf\n",kappa);
    fprintf(outfile,"   lambda: %lf\n",lambda);
  }
  fprintf(stdout,"x ranges from %6.3f to %6.3f by %5.3f.\n\n",lowbd,upbd,inc);
  fprintf(outfile,"    lowbd: %f\n",lowbd);
  fprintf(outfile,"     upbd: %f\n",upbd);
  fprintf(outfile,"      inc: %f\n",inc);
  fprintf(outfile,"  maxiter: %d\n",maxiter);
  fprintf(outfile,"  miniter: %d\n",miniter);
  fprintf(outfile," trunceps: %f\n",trunceps);
```

```
  fprintf(outfile," maxvalue: %f\n",maxvalue);
  fprintf(outfile," mindelta: %f\n",mindelta);
  fprintf(outfile,"minptsper: %f\n",minptsper);
  fprintf(outfile,"       T0: %e\n\n",T0);
  fprintf(outfile,"****\n");  /* Start-of-data flag */

/* Set fixed parameters */
  oopi = 0.3183098862;  /* 1/pi */
  topi = 0.6366197724;  /* 2/pi */
  twopi = 6.283185307;  /* 2*pi */
  T0tr = -log(T0);

/* Determine number of regressions---abort if none or too many */
  numpts = (upbd - lowbd)/inc;
  if ((numpts<1) || (numpts>1000)) {
    fprintf(stderr,"Number of regressor values, %d, out of bounds\n",numpts);
    exit(-1);
  }

/* Determine upper and lower limits of integration to be used,
   computed as a function of trunceps */
  T1den = -(1.0/sigma1)*log(sigma1*trunceps);
  if (beta1==0) T1num = -(2.0/sigma1)*log(0.5*sigma1*sigma1*trunceps);
  else          T1num = -(1.0/sigma1)*log(sigma1*trunceps);

  if (densflag) {
    fprintf(stdout,"Computed limits of integration: %le to %le\n",T0,T1num);
    if (maxvalue == (T1den = MIN(T1den,maxvalue)))
      fprintf(stdout,"Using: %le to %le\n\n",T0,T1den);
  }
  else {
    fprintf(stdout,"Computed limits of integration:\n");
    fprintf(stdout,"       %le to %le for numerator\n",T0,T1num);
    fprintf(stdout,"       %le to %le for denominator\n\n",T0,T1den);
    if ((maxvalue == (T1num = MIN(T1num,maxvalue))) ||
(maxvalue == (T1den = MIN(T1den,maxvalue)))) {
      fprintf(stdout,"Using: %le to %le for numerator\n",T0,T1num);
      fprintf(stdout,"       %le to %le for denominator\n\n",T0,T1den);
    }
  }

/* Main computation loop */

  for(i=0; i<=numpts; ++i) {

    x = lowbd + ((double) i)*inc;
    absx = fabs(x) + ((x==0)*0.1);

    /* Print column headings every 20th iteration */
    if (i == (i/20)*20) {
      fprintf(stdout,"\n  x-value ");
      if (!densflag) {
fprintf(stdout,"E(X2|X1=x)");
fprintf(stdout," numerator");
```

```
fprintf(stdout," num_iters");
fprintf(stdout," num_delta");
fprintf(stdout," denmnator");
fprintf(stdout," den_iters");
fprintf(stdout," den_delta\n\n");
      }
      else {
fprintf(stdout,"   density");
fprintf(stdout,"     iters");
fprintf(stdout,"     delta\n\n");
      }
    }

    /* Determine final stepsize to use in denominator */
    minperiod = twopi/absx;
    del = MIN((minperiod/minptsper),mindelta);
    del = MIN((T1den-T0)/miniter,del);
    deniters = 0.5 + (T1den-T0)/del;
    if (deniters>maxiter) {
      deniters = maxiter;
      fprintf(stdout,"Desired number of denominator iterations too large,");
      fprintf(stdout," using %d iterations\n",deniters);
    }
    dendel = (T1den-T0)/((double)deniters);

    /* Denominator numerical integration */
    den = simrule(deneval1,T0,T1den,deniters);

    if (!densflag) {       /* Computing regressions */

      /* Determine final stepsize to use in numerator */
      del = MIN((minperiod/minptsper),mindelta);
      del = MIN((T1num-T0)/miniter,del);
      numiters = 0.5 + (T1num-T0)/del;
      if (numiters>maxiter) {
numiters = maxiter;
fprintf(stdout,"Desired number of numerator iterations too large,");
fprintf(stdout," using %d iterations\n",numiters);
      }
      numdel = (T1num-T0)/((double)numiters);

      /* Numerator integral */
      if (beta1==0.0) num = simrule(numeval10a,0.0,T0tr,numiter01)
                 + simrule(numeval10b,1.0,T1num,numiters);
      else             num = simrule(numeval1,T0,T1num,numiters);

      quint = num/den;

      /* Determine regression and write to terminal and to output file */
      if (beta1==0) fac = -topi*sigma1*lambda*quint;
      else          fac = (lambda-beta1*kappa)*(x-mu1-sigma1*quint)/beta1;
      regressx = -topi*sigma1*k0 + kappa*(x-mu1) + fac;

      fprintf(stdout,"%10.2f %9.5f %9.5f %9d %9.5f %9.5f %9d %9.5f\n",
```

```
      x,regressx,num,numiters,numdel,den,deniters,dendel);
      fprintf(outfile,
      "%6.2f\t%13.8lf\t%13.8lf\t%7d\t%8.5f\t%13.8lf\t%7d\t%8.5f\n",
      x,regressx,num,numiters,numdel,den,deniters,dendel);
    }

    else {              /* Computing densities */

      density = oopi * den;
      fprintf(stdout,"%10.2f %9.5f %9d %9.5f\n",x,density,deniters,dendel);
      fprintf(outfile,
         "%6.2f\t%13.8lf\t%7d\t%8.5f\n",x,density,deniters,dendel);
   }
  }

  fclose(parmfile);
  fclose(outfile);
}

/* Insert the numerator and denominator integrand evaluation functions
   for the case alpha=1 */

#include "eval1.c"
```

6.4 testints.c

```
/*
 * Program name:
 *    testints
 *
 * Purpose:
 *    Computes the two numerical integrals and their quotient for one input
 *    value of the regressor, for the case alpha not equal one.  Command line
 *    arguments specify where the integral is to be truncated and the number
 *    of sample points in the Simpson's rule integrations. The values of alpha
 *    and beta1 are compiled into the code.  This routine is used to experiment
 *    with the effects of truncation and refinement on numerical accuracy, and
 *    allows, through change of the integrand evaluation routines,
 *    experimentation with changes of variables.
 *
 * Usage:
 *    testints  regressor  truncation  sample_points
 *
 *        regressor     - Value of the regressor (x) for which to compute
 *                        integrals
 *        truncation    - Upper limit of integration (T1) for the integrals
 *        sample_points - Number of steps (iters) in Simpson's rule
 *
 * Output:
 *    The approximate numerator integral, denominator integral, and their
 *    quotient are printed out.
 */
```

```
#include <math.h>
#include <stdio.h>

#define FUNC(x) ((*func)(x))

double  alpha = 1.5,    /* Constants compiled in code  */
        beta1 = 0.24;

double  ooa, abeta1, x;

main(argc,argv)
     int argc;
     char *argv[];
{
  double simrule(), numeval(), deneval(), atof(),
         T0, T1, integral, num, den, quint;
  int    iters, i;

  abeta1 = beta1*tan(alpha*1.570796327);
  ooa = 1.0/alpha;
  T0 = 1.0e-20;

/* Check command line for proper usage and get arguments */
  if (argc < 4) {
    fprintf(stderr,
    "Usage:  %s  regressor  truncation  sample_points\n",argv[0]);
    exit(-1);
  }
  x     = atof(argv[1]);
  T1    = atof(argv[2]);
  iters = atoi(argv[3]);

  fprintf(stdout,"\nalpha = %6.3f, beta1 = %6.3f\n",alpha,beta1);
  fprintf(stdout,
 "For x = %6.3f, T0 = %8.2e, T1 = %8.2e, steps = %d:\n",x,T0,T1,iters);

  den = simrule(deneval,T0,T1,iters);

  num = ooa*simrule(numeval,T0,T1,iters);

  quint = num/den;

  fprintf(stdout,"\nNumerator   = %21.15lf\n",num);
  fprintf(stdout,"Denominator = %21.15lf\n",den);
  fprintf(stdout,"Quotient    = %21.15lf\n\n",quint);
}


/* Insert the numerator and denominator integrand evaluation functions */

#include "eval.c"
```

6.5 testints1.c

```
/*
 * Program name:
 *    testints1
 *
 * Purpose:
 *    Computes the two numerical integrals and their quotient for one input
 *    value of the regressor, for the case alpha equal one.  Command line
 *    arguments specify where the integral is to be truncated and the number
 *    of sample points in the Simpson's rule integrations. The values of beta1,
 *    sigma1, and mu1 are compiled into the code.  This routine is used to
 *    experiment with the effects of truncation and refinement on numerical
 *    accuracy, and allows, through change of the integrand evaluation
 *    routines, experimentation with changes of variables.
 *
 * Usage:
 *    testints1  regressor  truncation  sample_points
 *
 *        regressor     - Value of the regressor (x) for which to compute
 *                        integrals
 *        truncation    - Upper limit of integration (T1) for the integrals
 *        sample_points - Number of steps (iters) in Simpson's rule
 *
 * Output:
 *    The approximate numerator integral, denominator integral, and their
 *    quotient are printed out.
 */

#include <math.h>
#include <stdio.h>

#define FUNC(x) ((*func)(x))

double  mu1    =  0.17650848,   /* Constants compiled in code */
        beta1  =  0.0,
        sigma1 =  1.0,
        T0     =  1.0e-20;

double  topi  =  0.6366197724, x;

main(argc,argv)
     int argc;
     char *argv[];
{
  double simrule(), numeval1(), numeval10a(), numeval10b(), deneval1(), atof(),
         T0tr, T1, integral, num, den, quint;
  int    iters, i;


/* Check command line for proper usage and get arguments */
  if (argc < 4) {
    fprintf(stderr,
         "Usage:  %s  regressor  truncation  sample_points\n",argv[0]);
```

```
    exit(-1);
  }
  x     = atof(argv[1]);
  T1    = atof(argv[2]);
  iters = atoi(argv[3]);

  fprintf(stdout,
"\nalpha = 1.0, beta1 = %6.3f, mu1 = %6.3f, sigma1 = %6.3f\n",
   beta1,mu1,sigma1);
  fprintf(stdout,
"For x = %6.3f, T0 = %8.2e, T1 = %8.2e, steps = %d:\n",x,T0,T1,iters);

  den = simrule(deneval1,T0,T1,iters);

  T0tr = -log(T0);

  if (beta1==0.0)  num = simrule(numeval10a,0.0,T0tr,iters)
                       + simrule(numeval10b,1.0,T1,iters);
  else             num = simrule(numeval1, T0,T1,iters);

  quint = num/den;

  fprintf(stdout,"\nNumerator   = %21.15lf\n",num);
  fprintf(stdout,"Denominator = %21.15lf\n",den);
  fprintf(stdout,"Quotient    = %21.15lf\n\n",quint);
}


/* Insert the numerator and denominator integrand evaluation functions */

#include "eval1.c"
```

6.6 eval.c

```
/* INTEGRAND EVALUATION FUNCTIONS FOR  0 < alpha < 1  and  1 < alpha < 2  */


/* Function to evaluate the numerator integrand, passed to the Simpson's rule
   function.  Requires global variables x, ooa, abeta1 */

double numeval(u)
     double u;
{
  double upowooa,num;

  upowooa = pow(u,ooa);
  num = exp(-u)*cos(x*upowooa - abeta1*u);

  return num;
}
```

```
/* Function to evaluate the denominator integrand, passed to the Simpson's rule
   function.  Requires global variables x, alpha, abeta1 */

double deneval(t)
     double t;
{
  double ttoalpha, den;

  ttoalpha =  pow(t,alpha);
  den  =  exp(-ttoalpha) * cos(x*t - abeta1*ttoalpha);

  return den;
}
```

6.7 eval1.c

```
/* INTEGRAND EVALUATION FUNCTIONS FOR alpha = 1 */


/* Function to evaluate the numerator u(x) in the case beta1 not equal zero,
   passed to the Simpson's rule function.  Requires global variables
   x, sigma1, mu1, beta1, topi */

double numeval1(t)
     double t;
{
  double num;

  num = exp(-sigma1*t)*sin(t*(x-mu1) + topi*beta1*sigma1*t*log(t));

  return num;
}



/* Function to evaluate the T0 to 1 part of the numerator v(x) in the case
   beta1 equal zero, passed to the Simpson's rule function.  Requires global
   variables x, sigma1, mu1 */

double numeval10a(t)
     double t;
{
  double num,emt;

  emt = exp(-t);
  num = exp(-sigma1*emt-t)*(1 - t)*cos(emt*(x-mu1));

  return num;
}



/* Function to evaluate the 1 to T1 part of the numerator v(x) in the case
```

```
   beta1 equal zero, passed to the Simpson's rule function.  Requires global
   variables x, sigma1, mu1 */

double numeval10b(t)
     double t;
{
  double num;

  num = exp(-sigma1*t)*(1+log(t))*cos(t*(x-mu1));

  return num;
}



/* Function to evaluate the denominator, passed to the Simpson's rule
   function.  Requires global variables x, sigma1, mu1, beta1, topi */

double deneval1(t)
     double t;
{
  double den;

  den  =  exp(-sigma1*t) * cos(t*(x-mu1) + topi*beta1*sigma1*t*log(t));

  return den;
}
```

6.8 simrule.c

```
/*
 * Function name:
 *    simrule
 *
 * Purpose:
 *    Perform numerical integration by Simpson's 1/3 rule
 *
 * Input arguments:
 *    func  - pointer to function (of one variable) to be integrated
 *    a     - variable of integration lower limit
 *    b     - variable of integration upper limit
 *    iters - numbers of sample points used to compute the integral (forced
 *            to be even)
 *
 * Returns:
 *    s     - Simpson's rule approximation to the integral
 */

#define FUNC(x) ((*func)(x))


double simrule(func,a,b,iters)
     double (*func)();
```

```
     double a,b;
     int iters;
{
  double x, sum, s, del;
  int    j;

  iters += iters & 1;                /* Force iters to be even */
  del = (b-a)/((double) iters);      /* Find stepsize */
  sum  = 0.5*FUNC(a);                /* Evaluate function at endpoints */
  sum += 0.5*FUNC(b);

  for (j=1; j<iters-2; j+=2) {
    x = a + j*del;                   /* Evaluate at odd points and add */
    sum += 2.0*FUNC(x);
    x += del;                        /* Evaluate at even points and add */
    sum += FUNC(x);
  }

  x = b - del;                       /* Evaluate at last odd point */
  sum += 2.0*FUNC(x);

  s = (2.0 * del / 3.0) * sum;       /* Normalize */
  return s;
}
```

6.9 makefile

```
# This makefile provides a mechanism for compiling the source programs on
# a UNIX operating system.  To compile a program, type
#                  make  [options]  name_of_program
# For example  "make reg1" creates (or updates) the object files reg1.o and
# simrule.o, and the C-code file eval1.c, and links them, creating the
# executable file reg1.  For more information, check the UNIX manual pages for
# make.

CFLAGS = -O
LIBS   = -lm

quint:      quint.o simrule.o eval.c
     cc -o quint $(CFLAGS) quint.o simrule.o $(LIBS)

reg:      reg.o
     cc -o reg $(CFLAGS) reg.o $(LIBS)

reg1:      reg1.o simrule.o eval1.c
     cc -o reg1 $(CFLAGS) reg1.o simrule.o $(LIBS)

testints:    testints.o simrule.o eval.c
     cc -o testints $(CFLAGS) testints.o simrule.o $(LIBS)

testints1:   testints1.o simrule.o eval1.c
     cc -o testints1 $(CFLAGS) testints1.o simrule.o $(LIBS)
```

References

[Har84] C.D. Hardin, Jr. Skewed stable variables and processes. Technical Report 79, Center for Stoch. Proc., University of North Carolina, Chapel Hill, 1984.

[JST91] C.D. Hardin Jr., G. Samorodnitsky, and M. S. Taqqu. Non-linear regression of stable random variables. To appear in *The Annals of Applied Probability*, 1991.

[ST89] G. Samorodnitsky and M.S. Taqqu. Conditional moments and linear regression for stable random variables. Preprint, 1989.

Clyde D. Hardin Jr.
The Analytic Sciences Corporation
55 Walkers Brook Drive
Reading, MA 01867

Gennady Samorodnitsky
Cornell University
School of ORIE
Ithaca, NY 14853

Murad S. Taqqu
Boston University
Department of Mathematics
Boston, MA 02215

A Characterization of the Asymptotic Behavior of Stationary Stable Processes *†‡

Joshua B. Levy
University of Lowell

Murad S. Taqqu
Boston University

Abstract

Let $X(t)$ be a stationary stable process and let $r(t)$ be the difference between the joint characteristic function of $(X(t), X(0))$ and the product of the characteristic functions of $X(t)$ and $X(0)$. The function $r(t)$ as $t \to \infty$ is a measure of asymptotic dependence. We analyze the behavior of $r(t)$ as $t \to \infty$ for various stable stochastic processes.

1 Introduction

The covariance function characterizes the dependence structure of a stationary Gaussian process. What can one use when the process $\{X(t), -\infty < t < \infty\}$ is a non-Gaussian α-stable process with $0 < \alpha < 2$? Such a process shares many common features with the Gaussian process, but it has infinite variance and hence the covariance is not defined. It is possible to use the covariation ([Mil78]), but the

*This research was supported at Boston University by the the ONR grant 90-J-1287.

†**AMS 1980 subject classifications**. Primary 60E07. Secondary 60G10.

‡**Keywords and phrases** Stable distributions, moving average, self-similarity, fractional Brownian motion, Log-fractional Lévy motion.

covariation is not symmetric and it is not always defined. We shall focus instead on the function

$$r(t) := r_X(\theta_1, \theta_2; t)$$

$$= E\exp\{i[\theta_1 X(t) + \theta_2 X(0)]\} - E\exp\{i\theta_1 X(t)\}E\exp\{i\theta_2 X(0)\} \quad (1.1)$$

where θ_1, θ_2 are real constants. We exclude the trivial case $\theta_1 = 0$ or $\theta_2 = 0$ since then $r(t) \equiv 0$. Unlike the covariation, $r(t)$ is always defined and is particularly useful as $t \to \infty$ for distinguishing between the asymptotic structures of the moving average, sub-Gaussian and real harmonizable processes. In the Gaussian case, $r(t)$ is asymptotically proportional to the covariance, if the covariance converges to zero.

THEOREM 1.1 *If $X(t)$ is a real stationary Gaussian process with mean $EX(t)$ and covariance* $\mathrm{Cov}(X(t), X(0)) = EX(t)X(0) - (EX(0))^2$, *then as $t \to \infty$*

$$r(t) \sim K(\theta_1, \theta_2)\mathrm{Cov}(X(t), X(0)) \quad \textit{if and only if} \quad \mathrm{Cov}(X(t), X(0)) \to 0$$

where

$$K(\theta_1, \theta_2) = \exp\{i(\theta_1 + \theta_2)EX(0) - \frac{1}{2}(\theta_1^2 + \theta_2^2)\mathrm{Var}X(0)\}(-\theta_1\theta_2).$$

PROOF: From (1.1), $r(t) = \exp\{i(\theta_1 + \theta_2)EX(0) - \frac{1}{2}(\theta_1^2 + \theta_2^2)\mathrm{Var}X(0)\}(e^{-I(t)} - 1)$ with $I(t) = \theta_1\theta_2 \, \mathrm{Cov}(X(t), X(0))$. The conclusion follows because $e^{-I(t)} - 1 \to 0$ if and only if $I(t) \to 0$. ■

The purpose of this paper is to find the behaviour of $r(t)$ as $t \to \infty$ for various real stationary non-Gaussian α-stable processes. We shall factor $E\exp\{i\theta_1 X(t)\}E\exp\{i\theta_2 X(0)\}$ in the right-hand side of (1.1) to get

$$r(t) = e^{-A_\alpha}(e^{-I(t)} - 1), \quad (1.2)$$

where

$$A_\alpha = -[\ln E\exp\{i\theta_1 X(t)\} + \ln E\exp\{i\theta_2 X(0)\}],$$

and

$$I(t) = -\ln E\exp\{i[\theta_1 X(t) + \theta_2 X(0)]\}$$
$$+ \ln E\exp\{i\theta_1 X(t)\} + \ln E\exp\{i\theta_2 X(0)\},$$

and then examine $I(t)$ as $t \to \infty$. (Here, $\ln z$ equals $\ln|z| + i\phi$ with $-\pi < \phi \leq \pi$ for any complex number z.) Note that

$$r(t) \sim -\exp\{-A_\alpha\}I(t) \text{ if and only if } I(t) \to 0. \tag{1.3}$$

We will need the expression of the finite-dimensional characteristic function of the α-stable process $\{X(t), t \in \mathcal{R}\}$. It is known (see [Har84]) that the stationary process $X(t)$ can be represented by the stable stochastic integral $\int_E f(t,x)M(dx) + \eta$ where

(i) $(E, \mathcal{E}, m)$ is a measure space with σ-finite measure m,

(ii) M is an independently scattered α-stable random measure with control measure m and skewness intensity $\beta(\cdot) : E \to [-1, 1]$,

(iii) $f(t,\cdot) \in L^\alpha(E, \mathcal{E}, m)$ if $\alpha \neq 1$ while, if $\alpha = 1$, $f(t,\cdot) \in L^1(E, \mathcal{E}, m)$ and $\beta(\cdot)f(t,\cdot)\ln|f(t,\cdot)| \in L^1(E, \mathcal{E}, m)$, and

(iv) η is a real constant.

For convenience we suppose $\eta = 0$. The finite-dimensional characteristic function of $\{X(t), t \in \mathcal{R}\}$ can then be expressed as

$$E\exp\{i\sum_{j=1}^{d}\theta_j X(t_j)\} = \exp\{-\int_E \xi(\sum_{j=1}^{d}\theta_j f(t_j, x), \beta(x))m(dx)\} \tag{1.4}$$

for $\theta_1, \ldots, \theta_d$, $t_1, \ldots, t_d$ real, $d \geq 1$, and

$$\xi(u(x), \beta(x)) = \begin{cases} |u(x)|^\alpha\{1 - i\beta(x)(\text{sign } u(x))\tan\frac{\pi\alpha}{2}\}, & \alpha \neq 1, \\ |u(x)|\{1 + i\frac{2}{\pi}\beta(x)(\text{sign } u(x))\ln|u(x)|\}, & \alpha = 1. \end{cases} \tag{1.5}$$

To simplify the notation we write $\xi(u) = \xi(u(\cdot), \beta(\cdot))$.

In the following section, we note that $r(t)$ tends to zero if the process is a moving average. We give the asymptotic forms of $r(t)$ for the Ornstein-Uhlenbeck process, the reverse Ornstein-Uhlenbeck process and the linear fractional Lévy motions and compare these to well-known results in the Gaussian case $\alpha = 2$. In Section 3, we show that $r(t)$ for the real stationary harmonizable process never converges and we provide a lower bound for the Cesaro mean $\frac{1}{2T}\int_{-T}^{T} r(t)dt$ as $T \to \infty$. The nature of this lower bound points to the difference between the cases $\alpha < 2$ and $\alpha = 2$. In the last section, we prove that for sub-Gaussian processes, $r(t)$ is always asymptotically positive when $0 < \alpha < 1$.

The different asymptotic behavior of $r(t)$ for different stable processes can be used as a criterion for distinguishing among them. Although other functionals can also serve this purpose, it is especially informative to use $\lim_{t\to\infty} r(t)$ since this is a measure of asymptotic dependence.

2 Moving Average Processes

Let M be an independently scattered α-stable random measure on $(\mathcal{R}, \mathcal{B})$ with Lebesgue control measure and skewness intensity $\beta(\cdot) : \mathcal{R} \to [-1, 1]$. Let f be a real-valued measurable function on $\mathcal{R}$ satisfying $\|f\|_\alpha^\alpha := \int_{\mathcal{R}} |f(x)|^\alpha dx < \infty$ for $0 < \alpha \leq 2$ and, if $\alpha = 1$, then also $\int_{\mathcal{R}} |f(x)(\ln|f(x)|)\beta(x)|dx < \infty$. The α-stable process defined by

$$X(t) = \int_{\mathcal{R}} f(t-x)M(dx) \ , \ -\infty < t < \infty,$$

is called a *moving average process.* We implicitly assume that X is stationary. Note that stationarity imposes a restriction on the spectral intensity $\beta(\cdot)$ because we must have

$$\int_{-\infty}^{\infty} \xi(\sum_{j=1}^{d} \theta_j f(t_j + h - x))dx = \int_{-\infty}^{\infty} \xi(\sum_{j=1}^{d} \theta_j f(t_j - x))dx \qquad (2.1)$$

for all real h. If both $\lim_{x\to-\infty} \beta(x)$ and $\lim_{x\to\infty} \beta(x)$ exist, then (2.1) implies $\lim_{x\to-\infty} \beta(x) = \lim_{x\to\infty} \beta(x)$. We do not know whether stationarity requires $\beta(x)$ to be constant a.e.-m.

THEOREM 2.1 *If* $\{X(t),\ -\infty < t < \infty\}$ *is an* α*-stable moving average process then*

$$\lim_{t\to\infty} r(t) = 0.$$

One can prove this result by establishing that moving average processes are mixing (see for example [CHW84]) and noting that mixing implies "$r(t) \to 0$".

If we know the kernel explicitly, we can obtain a sharper rate of decay for $r(t)$. For example, if $h \geq 0$ and $\{X(t),\ t \in \mathcal{R}\}$ is the h-increment process of Lévy stable motion, that is $X(t) = \int_t^{t+h} M(dx)$, then $f(s) = 1_{(-h,0]}(s)$ and, in this case, $r(t) \equiv 0$ for $t \geq h$.

For simplicity we assume throughout the rest of this section that $\beta(\cdot) \equiv \beta$ is a non-zero constant.

We shall now specify the rate of convergence of $r(t)$ to zero for two important moving average processes, the Ornstein-Uhlenbeck process and the 1-step increment of linear fractional Lévy motion.

Consider the Ornstein-Uhlenbeck process. It is defined for fixed $\lambda > 0$ and $0 < \alpha \leq 2$ by

$$X(t) = \int_{-\infty}^{t} e^{-\lambda(t-x)} M(dx),\ -\infty < t < \infty, \tag{2.2}$$

where M is an α-stable random measure with Lebesgue control measure and constant skewness intensity $\beta \neq 0$. The process is well-defined since $\int_{-\infty}^{t} |e^{-\lambda(t-x)}|^\alpha dx = \frac{1}{\lambda\alpha}$, is a moving average and hence is stationary. In addition, $X(t)$ is a Markov process and, for $\alpha = 2$, it is the only stationary Markov Gaussian process. Let $a^{<b>} = |a|^b \text{sign}\ a$.

THEOREM 2.2 *For the Ornstein-Uhlenbeck process* (2.2),

(i) $r(t) \equiv 0$ *if* $\alpha = 1, \beta = 0$, *and* sign $\theta_1\theta_2 = 1$.

(ii) If either $\alpha \neq 1$, *or* $\alpha = 1, \beta = 0$, sign $\theta_1\theta_2 = -1$, *then*

$$r(t) \sim e^{-A_\alpha}(-C)e^{-(\alpha\wedge 1)\lambda t}$$

as $t \to \infty$ *where*

$$A_\alpha = \frac{1}{\alpha\lambda}[\xi(\theta_1) + \xi(\theta_2)] \tag{2.3}$$

and

$$C = \begin{cases} -\frac{1}{\alpha\lambda}|\theta_1|^\alpha(1 - i\beta(\text{sign } \theta_1)\tan\frac{\pi\alpha}{2}), & 0 < \alpha < 1, \\ -\frac{2}{\lambda}|\theta_1|, & \alpha = 1,\ \beta = 0, \\ \frac{1}{\lambda}\theta_2^{<\alpha-1>}\theta_1(1 - i\beta(\text{sign } \theta_2)\tan\frac{\pi\alpha}{2}), & 1 < \alpha \le 2. \end{cases} \tag{2.4}$$

(iii) If $\alpha = 1$ *and* $\beta \neq 0$, *then*

$$r(t) \sim e^{-A_1}(i\beta\frac{2}{\pi}\theta_1)te^{-\lambda t}$$

as $t \to \infty$, *where*

$$A_1 = \frac{1}{\lambda}\{|\theta_1|[1+i\frac{2}{\pi}\beta(\text{sign } \theta_1)(\ln|\theta_1|-1)]+|\theta_2|[1+i\frac{2}{\pi}\beta(\text{sign } \theta_2)(\ln|\theta_2|-1)]\}. \tag{2.5}$$

To establish Theorem 2.2, we use the following majorizing inequality, established in Lemma 3.1 of [ALT90].

LEMMA 2.1 *Suppose* $\xi(\cdot)$ *is defined by* (1.5) *and* $r, s \in \mathcal{R}$. *There exist positive constants* J, K *and* L *such that*

$$|\xi(r+s) - \xi(r) - \xi(s)| \le \begin{cases} J|r|^\alpha, & 0 < \alpha < 1, \\ 2|r|, & \alpha = 1,\ \beta = 0, \\ K|r||s|^{\alpha-1} + L|r|^\alpha, & 1 < \alpha < 2, \\ 2|rs|, & \alpha = 2. \end{cases} \tag{2.6}$$

PROOF OF THEOREM 2.2: Setting

$$u_t(x) = \theta_1 e^{-\lambda(t-x)}1_{(-\infty,t]}(x), \quad v(x) = \theta_2 e^{\lambda x}1_{(-\infty,0]}(x), \tag{2.7}$$

we get

$$I(t) = \int_{-\infty}^{\infty}[\xi(u_t(x) + v(x)) - \xi(u_t(x)) - \xi(v(x))]dx. \tag{2.8}$$

(i) By (1.5), $\xi(\cdot) = |\cdot|$ when $\alpha = 1$, $\beta = 0$ and $\xi(u_t+v) = \xi(u_t)+\xi(v)$ if one also has sign $\theta_1\theta_2 = 1$. Hence $I(t) \equiv 0$ and, by (1.2), $r(t) \equiv 0$.

(ii) Suppose either $\alpha \neq 1$ or $\alpha = 1$, $\beta = 0$, sign $\theta_1\theta_2 = -1$. For real θ, we get from (1.4) and (2.2),

$$E\exp\{i\theta X(t)\} = \exp\{-\int_{-\infty}^{\infty} \xi(\theta f(t-x))dx\} = \exp\{-\frac{1}{\alpha\lambda}\xi(\theta)\},$$

so that (1.2) implies (2.3). Next, using (2.8) and (2.7) we decompose $I(t)$ as follows:

$$I(t) = I_1(t) + I_2(t) + I_3(t) \tag{2.9}$$

where $I_1(t) = \int_{-\infty}^{0}$, $I_2(t) = \int_0^t$, $I_3(t) = \int_t^\infty$. But $I_2(t) = 0$ since $v = 0$ on $(0,t]$, and $I_3(t) = 0$ since $u_t = 0 = v$ on (t,∞). Thus, $I(t) = I_1(t)$. It is convenient to write

$$I_1(t) = e^{-(\alpha\wedge 1)\lambda t}\int_{-\infty}^{0} G_t(x)dx \tag{2.10}$$

where

$$G_t(x) = e^{(\alpha\wedge 1)\lambda t}[\xi(u_t+v) - \xi(u_t) - \xi(v)]. \tag{2.11}$$

We will show that as $t \to \infty$, the integral $\int_{-\infty}^{0} G_t(x)dx$ is asymptotic to the constant C defined in relation (2.4).

For all $-\infty < x \leq 0$ we have, as $t \to \infty$,

$$|u_t+v|^\alpha - |u_t|^\alpha - |v|^\alpha = |\theta_2 e^{\lambda x}|^\alpha[|1+\frac{\theta_1 e^{-\lambda t}}{\theta_2}|^\alpha - 1] - |\theta_1 e^{-\lambda(t-x)}|^\alpha$$

$$\sim e^{\alpha\lambda x}(|\theta_2|^\alpha \alpha\frac{\theta_1}{\theta_2}e^{-\lambda t} - |\theta_1 e^{-\lambda t}|^\alpha)$$

$$= e^{\alpha\lambda x}(\alpha\theta_2^{<\alpha-1>}\theta_1 e^{-\lambda t} - |\theta_1|^\alpha e^{-\alpha\lambda t}) =: f_t(x). \tag{2.12}$$

Similarly, $(u_t+v)^{<\alpha>} - u_t^{<\alpha>} - v^{<\alpha>} \sim g_t$ where $g_t(x) := e^{\alpha\lambda x}[\alpha|\theta_2|^{\alpha-1}\theta_1 e^{-\lambda t} - \theta_1^{<\alpha>}e^{-\alpha\lambda t}]$. Hence,

$$\xi(u_t+v) - \xi(u_t) - \xi(v)$$

$$\sim \begin{cases} f_t - i\beta g_t \tan \frac{\pi\alpha}{2}, & \alpha \neq 1, \\ f_t, & \alpha = 1,\ \beta = 0,\ \text{sign } \theta_1\theta_2 = -1, \end{cases}$$

and, consequently, from (2.11),

$$\lim_{t\to\infty} G_t(x) = G_\infty(x)$$

$$:= \begin{cases} -|\theta_1|^\alpha(1 - i\beta(\text{sign } \theta_1)\tan \frac{\pi\alpha}{2})e^{\alpha\lambda x}, & 0 < \alpha < 1, \\ -2|\theta_1|e^{\lambda x}, & \alpha = 1,\ \beta = 0, \\ & \quad \text{sign } \theta_1\theta_2 = -1, \\ \alpha\theta_2^{<\alpha-1>}\theta_1(1 - i\beta(\text{sign } \theta_2)\tan \frac{\pi\alpha}{2})e^{\alpha\lambda x}, & 1 < \alpha \leq 2. \end{cases} \tag{2.13}$$

To establish $L^1(-\infty, 0)$-convergence as $t \to \infty$, note that by (2.7), $|u_t(x)| \leq |\theta_1|e^{-\lambda(t-x)}$ and $|v(x)| \leq |\theta_2|e^{\lambda x}$, so that by substituting $r = u_t$ and $s = v$ in Lemma 2.1 we get positive constants J, K, L, satisfying

$$\sup_{t>0} |G_t(x)| \leq \begin{cases} J|\theta_1|^\alpha e^{\alpha\lambda x}, & 0 < \alpha < 1, \\ 2|\theta_1|e^{\lambda x}, & \alpha = 1,\ \beta = 0, \\ (K|\theta_1||\theta_2|^{\alpha-1} + L|\theta_1|^\alpha)e^{\alpha\lambda x}, & 1 < \alpha < 2, \\ 2|\theta_1\theta_2|e^{2\lambda x}, & \alpha = 2. \end{cases}$$

These bounds belong to $L^1(-\infty, 0)$. From (2.10), (2.13), and the dominated convergence theorem we conclude

$$\lim_{t\to\infty} e^{(\alpha\wedge 1)\lambda t} I_1(t) = \lim_{t\to\infty} \int_{-\infty}^0 G_t(x)dx = C.$$

(iii) Suppose $\alpha = 1$ and $\beta \neq 0$. We get from (1.4) and (2.2),

$$E\exp\{i\theta X(t)\} = \exp\{-\int_{-\infty}^0 |\theta|e^{\lambda x}[1 + i\frac{2}{\pi}\beta(\text{sign } \theta)(\ln|\theta| + \lambda x)]dx\}$$

implying (2.5). As in case (ii), $I(t) = I_1(t) = e^{-\lambda t}\int_{-\infty}^0 G_t(x)dx$, where $G_t(x)$ is defined as in (2.11). Fix $-\infty < x < 0$. Then

$$G_t(x) = e^{\lambda t}\{|u_t+v|-|u_t|-|v|+i\frac{2}{\pi}\beta[(u_t+v)\ln|u_t+v|-u_t\ln|u_t|-v\ln|v|]\}.$$

It is easy to see that as $t\to\infty$, $(u_t+v)\ln|u_t+v|-v\ln|v| = o(te^{-\lambda t})$, but $-u_t\ln|u_t| \sim \theta_1\lambda t e^{-\lambda(t-x)}$. From (2.12), we have $|u_t+v|-|u_t|-|v| = o(te^{-\lambda t})$. Therefore,

$$G_t(x) \sim i\frac{2}{\pi}\beta\theta_1\lambda t e^{\lambda x} \text{ as } t\to\infty. \tag{2.14}$$

We now want to show that for some large enough t_0,

$$\sup_{t>t_0}|\frac{1}{t}G_t(x)| \in L^1(-\infty,0). \tag{2.15}$$

Since $||u_t+v|-|u_t|-|v|| \le 2|u_t| \le 2|\theta_1|e^{-\lambda(t-x)}$, we conclude that $|G_t(x)|$ is less than or equal to

$$e^{\lambda x}(2|\theta_1|+e^{\lambda t}|(\theta_2+\theta_1 e^{-\lambda t})\ln|\theta_2+\theta_1 e^{-\lambda t}|-\theta_2\ln|\theta_2|-\theta_1 e^{-\lambda t}\ln|\theta_1 e^{-\lambda t}||)$$

$$\le e^{\lambda x}(2|\theta_1|+|\theta_1|(1+|\ln|\theta_2||+|\ln|\theta_1 e^{-\lambda t}||+|\ln|\theta_2+\theta_1 e^{-\lambda t}||)),$$

by using the inequality

$$|(r+s)\ln|r+s|-r\ln|r|-s\ln|s|| \le |s|(1+|\ln|r||+|\ln|s||+|\ln|r+s||),$$

valid when r and $r+s$ have the same sign. Hence (2.15) holds and, by (2.14), as $t\to\infty$,

$$I(t) \sim te^{-\lambda t}\int_{-\infty}^0 i\frac{2}{\pi}\beta\theta_1\lambda e^{\lambda x}dx = i\frac{2}{\pi}\beta\theta_1 te^{-\lambda t}. \quad ∎$$

Remarks

- The rate of decay of $r(t)$ is $e^{-\lambda t}$ when $\alpha > 1$ but is $e^{-\alpha\lambda t}$ when $\alpha < 1$. When $\alpha = 1$, the rate of decay is either 0, $e^{-\lambda t}$ or $te^{-\lambda t}$.

- In the Gaussian case $\alpha = 2$, we get

$$r(t) \sim \exp\{-\frac{1}{2\lambda}(\theta_1^2 + \theta_2^2)\} \left(-\frac{\theta_1\theta_2}{\lambda}\right) e^{-\lambda t}$$

as $t \to \infty$. This relation follows also directly from Theorem 1.1 because the covariance of the Gaussian Ornstein-Uhlenbeck process equals, for $s \le t$,

$$EX(s)X(t) = \int_{-\infty}^{s} e^{-\lambda(s-x)} e^{-\lambda(t-x)} 2dx = \frac{1}{\lambda} e^{-\lambda(t-s)}.$$

(The "$2dx$" is a consequence of $M(A) \sim N(0, 2|A|)$.)

Now consider the reverse (or fully anticipating) Ornstein-Uhlenbeck process, defined as

$$X(t) = \int_{t}^{\infty} e^{\lambda(t-x)} M(dx), \quad -\infty < t < \infty,$$

a process which is also stationary and Markov. Here, $f(x) = e^{\lambda x} 1_{(-\infty,0)}(x)$. Although this process coincides with the Ornstein-Uhlenbeck process if $\alpha = 2$, it is a different one when $\alpha < 2$ because their $r(t)$'s are asymptotically different. If X_1 and X_2 denote respectively the Ornstein-Uhlenbeck and the reverse Ornstein-Uhlenbeck processes, then $\{X_1(t),\ t \in \mathcal{R}\}$ and $\{X_2(-t),\ t \in \mathcal{R}\}$ have identical finite-dimensional distributions and, consequently, $r_{X_2}(\theta_1, \theta_2; t) = r_{X_1}(\theta_2, \theta_1; t)$. Hence,

COROLLARY 2.1 *For the reverse Ornstein-Uhlenbeck process,* $\lim_{t\to\infty} r(t)$ *is as in Theoreom 2.2, but with* θ_1 *and* θ_2 *interchanged.*

We now study moving averages whose $r(t)$ decreases like a power function. Consider the *1-step increment of linear fractional Lévy motion* ([ALT90]). It is a stationary process, defined for real constants a and b, $0 < \alpha \le 2$, and $0 < H < 1$, $H \neq \frac{1}{\alpha}$ by

$$X_{\alpha,H}(t) \equiv X_{\alpha,H}(a, b; t) = \int_{-\infty}^{\infty} f_{\alpha,H}(a, b; t - x) M(dx)$$

where

$$f_{\alpha,H}(a,b;s) = a[(s+1)_+^{H-\frac{1}{\alpha}} - (s)_+^{H-\frac{1}{\alpha}}] + b[(s+1)_-^{H-\frac{1}{\alpha}} - (s)_-^{H-\frac{1}{\alpha}}],$$

and M is an α-stable random measure with Lebesgue control measure and skewness intensity $\beta(\cdot)$ that is constant when $\alpha \neq 1$ and equal to zero if $\alpha = 1$. Alternatively, $X_{\alpha,H}(t) = Y_{\alpha,H}(t+1) - Y_{\alpha,H}(t)$, where $Y_{\alpha,H}$ is the linear fractional Lévy motion, defined as

$$Y_{\alpha,H}(t) \equiv Y_{\alpha,H}(a,b;t)$$

$$= \int_{-\infty}^{+\infty} \{a[(t-x)_+^{H-\frac{1}{\alpha}} - (-x)_+^{H-\frac{1}{\alpha}}] + b[(t-x)_-^{H-\frac{1}{\alpha}} - (-x)_-^{H-\frac{1}{\alpha}}]\} M(dx).$$

The process $Y_{\alpha,H}$ has stationary increments and is H-self-similar, that is, for all $c > 0$, the processes $\{Y_{\alpha,H}(ct),\ t \in \mathcal{R}\}$ and $\{c^H Y_{\alpha,H}(t),\ t \in \mathcal{R}\}$ have the same finite-dimensional distributions. (H is called the self-similarity parameter.)

The stationary process $X_{\alpha,H}(t)$ exhibits long-range dependence. Long-range dependence, in the Gaussian case $\alpha = 2$, occurs when the covariance decays to zero like a power function. The process $X_{\alpha,H}(t)$ exhibits long-range dependence when $\alpha = 2$, because its covariance is asymptotically proportional to t^{2H-2} as $t \to \infty$ (see [MV68]). To extend the notion of long-range dependence to $0 < \alpha < 2$, we must find an adequate replacement for the covariance. We propose to use $r(t)$ instead. It turns out (see [ALT90],) that there are complex-valued constants B and C depending on θ_1, θ_2, a, b and H such that

$$r(t) \sim \begin{cases} Bt^{\alpha H-\alpha} & \text{if either } 0 < \alpha < 1,\ 0 < H < 1, \text{or} \\ & \quad 1 < \alpha < 2,\ 1 - 1/(\alpha(\alpha-1)) < H < 1,\ H \neq \frac{1}{\alpha}, \\ Ct^{H-(1/\alpha)-1} & \text{if } 1 < \alpha < 2 \text{ and } 0 < H < 1 - 1/(\alpha(\alpha-1)). \end{cases} \tag{2.16}$$

Observe that $r(t)$ also decays like a power function. The exponent of decay is similar to the one in the case $\alpha = 2$ when H is high enough, but has a different expression when H is smaller than the cutoff point $1/(\alpha(\alpha-1))$.

The constants B and C are known explicitly: see [ALT90]. Because they depend on a and b in a complicated fashion, one can use (2.16) to establish the fact that the processes $X_{\alpha,H}(t)$ (and $Y_{\alpha,H}(t)$) are different for different a and b (see [CM89] and [ST89] for proofs that do not use $r(t)$). The constants B and C are real-valued in the symmetric case $\beta = 0$.

A process related to linear fractional Lévy motion $Y_{\alpha,H}(t)$ is the *Log-fractional Lévy motion* (see [KMV88]). It is defined only for $1 < \alpha < 2$, has stationary increments, and is H-self-similar with $H = 1/\alpha$. Its 1-step increment forms a stationary moving average process given by

$$X_{\alpha,\frac{1}{\alpha}}(t) = \int_{-\infty}^{\infty} [\ln|t+1-x| - \ln|t-x|] M(dx)$$

(where M is an α-stable random measure with Lebesgue control measure and constant skewness intensity $\beta \in [-1,1]$.) Since $H = 1/\alpha > 1 - 1/[\alpha(\alpha-1)]$, it is not surprising that

$$r(t) \sim Dt^{1-\alpha}$$

as $t \to \infty$ for some complex-valued constant D (see [ALT90]. The rate of decay is $1-\alpha$ and equals the rate of decay in (2.16) with $H = 1/\alpha$.

3 Real harmonizable processes

We have seen that $\lim_{t\to\infty} r(t) = 0$ for moving average processes. In contrast, $r(t)$ never converges to zero when the process $X(t)$ is a real symmetric α-stable harmonizable process with $0 < \alpha < 2$.

Let $0 < \alpha < 2$ and let M be a complex symmetric α-stable ($S\alpha S$) random measure with finite control measure m on $(\mathcal{R}, \mathcal{B})$. The process

$$X(t) = \operatorname{Re} \int_{-\infty}^{+\infty} e^{itx} M(dx), \quad -\infty < t < \infty, \tag{3.1}$$

is called real $S\alpha S$ harmonizable. If X is stationary then it can always be represented as (3.1) with M rotationally invariant (=isotropic). (This means that for any $\phi \in (0, 2\pi]$, $e^{i\phi}M(B) \stackrel{d}{=} M(B)$ for all $B \in \mathcal{B}$.) The characteristic function of a real stationary $S\alpha S$ harmonizable process $X(t)$ is then

$$E\exp\{i\sum_{j=1}^{d}\theta_j X(t_j)\} = \exp\{-c_0\int_{-\infty}^{+\infty}|\sum_{j=1}^{d}\theta_j e^{it_j x}|^\alpha m(dx)\}$$

where $c_0 = \frac{1}{2\pi}\int_0^{2\pi}|\cos\phi|^\alpha d\phi$ (see Theorem 5.5 of [Cam83]). Therefore, $r(t) = e^{-A_\alpha}(e^{-I(t)} - 1)$ where

$$A_\alpha = c_0(|\theta_1|^\alpha + |\theta_2|^\alpha)m(\mathcal{R}) \tag{3.2}$$

and

$$I(t) = c_0\int_{-\infty}^{+\infty}(|\theta_1 e^{itx} + \theta_2|^\alpha - |\theta_1|^\alpha - |\theta_2|^\alpha)m(dx). \tag{3.3}$$

Note that $r(t)$ does not tend to zero as $t \to \infty$, because $I(t)$ does not tend to zero. It is convenient to focus here on the Cesaro mean of $r(t)$.

THEOREM 3.1 *For any real stationary $S\alpha S$ harmonizable process*

$$\liminf_{T\to\infty}\frac{1}{2T}\int_{-T}^{T}r(t)dt \geq e^{-A_\alpha}c_0(m\{0\}F_0 + \frac{1}{2\pi}m(\mathcal{R}\backslash\{0\})F_1) \tag{3.4}$$

if $0 < \alpha < 2$, where A_α is given by (3.2),

$$F_0 = |\theta_1|^\alpha + |\theta_2|^\alpha - |\theta_1 + \theta_2|^\alpha,$$

and

$$F_1 = \int_0^{2\pi}(|\theta_1|^\alpha + |\theta_2|^\alpha - |\theta_1 e^{it} + \theta_2|^\alpha)dt > 0.$$

PROOF: Since $e^{-I(t)} - 1 \geq -I(t)$, we have

$$\liminf_{T\to\infty}\frac{1}{2T}\int_{-T}^{T}r(t)dt \geq e^{-A_\alpha}\lim_{T\to\infty}\frac{1}{T}\int_0^T(-I(t))dt.$$

Define $K(x) = \max\{k : k = 0, 1, 2, \ldots, \; 2\pi k \le x, \; 2\pi(k+1) > x\}$ and write

$$\frac{1}{T}\int_0^T (-I(t))dt = J_1(T) + J_2(T) + J_3(T)$$

where

$$J_1(T) = \frac{c_0}{T}\int_0^T (|\theta_1|^\alpha + |\theta_2|^\alpha - |\theta_1 + \theta_2|^\alpha) m(\{0\})dt = c_0 m(\{0\})F_0,$$

$$J_2(T) = c_0 \int_{\{x\neq 0\}} \frac{1}{Tx} \sum_{k=1}^{K(Tx)} \int_{2\pi(k-1)}^{2\pi k} (|\theta_1|^\alpha + |\theta_2|^\alpha - |\theta_1 e^{it} + \theta_2|^\alpha) dt m(dx),$$

$$J_3(T) = c_0 \int_{\{x\neq 0\}} \frac{1}{Tx} \int_{2\pi K(Tx)}^{Tx} (|\theta_1|^\alpha + |\theta_2|^\alpha - |\theta_1 e^{it} + \theta_2|^\alpha) dt m(dx).$$

As $T \to \infty$,

$$J_2(T) = c_0 \int_{\{x\neq 0\}} \frac{K(Tx)}{Tx} \int_0^{2\pi} (|\theta_1|^\alpha + |\theta_2|^\alpha - |\theta_1 e^{it} + \theta_2|^\alpha) dt m(dx)$$

$$\to \frac{c_0}{2\pi} m(\mathcal{R}\backslash\{0\})F_1,$$

because for $x \neq 0$, $2\pi K(Tx) \le Tx < 2\pi(K(Tx)+1)$ implies $\lim_{T\to\infty} \frac{K(Tx)}{Tx} = \frac{1}{2\pi}$ and the dominated convergence theorem applies. Similarly, $\lim_{T\to\infty} J_3(T) = 0$.

We now prove that the constant F_1 is positive. By Jensen's inequality,

$$\frac{1}{2\pi}\int_0^{2\pi} (\theta_1^2 + \theta_2^2 + 2\theta_1\theta_2 \cos t)^{\alpha/2} dt < \left[\frac{1}{2\pi}\int_0^{2\pi} (\theta_1^2 + \theta_2^2 + 2\theta_1\theta_2 \cos t)dt\right]^{\alpha/2}$$

$$= (\theta_1^2 + \theta_2^2)^{\alpha/2}.$$

Hence $\frac{1}{2\pi} F_1 > (\theta_1^2)^{\alpha/2} + (\theta_2^2)^{\alpha/2} - (\theta_1^2 + \theta_2^2)^{\alpha/2} > 0$. This concludes the proof of the theorem. ∎

Remarks. (1) If $m(\{0\}) = 0$, then for large T, the Cesaro mean is bounded away from zero.

(2) The theorem holds also for $\alpha = 2$, but in that case $F_1 = 0$ since $\int_0^{2\pi} (\theta_1^2 + \theta_2^2 - |\theta_1 e^{it} + \theta_2|^2) dt = -2\theta_1\theta_2 \int_0^{2\pi} \cos t dt = 0$. If also

$\mathrm{Cov}(X(t), X(0)) \to 0$ as $t \to \infty$, then $r(t) \to 0$ as $t \to \infty$, $m\{0\} = 0$, and the lower bound in (3.4) is attained and is zero. This contrasts with the case $\alpha < 2$, where the lower bound can be positive.

4 Sub-Gaussian processes

Consider a mean-zero Gaussian process $\{G(t), \ -\infty < t < \infty\}$. Let $0 < \alpha < 2$ and A be an $\alpha/2$-stable random variable totally skewed to the right, so that the Laplace transform of A satisfies $E\exp\{-\lambda A\} = \exp\{-\lambda^{\alpha/2}\}$ for $\lambda \geq 0$. Assume that A is independent of the process $G(\cdot)$. Then the process

$$X(t) = A^{1/2}G(t), \ -\infty < t < \infty$$

is called a sub-Gaussian process with governing Gaussian process $G(\cdot)$. It is easy to see by conditioning on A that the characteristic function of $(X(t_1), \ldots, X(t_d))$ is

$$E\exp\{i\sum_{j=1}^{d}\theta_j X(t_j)\} = \exp\{-(\frac{1}{2}\mathrm{Var}\sum_{j=1}^{d}\theta_j G(t_j))^{\alpha/2}\}. \tag{4.1}$$

Therefore, the process $\{X(t), \ -\infty < t < \infty\}$ is $S\alpha S$ and is stationary if the Gaussian process $G(\cdot)$ is stationary.

The dependence structure of the stationary sub-Gaussian process, and thus the asymptotic behavior of $r(t)$ as $t \to \infty$, is controlled by the *correlation* $\rho(t) = EG(t)G(0)/EG^2(0)$ of the stationary mean-zero Gaussian process G. We shall examine the case where $\rho(t)$ has a limit. Then $\lim_{t\to\infty} r(t)$ also exists, but unlike the moving average processes, it is not necessarily zero. In fact, $\lim_{t\to\infty} r(t) > 0$ if $\alpha < 1$.

THEOREM 4.1 *Let $X(t) = A^{1/\alpha}G(t)$, $-\infty < t < \infty$, be a stationary sub-Gaussian process, governed by a stationary mean-zero Gaussian process G with variance σ^2 and correlation function $\rho(t)$. If*

$\lim_{t\to\infty} \rho(t) = \rho_\infty$, *then*

$$r_\infty = \lim_{t\to\infty} r(t) = e^{-A_\alpha}(e^{-D} - 1)$$

where

$$A_\alpha = \left(\frac{\sigma^2}{2}\right)^{\alpha/2} (|\theta_1|^\alpha + |\theta_2|^\alpha)$$

and

$$D = \left(\frac{\sigma^2}{2}\right)^{\alpha/2} [(\theta_1^2 + \theta_2^2 + 2\theta_1\theta_2\rho_\infty)^{\alpha/2} - |\theta_1|^\alpha - |\theta_2|^\alpha].$$

In particular, the sign of r_∞ is as follows:

- $0 < \alpha < 1 \Rightarrow r_\infty > 0$,
- $\alpha = 1 \Rightarrow r_\infty > 0$ *if and only if* $\frac{\theta_1\theta_2}{|\theta_1\theta_2|}\rho_\infty < 1$, *and* $r_\infty = 0$ *if and only if there is equality,*
- $1 < \alpha < 2 \Rightarrow r_\infty$ *has the same sign as* $\frac{1}{2}[(|\theta_1|^\alpha + |\theta_2|^\alpha)^{2/\alpha} - \theta_1^2 - \theta_2^2] - \theta_1\theta_2\rho_\infty$.

PROOF: The form of the constant A_α follows directly from (4.1) and (1.2). Similarly,

$$I(t) = (\frac{\sigma^2}{2})^{\alpha/2}[(\theta_1^2 + \theta_2^2 + 2\theta_1\theta_2\rho(t))^{\alpha/2} - |\theta_1|^\alpha - |\theta_2|^\alpha]$$

implying $\lim_{t\to\infty} I(t) = D$.

We focus now on the sign of D (since r_∞ and D have opposite signs). If $0 < \alpha < 1$, then $(\theta_1^2 + \theta_2^2 + 2\theta_1\theta_2)^{\alpha/2} \leq (|\theta_1| + |\theta_2|)^{2\alpha/2} < |\theta_1|^\alpha + |\theta_2|^\alpha$, and hence $D < 0$. Suppose now $1 \leq \alpha < 2$. Then $D < 0$ if and only if $(\theta_1^2 + \theta_2^2 + 2\theta_1\theta_2\rho_\infty)^{\alpha/2} < |\theta_1|^\alpha + |\theta_2|^\alpha$, that is, if and only if $\theta_1\theta_2\rho_\infty < \frac{1}{2}[(|\theta_1|^\alpha + |\theta_2|^\alpha)^{2/\alpha} - \theta_1^2 - \theta_2^2] \equiv g(\theta_1, \theta_2)$ and $D = 0$, if there is equality. Finally, suppose $\alpha = 1$. Then $g(\theta_1, \theta_2) = |\theta_1\theta_2|$ and hence $D < 0$ if and only if $\frac{\theta_1\theta_2}{|\theta_1\theta_2|}\rho_\infty < 1$, with equality if $D = 0$. In this case, we cannot have $D > 0$ because $|\rho_\infty| \leq 1$. ∎

Remark. Suppose $\rho_\infty = 0$. Then $r_\infty > 0$ if $0 < \alpha < 2$.

Acknowledgments. We thank Piotr Kokoszka for his comments on the paper.

References

[ALT90] A. Astrauskas, J.B. Levy, and M. S. Taqqu. The asymptotic dependence structure of the linear fractional Lévy motion. To appear, *Lietuvos Matematikos Rinkinys (Lithuanian Mathematical Journal)*, 1990.

[Cam83] S. Cambanis. Complex symmetric stable variables and processes. In P.K. Sen, editor, *Contribution to Statistics, Essays in Honor of Norman L. Johnson*, pages 63–79. North-Holland, 1983.

[CHW84] S. Cambanis, C.D. Hardin, Jr., and A. Weron. Ergodic properties of stationary stable processes. Technical Report 59, Center for Stoch. Proc., University of North Carolina, Chapel Hill, 1984.

[CM89] S. Cambanis and M. Maejima. Two classes of self-similar stable processes with stationary increments. *Stoch. Proc. Appl.*, 32:305–329, 1989.

[Har84] C.D. Hardin, Jr. Skewed stable variables and processes. Technical Report 79, Center for Stoch. Proc., University of North Carolina, Chapel Hill, 1984.

[KMV88] Y. Kasahara, M. Maejima, and W. Vervaat. Log-fractional stable processes. *Stoch. Proc. Appl.*, 30:329–339, 1988.

[Mil78] G. Miller. Properties of certain symmetric stable distributions. *J. Mult. Anal.*, 8:346–360, 1978.

[MV68] B.B. Mandelbrot and J.W. Van Ness. Fractional Brownian motions, fractional noises and applications. *SIAM Review*, 10:422–437, 1968.

[ST89] G. Samorodnitsky and M.S. Taqqu. The various linear fractional Lévy motions. In K.B. Athreya T.W. Anderson and D.L. Iglehart, editors, *Probability, Statistics and Mathematics: Papers in Honor of Samuel Karlin*, pages 261–270, Boston, 1989. Academic Press.

Joshua B. Levy
University of Lowell
College of Management Science
Lowell, MA 01854, USA

Murad S. Taqqu
Boston University
Department of Mathematics
111 Cummington Street
Boston, MA 02215, USA

AN EXTREMAL PROBLEM IN H^p OF THE UPPER HALF PLANE WITH APPLICATION TO PREDICTION OF STOCHASTIC PROCESSES

BALRAM S. RAJPUT   KAVI RAMA-MURTHY   CARL SUNDBERG

Abstract. Let $1 \le p \le 2$ and let $\mathbf{H}^p$ be the Hardy space of the upper half plane. Let $\varphi \in \mathbf{H}^p$ and $T > 0$ be such that $e^{-iTz}\varphi(z) \notin \mathbf{H}^p$. Under certain conditions, we provide an explicit formula for the best approximation $\mathbf{P}_T(\varphi)$ of $e^{-Tz}\varphi(z)$ in $\mathbf{H}^p$. Using this formula for $\mathbf{P}_T(\varphi)$, we derive a formula for the best linear predictor of a continuous parameter "L^p-representable" complex regular stochastic process. The class of processes to which this formula applies includes the processes which are regular and are Fourier transforms of α-stable and α-semi-stable random measures, $1 \le \alpha \le 2$. The main tools used for the proofs depend heavily on the ideas and results from the theory of H^p spaces.

0. Introduction

This paper is composed of two parts. The first part is devoted to study an extremal problem in the Hardy space of the upper half plane and the second to study the linear prediction problem of certain continuous parameter stochastic processes. The first part is completely independent of the second; however, the second part depends, in a crucial way, on the results obtained in the first.

Let $1 \le p < \infty$, and let $\mathbf{H}^p$ be the Hardy space of the upper half plane $\mathcal{H}$. Let $\varphi \in \mathbf{H}^p$ and $T > 0$ be such that $e_{-T}\varphi \notin \mathbf{H}^p$, where $e_{-T}(z) = e^{-iTz}$, for every z in the complex plain $\mathbf{C}$. The main problem of interest in the first part of the paper is to determine the "best approximation" $\mathbf{P}(e_{-T}\varphi) \equiv \mathbf{P}_T(\varphi)$ of $e_{-T}\varphi$ in $\mathbf{H}^p$. This extremal problem is the analog in the upper half plane of the "dual extremal problem" in the H^p-space of the disc for the rational kernel $k(z) = z^{-(n+1)}\sum_{j=0}^{n} a_j z^{n-j} = \sum_{j=0}^{n} a_j z^{-(j+1)}, a_j \epsilon \mathbf{C}$. Although a solution to this problem in the disc has been known for quite some time (Duren [6, p.140]), the corresponding problem in the upper half plane was open.

This research is partially supported by Tennessee Science Alliance and AFSOR Grant No 90-0168.
AMS 1980 Subject Classification: Primary 30D55, 60G25, 62M20; Secondary 60E07, 60G10.
Key Words and Phrases: Best approximation, best linear predictors, stable and infinitely divisible processes.

In our *main* result of the first part of the paper (Theorem 1.2.1), we provide, under certain conditions on φ, an explicit formula for the "best approximation" $\mathbf{P}_T(\varphi)$ of $e_{-T}\varphi$ in $\mathbf{H}^p$, $1 \leq p \leq 2$; we also provide, in this theorem, explicit formulae for the related "extremal kernel" and the "minimizing interpolating function". Further, we give a characterization of the "extremal kernel" of $e_{-T}\varphi$ (Theorem 1.2.3); this result holds without any restrictive hypotheses on φ. In addition to these, we also prove two other related results.

Our formula for $\mathbf{P}_T(\varphi)$ is the counterpart in $\mathcal{H}$ of the solution of the "dual extremal problem" in the disc for the rational kernel k mentioned above. This and other formulae derived in Theorem 1.2.1 are reminiscent of the classical works of Landau, F. Riesz and Fejér which are devoted to solve certain "coefficient problems" and "minimum interpolation problems". We refer the reader to Duren [6, pp. 129-146] for the discussion of these and other related work and for a historical account of the theory of extremal problems in the disc from its evolutionary stages to its contemporary development.

It is well known that the linear space of a zero mean second order mean square continuous stationary process can be "viewed or represented" as an L^2-space of complex functions relative to a suitable finite measure on the real line $\mathbf{R}$. This fact plays an important role in the classical Wiener-Kolmogorov theory of best least square linear prediction of such a process. A similar situation prevails with regard to the linear space and the linear prediction of the α-stable process considered in [3]; except, in this case, the L^2-space is replaced by a suitable L^α-space. Motivated by these, we study, in the second part of this paper , the linear prediction for a class of processes whose linear space can be "represented" as a subspace of an L^p-space, $1 \leq p < \infty$, of complex functions relative to a suitable finite measure on $\mathbf{R}$. We refer to these processes here as "L^p-representable" processes.

Our *main* result of the second part of the paper is Theorem 2.2.3, this provides, under certain conditions, an explicit formula for the best linear predictor for a complex continuous parameter "L^p-representable" harmonizable and regular process X, $1 \leq p \leq 2$. The class of processes to which this formula applies includes the centered second order mean square continuous regular stationary processes and the regular Fourier transforms of α-stable and (r, α)-semi-stable random measures, $1 \leq \alpha < 2$ and $0 < r < 1$. In the special case when X is Gaussian, our formula for the linear predictor coincides with the formula of the Gaussian predictor due to Kolmogorov [11-13] Krein [16] and Weiner [29] (see Gihman and Skorohod [8, p. 307] or Dym and McKean [7, p. 88], for the formula of this Gaussian predictor). This result, in particular, confirms a conjecture of Cambanis and Soltani [3] and complements their work on linear prediction for a discrete parameter regular process which is a Fourier transform of an α-stable random measure, $1 < \alpha < 2$. Besides our main theorem, we formulate another result (Theorem 2.2.1), this provides a characterization of regularity and singularity of an "L^p-representable" harmonizable process, $1 \leq p < \infty$; this also provides a Wold-Cramér (W-C) type decomposition for such a process. We hasten to point out, however, that, except for the case $p = 1$ and for the uniqueness part in the W-C decomposition, the proof of this result is due to Cambanis and Soltani [3].

The methods of proof in the first part depend heavily on ideas and results from the theory of H^p-spaces and, in particular, on the interplay between complex function theory in the upper half plane and the Fourier transform on the real line. Some of these ideas are also needed in the second part; but, more importantly, the results in the second part are proved using the results of the first part (e.g., the formula for $\mathbf{P}_T(\varphi)$) and the isomorphism between the linear space of the "L^p-representable" process and the representing L^p-function space.

We close this section by pointing out that, for technical reasons involving the definition of the Fourier transform, we restrict ourselves to the case $1 \leq p \leq 2$.

I. An Extremal Problem in the $\mathbf{H}^p$-space of the Upper Half Plane

1.1. Preliminaries and Notations

Let $1 \leq p < \infty$ and let $L^p \equiv L^p(dx)$ be the usual Banach space of complex functions on $\mathbf{R}$ with norm $\|f\|_p = (\int_{\mathbf{R}} |f(x)|^p dx)^{1/p}$. Let $\mathbf{H}^p$, as in Section 0, be the Hardy space of analytic functions in the upper half plane $\mathcal{H}$. Then, as is well known, $\mathbf{H}^p$ is a closed subspace of L^p, and, for every $k \in L^p$, there exists a unique element $\mathbf{P}(k) \in \mathbf{H}^p$ which is the *best approximation* of k in $\mathbf{H}^p$; equivalently, there is a unique element $\mathbf{K} \equiv \mathbf{K}(k)$ in the coset $k + \mathbf{H}^p$ of the minimal L^p-norm [9, p. 134]. Thus, $\mathbf{P}(k)$ and $\mathbf{K}$ satisfy

$$\|\mathbf{K}\|_p = \inf_{f \in \mathbf{H}^p} \|k - f\| = \|k - \mathbf{P}(k)\|_p \ . \tag{1.1}$$

Using standard duality arguments, it follows that there exists a unique element $\mathbf{F} \in \mathbf{H}^q$, $\|\mathbf{F}\|_q = 1$ satisfying

$$\begin{aligned} \inf_{f \in \mathbf{H}^p} \|k - f\| &= \sup_{g \in \mathbf{H}^q} \{ \int k(x) g(x) dx : \|g\|_q \leq 1 \} \\ &= \int_{\mathbf{R}} \mathbf{K}(x) \mathbf{F}(x) dx \ , \end{aligned} \tag{1.2}$$

where q is the conjugate exponent of p [9, p.134].

The original *extremal problem* asks for determining the sup in (1.2) and the function $\mathbf{F}$, and the *dual extremal problem* is usually concerned with determining the inf in (1.1) and the function $\mathbf{K}$ (equivalently, $\mathbf{P}(k)$). The functions $\mathbf{K}$ and $\mathbf{F}$ are, respectively, referred to as the *extremal kernel* and the *dual extremal function* of k; and

$$\delta(k) = \|k - \mathbf{P}(k)\|_p$$

is referred to as the (p-th mean) *error* in the approximation of k by $\mathbf{P}(k)$.

As noted in the Introduction, we are interested here in finding a suitable formula for the best approximation $\mathbf{P}_T(\varphi)$ (equivalently, for the extremal kernel $\mathbf{K}_T \equiv \mathbf{K}_T(\varphi) \equiv \mathbf{K}(e_{-T}\varphi)$), where $\varphi \in \mathbf{H}^p$ and $T > 0$ is such that $e_{-T}\varphi \notin \mathbf{H}^p$. In the case when $1 \leq p \leq 2$; one can define another related function which will be called the *minimizing interpolating function* relative to $e_{-T}\varphi$. This function is the analog (in the upper half plane) of the function which solves the "minimal interpolation problem" at the origin (see [6, p.140] , [9, p.175]). It is defined as the unique function $\mathbf{\Psi}_T \equiv \mathbf{\Psi}_T(\varphi) \equiv \mathbf{\Psi}(e_{-T}\varphi)$ in $\mathbf{H}^p$ which satisfies

$$\|\mathbf{\Psi}_T\|_p = \inf\{\|\psi\| : \psi \in \mathbf{H}^p, \hat{\psi} = \hat{\varphi} \quad \text{a.e. on} \quad [0,T]\} , \tag{1.3}$$

where, throughout, $\hat{f}$ denotes the Fourier transform of a function $f \in L^p$. The existence and uniqueness of $\mathbf{\Psi}_T$ can be proved using the existence and uniqueness of $\mathbf{K}_T$ and the equality

$$e_{-T}\varphi + \mathbf{H}^p = \{e_{-T}\psi : \psi \in \mathbf{H}^p, \hat{\psi} = \hat{\varphi} \quad \text{a.e. on} \quad [0,T]\} .$$

It follows, in fact, that $\mathbf{K}_T = e_{-T}\mathbf{\Psi}_T$. The three functions $\mathbf{K}_T$, $\mathbf{P}_T(\varphi)$ and $\mathbf{\Psi}_T$ are related to each other as follows:

$$\mathbf{K}_T = e_{-T}\varphi - \mathbf{P}(e_{-T}\varphi) \, , \mathbf{\Psi}_T = e_T\boldsymbol{K}_T = \varphi - e_T\boldsymbol{P}_T(\varphi) \, ; \tag{1.4}$$

for future reference, we also record the following obvious facts:

$$e_T\mathbf{K}_T \in \mathbf{H}^p \quad \text{and} \quad \widehat{[e_T\mathbf{K}_T]} = \hat{\varphi} \quad \text{a.e. on} \quad [0,T] \, . \tag{1.5}$$

All the facts noted above are standard in the theory of $\mathbf{H}^p$-spaces, and can be found in [6, 9, 14]. Besides these, we shall need several other basic facts (BF's) from the complex function theory and the theory of H^p-spaces. For convenience and for ready reference, we have collected some of these facts in an Appendix (Subsection 1.4). We close this subsection by introducing a few more notations: For a function g, $supp(g)$ will denote the support of g. For a function $f \in \mathbf{H}^2$ and $T > 0$, $\mathcal{P}_T(f)$ will denote the $\mathbf{H}^2$-function defined by

$$\mathcal{P}_T(f)(z) = \frac{1}{\sqrt{2\pi}} \int_0^T e^{itz} \hat{f}(t) dt, z \in \mathbf{C} , \tag{1.6}$$

(see [14, p. 177]). These and other notations introduced above will remain fixed throughout the paper, unless stated otherwise.

1.2. Statements and Proofs of Results

The main result of this part of the paper is the following theorem.

Theorem 1.2.1. *Let $T > 0$ and $1 \leq p \leq 2$; and let φ be an $\mathbf{H}^p$-function. Assume that 0 belongs to $supp(\hat{\varphi})$; and if $p < 2$, assume also that φ and $\mathcal{P}_T(\varphi^{p/2})$ have no zeros in $\mathcal{H}$, where $\varphi^{p/2}$ is any analytic branch of the $p/2$-th root of φ. Then $\mathbf{P}(e_{-T}\varphi)$, the best approximation of $e_{-T}\varphi$ in $\mathbf{H}^p$, is given by the formula*

$$\mathbf{P}(e_{-T}\varphi) = e_{-T}\varphi - e_{-T}(\mathcal{P}_T(\varphi^{p/2}))^{2/p} , \tag{1.7}$$

where $(\mathcal{P}_T(\varphi^{p/2}))^{2/p}$ is that analytic branch of the $2/p$-th root of $\mathcal{P}_T(\varphi^{p/2})$ which satisfies (and is uniquely determined by) the condition

$$\lim_{y \uparrow \infty} \frac{(\mathcal{P}_T(\varphi^{p/2}))^{2/p}(iy)}{\varphi(iy)} = 1, \tag{1.8}$$

and the (p-th mean) error $\delta(T)(\equiv \delta(e_{-T}\varphi))$ in the approximation is given by the formula

$$\delta(T)^p = \int_{\mathbf{R}} |\mathcal{P}_T(\varphi^{p/2})|^2(x)dx = \int_0^T |\widehat{\varphi^{p/2}}(t)|^2 dt. \tag{1.9}$$

Further, the corresponding extremal kernel and the minimizing interpolating function are given, respectively, by

$$\mathbf{K}_T = e_{-T}(\mathcal{P}_T(\varphi^{p/2}))^{2/p} \quad and \quad \mathbf{\Psi}_T = (\mathcal{P}_T(\varphi^{p/2}))^{2/p}. \tag{1.10}$$

Remark 1.2.2. (a). The above result is the analog, in the upper half plane, of the solution for the dual extremal problem in H^p-space of the disc for the rational kernel k mentioned in the Introduction [6, p. 140].

(b). The necessity of requiring that $p \leq 2$ is explained in Section 0. The hypothesis that $0 \in supp(\hat{\varphi})$ is natural because it ensures that $e_{-T}\varphi \notin \mathbf{H}^p$, for any $T > 0$ (see BF-4). The hypotheses that φ and $\mathcal{P}_T(\varphi^{p/2})$ be non-vanishing on $\mathcal{H}$ are restrictive but essential, since one needs to know suitable analytic roots of these functions in $\mathcal{H}$. It seems that, when $p < 2$, a suitable computable formula for the function $\mathbf{P}_T(\varphi)$ (and, hence, also for $\mathbf{K}_T$ and $\mathbf{\Psi}_T$) cannot be found without these two hypotheses. (Note that a similar situation prevails with regard to the above noted solution for the dual extremal problem in the disc [6,p.141]). Since these two hypotheses are essential, we provide, in Proposition 1.2.5, a criterion (in terms of the Fourier transform) for concluding that an $\mathbf{H}^p$-function has no zeros in $\mathcal{H}$. This result is the analog in $\mathcal{H}$ of the classical Eneström-Kakeya theorem in the unit disc [6, p.141].

(c). In the case when $p = 2$, the above theorem asserts that if $0 \in supp(\hat{\varphi})$, then $\mathbf{P}_T(\varphi) = e_{-T}\varphi - e_{-T}(\mathcal{P}_T(\varphi))$. Theorem 1.2.1 is a very inefficient way to obtain this *well known* result; we include here a simple proof of this: In view of (1.3), it is enough to show that $\mathbf{\Psi}_T = \mathcal{P}_T(\varphi)$. Note that $\mathcal{P}_T(\varphi) \in \mathbf{H}^2$ (recall (1.9) and use Plancherel formula) and that $[\mathcal{P}_T(\varphi)]\hat{} = I_{[0,T]}\hat{\varphi} = \hat{\varphi}\, on\, [0, T]$; here, and throughout the paper, I_A will denote the indicator function of the set A. Hence $\mathcal{P}_T(\varphi)$ belongs to the set of the functions on the right side of (1.3). Further, for any $\psi \in \mathbf{H}^2$ with $\hat{\psi} = \hat{\varphi}$ a.e. on [0,T], we have $\|\mathcal{P}_T(\varphi)\|_2^2 = \|[\mathcal{P}_T(\varphi)\hat{]}\|_2^2 = \int_0^T |\hat{\varphi}(t)|^2 dt = \int_0^T |\hat{\psi}(t)|^2 dt \leq \int_0^\infty |\hat{\psi}(t)|^2 dt = \|\psi\|_2^2$. Thus, it must be true that $\mathbf{\Psi}_T = \mathcal{P}_T(\varphi)$.

(d). Let the function φ in Theorem 1.2.1 be such that $\varphi(iy) > 0$, for all $y > 0$ (this is the case, for example, if φ is outer and $\varphi(x) = \varphi(-x)$ a.e. on $\mathbf{R}$ (see BF-2)). If $\mathcal{P}_T(\varphi^{p/2})(iy) > 0$, for all $y > 0$, then, it follows, from (1.8), that the unique $2/p$-th root used in Theorem 1.2.1 is one which satisfies $(\mathcal{P}_T(\varphi^{p/2}))^{2/p}(iy) > 0$, for $y > 0$. This observation may be of some value while applying Theorem 1.2.1 to concrete functions φ.

Our next result (Theorem 1.2.3) provides a characterization of the extremal kernel $\mathbf{K}_T$ for $e_{-T}\varphi, \varphi \in \mathbf{H}^p, 1 \leq p \leq 2$. This is an analog of a similar characterization of the extremal kernel in the disc for the rational kernel k mentioned in the Introduction [6, p.138-143]. This result plays a crucial role for the proof of our main result (Theorem 1.2.1); but we also feel that it may be of independent theoretical interest. Note also that this provides a formula for $\mathbf{K}_T$ (hence, also, for $\mathbf{P}_T(\varphi)$) without any restrictive hypotheses on φ. However, since, unlike formula (1.7), the parameters involved in this formula are hard to compute for given concrete functions φ, it is not very useful for purposes of finding explicit form of $\mathbf{P}_T(\varphi)$. We are now ready to state Theorem 1.2.3; we will need a few more notations, these will be used throughout the paper. For a set $A \subseteq \mathbf{C}, \bar{A}$ will denote the closure of A; and $\mathbf{N}$ will denote the set of non-negative integers.

Theorem 1.2.3. *Let $T > 0, 1 \leq p \leq 2$; and let $\varphi \in \mathbf{H}^p$. Assume that $0 \in supp(\hat{\varphi})$; then a function $K \in L^p$ is the extremal kernel $\mathbf{K}_T$ of $e_{-T}\varphi$ if and only if*

$$e_T K \in \mathbf{H}^p, [e_T K\hat{]} = \hat{\varphi} \quad a.e.\ on \quad [0,T]\ , \tag{1.11}$$

and $e_T K$ admits the representation

$$(e_T K)(z) = A\, e^{i(T-\lambda)z/p} z^{2N/p} e^{a\, z/p} \prod_n \left(\frac{\bar{\beta}_n}{\beta_n}\right)\left(\frac{z-\beta_n}{z-\bar{\beta}_n}\right)$$

$$\times \prod_n \left(1-\frac{z}{\bar{\beta}_n}\right)^{2/p} \exp\left\{\frac{z}{p}\left(\frac{1}{\beta_n}+\frac{1}{\bar{\beta}_n}\right)\right\} \times \prod_n \left(1-\frac{z}{\bar{\gamma}_n}\right)^{2/p} \exp\left\{\frac{z}{p}\left(\frac{1}{\gamma_n}+\frac{1}{\bar{\gamma}_n}\right)\right\}, \tag{1.12}$$

for all $z \in \mathcal{H}$, where $\{\beta_n\} \subseteq \mathcal{H}$, $\{\gamma_n\} \subseteq \overline{\mathcal{H}}$ and satisfy

$$\sum_n |\beta_n|^{-2}(1+\Im(\beta_n)) < \infty, \sum_n |\gamma_n|^{-2}(1+\Im(\gamma_n)) < \infty\ , \tag{1.13}$$

and $A \in \mathbf{C}$, $N \in \mathbf{N}, \lambda \geq 0$ and $a \in \mathbf{R}$.

The analytic branches $\left(1-\frac{z}{\bar{\beta}_n}\right)^{2/p}$ and $\left(1-\frac{z}{\bar{\gamma}_n}\right)^{2/p}$ on $\mathcal{H}$ are chosen so that they are close to 1 when z is close to zero; the choice of the analytic branch $z^{2N/p}$ is arbitrary; the products in (1.12) converge uniformly on compact subsets of $\mathcal{H}$.

Further, in representation (1.12), the parameters N, a, λ, β_n's and γ_n's are unique; and the constant A is unique modulo the choice of the analytic root $z^{2N/p}$.

The following result provides information about continuity (in T) of the functions $\mathbf{P}_T(\varphi)$ and $\mathbf{\Psi}_T$. The proof of the continuity of $\mathbf{P}.(\varphi)$ (and, hence, also of $\mathbf{\Psi}.$) on $(0,\infty)$ and the indicated limits at 0, in the case when $1 < p$, follows trivially using the well-known fact that the map $L^p \ni k \longmapsto \mathbf{P}(k) \in \mathbf{H}^p$ is continuous. In this sense, these facts are not new, the remaining facts, however, in this proposition appear new.

Proposition 1.2.4. *Let $1 \leq p < \infty$ and $\varphi \in \mathbf{H}^p$, then the map $T \longmapsto \mathbf{P}_T(\varphi)$ is continuous from $(0,\infty)$ into $\mathbf{H}^p$, and $\lim_{T\downarrow 0} \mathbf{P}_T(\varphi) = \varphi$. Further, if $1 \leq p \leq 2$, then the map $T \longmapsto \mathbf{\Psi}_T(\varphi)$ is continuous from $(0,\infty)$ into $\mathbf{H}^p$, and one also has $\lim_{T\uparrow\infty} \mathbf{\Psi}_T(\varphi) = \varphi$, $\lim_{T\uparrow\infty} \mathbf{P}_T(\varphi) = 0$ and $\lim_{T\downarrow 0} \mathbf{\Psi}_T(\varphi) = 0$. (All the four limits here are in L^p).*

Proposition 1.2.5. *Let $1 \leq p \leq 2$ and let $\psi \in \mathbf{H}^p, \psi$ is not equal to 0. Suppose $\hat{\psi} \geq 0$ and that it is monotone decreasing on $[0, \infty)$, then ψ has no zeros in $\mathcal{H}$.*

For the proof of Theorem 1.2.1, besides using Theorem 1.2.3, we shall also need two other results (Lemmas 1.2.6 and 1.2.7); and, for the proof of Theorem 1.2.3, we shall need Lemma 1.2.8. All these three lemmas are technical in nature. The proof of Proposition 1.2.4 depends heavily on Lemma 1.2.9; this last lemma asserts that the map $L^p \ni k \longmapsto \mathbf{P}(k) \in \mathbf{H}^p, 1 \leq p < \infty$, is continuous. This result is well known for $1 < p < \infty$ [24, pp. 368 and 388]. But, for the case $p = 1$, we are not able to find any reference and we wonder if this result is new. At any rate, we have included a proof for this case; and, for purposes of ready reference, we have stated it for all $1 \leq p < \infty$.

Lemma 1.2.6. *Let $1 \leq p \leq 2$; and let $f, g \in \mathbf{H}^2$ be such that $0 \in supp(\hat{f})$ and $\hat{f} = \hat{g}$ a.e. on $[0, T]$. Then for any analytic branch $g^{2/p}$ of the $2/p$-th root of g on $\mathcal{H}$, one can choose an analytic branch $f^{2/p}$ of the $2/p$-th root of f on $\mathcal{H}$ (if $p < 2$, it is assumed, in addition, that f and g are non-vanishing on $\mathcal{H}$) so that*

$$\widehat{f^{2/p}} = \widehat{g^{2/p}} \quad a.e. \text{ on } \quad [0, T]; \tag{1.14}$$

this analytic branch, in addition, satisfies (and is uniquely determined by) the condition

$$\lim_{y \uparrow \infty} \frac{f^{2/p}(iy)}{g^{2/p}(iy)} = 1. \tag{1.15}$$

(Note that this lemma is of interest only in the case when $1 \leq p < 2$; for $p = 2$, the only non-obvious conclusion is that $\lim_{y \uparrow \infty} \frac{f(iy)}{g(iy)} = 1$).

Lemma 1.2.7. *Let N be a non-negative integer and a, $b \in \mathbf{R}$; and let $\{\zeta_n\} \subseteq \overline{\mathcal{H}}, \zeta_n \neq 0$ satisfying $\sum_n (1 + \Im(\zeta_n))|\zeta_n|^{-2} < \infty$. Finally, let f be the entire function defined by*

$$f(z) = z^N e^{(a+ib)z} \prod_n \left(1 - \frac{z}{\bar{\zeta}_n}\right) \exp\left\{\frac{z}{2}\left(\frac{1}{\zeta_n} + \frac{1}{\bar{\zeta}_n}\right)\right\}; \tag{1.16}$$

(see BF-9). If $f \in L^2 \bigcap L^\infty, 0 \in supp(\hat{f})$ and $supp(\hat{f}) \subseteq [0,T]$, for some $T > 0$, then $b \leq T/2$. (Here and in the proof of Lemma 1.2.7, we denote $f/\mathbf{R}$ also by f).

Lemma 1.2.8. *Let $1 \leq p < \infty$ and let f be a non-vanishing $\mathbf{H}^p$-function. Let μ be the singular measure in the canonical factorization of f (see BF-2). If x_0 is an isolated point of $supp(\mu)$, the support of the measure μ, then*

$$|f(x_0 + iy)| \leq C(f) \exp\left(-\frac{\mu\{x_0\}}{2\pi y}\right) , \tag{1.17}$$

for all positive small y, where $C(f)$ is a positive constant.

Lemma 1.2.9. *Let $1 \leq p < \infty$; then the map $L^p \ni k \longmapsto \mathbf{P}(k) \in \mathbf{H}^p$ is continuous.*

We are now ready to prove our results.

Proof of Theorem 1.2.1. Let $f = \varphi^{p/2}$; then $f \in \mathbf{H}^2$ and $\hat{f} \in L^2$. Therefore, using BF-6(a), we have that $\mathcal{P}_T(f)(z) = \frac{1}{\sqrt{2\pi}} \int_0^T e^{itz} \hat{f}(t)dt$, $z \in \mathbf{C}$, is an entire function of order 1. Therefore, since by hypothesis $\mathcal{P}_T(f)$ has no zeros in $\mathcal{H}$, by BF-7, we can find $C \in \mathbf{C}, N \in \mathbf{N}, a, b \in \mathbf{R}$ and $\zeta_n \in \overline{\mathcal{H}}$ satisfying $\sum_n |\zeta_n|^{-2} < \infty$ so that

$$\mathcal{P}_T(f)(z) = C z^N e^{az} e^{ibz} \prod_n \left(1 - \frac{z}{\overline{\zeta}_n}\right) \exp\left(\frac{z}{\overline{\zeta}_n}\right) , \tag{1.18}$$

for all $z \in \mathbf{C}$, where the product converges uniformly on compact subsets of $\mathbf{C}$.

Since $[\mathcal{P}_T(f)\hat{]} = \hat{f} I_{[0,T]} \in L^2, \mathcal{P}_T(f) \in \mathbf{H}^2$; therefore, $supp([e_{-T}\mathcal{P}_T(f)\hat{]}) \subseteq [-T, 0]$ (see BF-3) and $e_{-T}\mathcal{P}_T(f) \in L^2$. Thus, using the analog of BF-3 for the lower half plane, we have that $e_{-T}\mathcal{P}_T(f)$ belongs to the $\mathbf{H}^2$-space of the lower half plane. Hence, form (1.18) and BF-1 (e.g., (0-2)), we must have $\sum_n \frac{\Im(\zeta_n)}{|\zeta_n|^2} < \infty$. Therefore, from (1.18), we have

$$\mathcal{P}_T(f)(z) = C\, z^N e^{(a+ib)z} \prod_n \left(1 - \frac{z}{\overline{\zeta}_n}\right) \exp\left\{\frac{z}{2}\left(\frac{1}{\overline{\zeta}_n} + \frac{1}{\zeta_n}\right) + \frac{z}{2}\left(\frac{1}{\overline{\zeta}_n} - \frac{1}{\zeta_n}\right)\right\}$$

$$= C\, z^N e^{az} e^{i(b+b_1)z} \prod_n \left(1 - \frac{z}{\overline{\zeta}_n}\right) \exp\, \frac{z}{2}\left(\frac{1}{\overline{\zeta}_n} + \frac{1}{\zeta_n}\right) , \tag{1.19}$$

for all $z \in \mathbf{C}$, where $b_1 = \sum_n \frac{\Im(\zeta_n)}{|\zeta_n|^2} < \infty$. (Note that the product $\prod_n(1-z/\bar{\zeta}_n) \exp \frac{z}{2}\left(\frac{1}{\bar{\zeta}_n}+\frac{1}{\zeta_n}\right)$ converges uniformly on compact subsets of $\mathbf{C}$ (see BF-9)). Let $(1-z/\bar{\zeta}_n)^{2/p}$ be the analytic branch of the 2/p-th root of $(1-z/\bar{\zeta}_n)$ in $\mathcal{H}$ which is close to 1 as z is close to 0; then, using (1.19), BF-8, BF-9 and the fact $\sum_n(1+\Im(\zeta_n))|\zeta_n|^{-2} < \infty$, it follows that the function

$$C^{2/p}\, z^{2N/p}\, e^{2az/p}\, e^{i2z(b+b_1)/p} \prod_n \left(1-\frac{z}{\bar{\zeta}_n}\right)^{2/p} \exp\left\{\frac{z}{p}\left(\frac{1}{\bar{\zeta}_n}+\frac{1}{\zeta_n}\right)\right\} \tag{1.20}$$

defines an analytic branch of the 2/p-th root of $\mathcal{P}_T(f)$ in $\mathcal{H}$, where $C^{2/p}$ is any 2/p-th root of C, $z^{2N/p}$ is any analytic branch of the 2/p-th root of z^N in $\mathcal{H}$ and the product converges uniformly on compact subsets of $\mathcal{H}$. Denote this branch by Ψ.

Since, as noted above, $\mathcal{P}_T(f) = \mathcal{P}_T(\varphi^{p/2}) \in \mathbf{H}^2$, we have that Ψ belongs to $\mathbf{H}^p$. Further, since $[\mathcal{P}_T(\varphi^{p/2})\hat{]} = [\varphi^{p/2}\hat{]}$ on [0,T] and since $0 \in supp(\hat{\varphi})$, we have $0 \in supp([\mathcal{P}_T(\varphi^{p/2})\hat{]})$ (use BF-4 and the canonical representation BF-2 for this). Therefore, using Lemma 1.2.6, we can choose an analytic branch $(\mathcal{P}_T(\varphi^{2/p}))^{2/p}$ of the 2/p-th root of $\mathcal{P}_T(\varphi^{p/2})$ on $\mathcal{H}$ which satisfies (1.8) and $[(\mathcal{P}_T(\varphi^{p/2}))^{2/p}\hat{]} = \hat{\varphi}$ a.e. on [0,T]. Therefore, observing that $(\mathcal{P}_T(\varphi^{p/2}))^{2/p}$ is a constant multiple of the expression given in (1.20) and noting that Ψ has no zeros in $\mathcal{H}$ and the fact $\sum_n(1+\Im(\zeta_n))|\zeta_n|^{-2} < \infty$, we will have, by the if part of Theorem 1.2.3, that $(\mathcal{P}_T(\varphi^{p/2}))^{2/p} = e_T\mathbf{K}_T$, provided we can show that $2(b+b_1) = T-\lambda$, for some $\lambda \geq 0$; equivalently, $b+b_1 \leq T/2$. But this follows from Lemma 1.2.7 and (1.19). (Note that, since $|\mathcal{P}_T(\varphi^{p/2})(z)| = |\frac{1}{\sqrt{2\pi}}\int_0^T e^{itz}\widehat{\varphi^{p/2}}(t)dt| \leq Const.\|\varphi^{p/2}\|_2$, for all $z \in \mathcal{H}, \mathcal{P}_T(\varphi^{p/2}) \in L^\infty$). Thus, $\mathbf{K}_T = e_{-T}(\mathcal{P}_T(\varphi^{p/2}))^{2/p}$; we have already noted that (1.8) holds. Now, using (1.4), we get (1.7) and (1.10). Finally, (1.9) follows from (1.7) and the definition of $\delta(T)$. ∎

Proof of Theorem 1.2.3. We first prove the only if part; so we let $K = \mathbf{K}_T$. Then, (1.11) is precisely (1.5). Now we proceed to establish the representation (1.12). Since

$e_T\mathbf{K}_T \in \mathbf{H}^p$ and $\mathbf{F}_T$, the dual extremal function of $e_{-T}\varphi$, belongs to $\mathbf{H}^q$, where q is the conjugate exponent of p, we have that $e_T\mathbf{F}_T\mathbf{K}_T \in \mathbf{H}^1$. Thus, by BF-3, $supp([\mathbf{F}_T\mathbf{K}_T\hat{]}) \subseteq [-T,\infty)$. Therefore, since $\mathbf{F}_T\mathbf{K}_T$ is a.e. real on $\mathbf{R}$ by BF-10, $supp([\mathbf{F}_T\mathbf{K}_T\hat{]}) \subseteq [-T,T]$; and, hence, $supp([e_T\mathbf{F}_T\mathbf{K}_T\hat{]}) \subseteq [0,2T]$. Therefore, by BF-6, $e_T\mathbf{F}_T\mathbf{K}_T$, and, hence, $\mathbf{F}_T\mathbf{K}_T$ extends to an entire function of order 1; we denote this extension of $\mathbf{F}_T\mathbf{K}_T$ by R_T. Since $\mathbf{F}_T\mathbf{K}_T \geq 0$ a.e. on $\mathbf{R}$, we have $R_T(x) \geq 0$, for all $x \in \mathbf{R}$; this implies that all real roots of R_T must be of even order. Further, using the fact that R_T is real on $\mathbf{R}$, it follows that if z is a root of R_T then so is $\bar{z}$. Using these facts and BF-7, it follows that there exists a sequence $\{\alpha_n\} \subseteq \overline{\mathcal{H}}$ such that

$$\sum_n |\alpha_n|^{-2} < \infty, \tag{1.21}$$

and that

$$R_T(z) = Cz^{2N}e^{az}\prod_n \left(1-\frac{z}{\alpha_n}\right)\left(1-\frac{z}{\bar{\alpha}_n}\right)\exp\left\{z\left(\frac{1}{\alpha_n}+\frac{1}{\bar{\alpha}_n}\right)\right\}, \tag{1.22}$$

for all $z \in \mathbf{C}$, where $C > 0, a \in \mathbf{R}, N \in \mathbf{N}$, and the product converges uniformly on compact subsets of $\mathbf{C}$. Clearly, from (1.22), the zeros of the $\mathbf{H}^1$-function e_TR_T in $\mathcal{H}$ are those α_n's which satisfy $\Im(\alpha_n) > 0$; therefore, by (1.21) and BF-1 (e.g., (0-2)), we have

$$\sum_n (1+\Im(\alpha_n))\,|\alpha_n|^{-2} < \infty\,. \tag{1.23}$$

Now, from BF-10, we have $|\mathbf{K}_T| = R_T$ a.e. on $\mathbf{R}$, if $p=1$, and

$$|\mathbf{K}_T(x)| = \frac{R_T(x)}{|\mathbf{F}_T(x)|} = \left(\frac{\|\mathbf{K}_T\|_p}{|\mathbf{K}_T(x)|}\right)^{p/q} R_T(x) \quad \text{a.e. on} \quad \mathbf{R}\,,$$

if $1 < p \leq 2$. Therefore, using (1.22), we have

$$|\mathbf{K}_T(x)|^p = C\|\mathbf{K}_T\|_p^{p-1}x^{2N}e^{ax}\prod_n \left|\left(1-\frac{x}{\alpha_n}\right)\right|^2 \\ \times \exp\left\{x\left(\frac{1}{\alpha_n}+\frac{1}{\bar{\alpha}_n}\right)\right\} \quad \text{a.e. on} \quad \mathbf{R}\,, \tag{1.24}$$

in both cases; i.e., for $p = 1$ and $1 < p \leq 2$.

The next step in establishing the representation (1.12) is to show that

$$\Psi(z) \equiv e^{iTz+az} z^{2N} \prod_n \left(1 - \frac{z}{\bar{\alpha}_n}\right)^2 \exp\left\{z\left(\frac{1}{\alpha_n} + \frac{1}{\bar{\alpha}_n}\right)\right\}, \tag{1.25}$$

$z \in \mathbf{C}$, is an $\mathbf{H}^1$-function and that $\frac{C}{C_0} e_{-\lambda}\Psi$ is outer, for some $C_0 \in \mathbf{C}$ with $|C_0| = 1$ and $\lambda \geq 0$. (Note that, in view of (1.23) and BF-9, the product in (1.25) converges uniformly on compact subsets of $\mathbf{C}$ and Ψ is entire). From (1.22) and (1.25), we have

$$\frac{e_T R_T}{\Psi}(z) = C \prod_n \frac{(1 - z/\alpha_n)}{(1 - z/\bar{\alpha}_n)} = C \prod_n \left(\frac{\bar{\alpha}_n}{\alpha_n}\right)\left(\frac{z - \alpha_n}{z - \bar{\alpha}_n}\right) = CB(z), \tag{1.26}$$

for all $z \in \mathcal{H}$, where $B(z) = \prod_n \left(\frac{\bar{\alpha}_n}{\alpha_n}\right)\left(\frac{z-\alpha_n}{z-\bar{\alpha}_n}\right)$. From (1.22), clearly B is the Blaschke product of the $\mathbf{H}^1$-function $e_T R_T$; and, from (1.26), $e_T R_T = CB\Psi$ on $\mathcal{H}$. Thus, since Ψ is analytic on $\mathcal{H}$ and $e_T R_T$ has no singular factor (being entire, see BF-2), it follows that $\frac{C}{C_0} e_{-\lambda}\Psi$ is outer for some $C_0 \in \mathbf{C}$ with $|C_0| = 1$ and $\lambda \geq 0$.

Let $\left(\frac{C}{C_0} e_{-\lambda}\Psi\right)^{1/p}$ be the unique analytic branch on $\mathcal{H}$ of the 1/p-th root of $\frac{C}{C_0} e_{-\lambda}\Psi$ which makes this branch an outer function. We now observe, from (1.24) and (1.25), that

$$|(e_T \mathbf{K}_T)(x)| = \|\mathbf{K}_T\|_p^{p-1} \left|\left(\frac{C}{C_0} e_{-\lambda}\Psi\right)^{1/p}(x)\right| \quad \text{a.e. on} \quad \mathbf{R};$$

therefore, since an outer factor of an $\mathbf{H}^p$-function is uniquely determined by the absolute value of its boundary function, it follows that $\|\mathbf{K}_T\|_p^{p-1}\left(\frac{C}{C_0} e_{-\lambda}\Psi\right)^{1/p}$ is the outer factor of the $\mathbf{H}^p$-function $e_T \mathbf{K}_T$. Using (1.25), BF-8 and BF-9, we have

$$\|\mathbf{K}_T\|_p^{p-1}\left(\frac{C}{C_0} e_{-\lambda}\Psi\right)^{1/p} = A_0 e^{i(T-\lambda)z/p} z^{2N/p} e^{az/p} \prod_n \left(1 - \frac{z}{\bar{\alpha}_n}\right)^{2/p}$$

$$\times \exp\left\{\frac{z}{p}\left(\frac{1}{\alpha_n} + \frac{1}{\bar{\alpha}_n}\right)\right\}, \tag{1.27}$$

for all $z \in \mathcal{H}$, where $A_0 \in \mathbf{C}, |A_0| = \|\mathbf{K}_T\|_p^{\frac{p-1}{p}} C^{1/p}$, $\left(1 - \frac{z}{\bar{\alpha}_n}\right)^{2/p}$ is that analytic branch of 1/p-th root of $\left(1 - \frac{z}{\bar{\alpha}_n}\right)^2$ which is close to 1 when z is close to zero and $z^{2N/p}$ is any analytic branch of the 1/p-th root of z^{2N}. (Note also that, by BF-9, the product in (1.27) converges uniformly on compact subsets of $\mathcal{H}$). As shown above $[e_T\mathbf{K}_T\hat{]} = \hat{\varphi}$ a.e. on [0,T], and, since $0 \in supp(\hat{\varphi})$, we have $0 \in supp([e_T\mathbf{K}_T\hat{]})$. Therefore, by BF-4, $e_T\mathbf{K}_T$ cannot have an e_τ factor $\tau > 0$. Further, since $e_T\mathbf{K}_T\mathbf{F}_T$ extends to the entire function $e_T R_T, e_T\mathbf{K}_T\mathbf{F}_T$ and hence $e_T\mathbf{K}_T$ (use the canonical representation) cannot have a singular factor in its canonical representation. Therefore, writing $\{\alpha_n\} = \{\beta_n\} \bigcup \{\gamma_n\}$, where β_n's are zeros of $e_T\mathbf{K}_T$, it follows, using (1.23), (1.27) and the canonical representation result again, that $e_T\mathbf{K}_T$ has the representation (1.12) and β_n's and γ_n's satisfy (1.13); the constant A in (1.12) is given by $A = A_0 w$, for a suitable w with $|w| = 1$. This completes the proof of the only if part.

Now we prove the if part. First note that, since $e_T K \in \mathbf{H}^p$, we have, from (1.12), that

$$h(z) \equiv e^{i(T-\lambda)z/p} z^{2N/p} e^{az/p} \prod_n \left(1 - \frac{z}{\bar{\beta}_n}\right)^{2/p} \exp\left\{\frac{z}{p}\left(\frac{1}{\beta_n} + \frac{1}{\bar{\beta}_n}\right)\right\}$$
$$\times \prod_n \left(1 - \frac{z}{\bar{\gamma}_n}\right)^{2/p} \exp\left\{\frac{z}{p}\left(\frac{1}{\gamma_n} + \frac{1}{\bar{\gamma}_n}\right)\right\}$$

is an $\mathbf{H}^p$-function. Further, since $|h(z)|^p$ agrees on $\mathcal{H}$ with the absolute value of the entire function

$$g(z) = e^{i(T-\lambda)z} z^{2N} e^{az} \prod_n \left(1 - \frac{z}{\bar{\beta}_n}\right)^2 \exp\left\{z\left(\frac{1}{\beta_n} + \frac{1}{\bar{\beta}_n}\right)\right\}$$
$$\times \prod_n \left(1 - \frac{z}{\bar{\gamma}_n}\right)^2 \exp\left\{z\left(\frac{1}{\gamma_n} + \frac{1}{\bar{\gamma}_n}\right)\right\},$$

it follows, using (1.13) and (BF-1), that

$$g(z) \times \prod_n \left(\frac{\bar{\beta}_n}{\beta_n}\right)\left(\frac{z - \beta_n}{z - \bar{\beta}_n}\right) \times \prod_n \left(\frac{\bar{\gamma}_n}{\gamma_n}\right)\left(\frac{z - \gamma_n}{z - \bar{\gamma}_n}\right), \tag{1.28}$$

$z \in \mathcal{H}$, is an $\mathbf{H}^1$-function. (Note that, by BF-9 and (1.13), the product in g converges uniformly on compact subsets of $\mathbf{C}$). Noting that

$$\left(\frac{\bar{\beta}_n}{\beta_n}\right)\left(1-\frac{z}{\bar{\beta}_n}\right)\left(\frac{z-\beta_n}{z-\bar{\beta}_n}\right)=\left(1-\frac{z}{\beta_n}\right)\ , \ z \in \mathcal{H}, \tag{1.29}$$

and an equality similar to (1.29) for γ_n factors, we observe that the expression in (1.28) is the restriction (to $\mathcal{H}$) of the entire function $e_{T-\lambda}R$ where R is the entire function defined by

$$R(z) = z^{2N} e^{az} \prod_n \left(1-\frac{z}{\beta_n}\right)\left(1-\frac{z}{\bar{\beta}_n}\right) \exp\left\{z\left(\frac{1}{\beta_n}+\frac{1}{\bar{\beta}_n}\right)\right\}$$

$$\times \prod_n \left(1-\frac{z}{\gamma_n}\right)\left(1-\frac{z}{\bar{\gamma}_n}\right) \exp\left\{z\left(\frac{1}{\gamma_n}+\frac{1}{\bar{\gamma}_n}\right)\right\}\ , \tag{1.30}$$

(see BF-9). From (1.12) and (1.30), we observe that $F \equiv \frac{R}{K}$ is analytic on $\mathcal{H}$. We also set $F(x) = \frac{R(x)}{K(x)}$, whenever $K(x) \neq 0, x \in \mathbf{R}$. Using (1.12) and (1.30) again, we have

$$|K(z)|^p = |A|^p \ |e^{-iTz+\frac{i}{p}(T-\lambda)z}|^p \ |R(z) \prod_n \left(\frac{\bar{\beta}_n}{\beta_n}\right)\left(\frac{z-\beta_n}{z-\bar{\beta}_n}\right)|^p,$$

for all $z \in \mathcal{H}$. Therefore, since the absolute value of the Blaschke product is 1 a.e. on $\mathbf{R}$, $|K(x)|^p = |A|^p|R(x)|$ a.e. on $\mathbf{R}$. Hence, since $K \neq 0$ a.e. on $\mathbf{R}$ and $F(x) = \frac{R(x)}{K(x)}$, whenever $K(x) \neq 0$, we have $|F(x)| = |A|^{-p}|K(x)|^{p-1}$ a.e. on $\mathbf{R}$. Thus $|F|^q = |A|^{-pq}|K|^p$ a.e. on $\mathbf{R}$, if $1 < p \leq 2$, and $|F| = |A|^{-1}$ a.e. on $\mathbf{R}$, if $p = 1$; also recall that $KF = R \geq 0$ a.e. on $\mathbf{R}$. Further, using (1.11), $[e_T K - \varphi]\hat{} = 0$ a.e. on $(-\infty, T]$; hence $[K - e_{-T}\varphi]\hat{} = 0$ a.e. on $(-\infty, 0]$; i.e., $K - e_{-T}\varphi \in \mathbf{H}^p$ (see BF-3). Thus, in view of BF-10, if we can show that $F \in \mathbf{H}^q$, it will follow that $K = \mathbf{K}_T$. We show this in the following.

We noted above that $e_{T-\lambda}R$ is an $\mathbf{H}^1$-function; hence, since $\lambda \geq 0, e_T R = e_\lambda(e_{T-\lambda}R)$ is also $\mathbf{H}^1$. Therefore, since $F = \frac{e_T R}{e_T K}, e_T K \in \mathbf{H}^p$ and $F \in L^q$, it follows, from BF-5, that F will be in $\mathbf{H}^q$, provided it can be shown that $e_T K$ has no e_τ factor $\tau > 0$ and no singular S_μ factor in its canonical representation. That $e_T K$ cannot have an e_τ factor follows from BF-4, since $0 \in supp(\hat{\varphi})$ and $[e_T \hat{K]} = \hat{\varphi}$ (see (1.11)). Let μ be the singular measure appearing in the singular factor of $e_T K$ (see (0-4)), we will show that $supp(\mu)$ is empty; this will complete the proof of the if part. First observe that, since by (1.13) $\sum_n |\beta_n|^{-2}$ and $\sum_n |\gamma_n|^{-2}$ are finite, for any $x \in \mathbf{R}$ which is different from γ_n's, we can choose an open *nbd.* $\triangle$ of x which contains no β_n and γ_n. Then it is easy to show, using (1.13), BF-8 and BF-9, that the terms in the three products in (1.12) extends analytically to $\mathcal{H} \bigcup \triangle$ and that these products converge uniformly on compact subsets of $\triangle$. Thus K extends analytically in a *nbd.* of x; therefore, by BF-2, x cannot belong to $supp(\mu)$. Hence $supp(\mu) \subseteq \{\gamma_n : \gamma_n \in \mathbf{R}\}$; however, if $x_0 = \gamma_{n_0}$, for some n_o, then, from (1.12) and arguing as above, we have that $K(x_0 + iy) = k_0(x_0 + iy)\left(1 - \frac{x_0+iy}{x_0}\right)^{2/p}, y > 0$, where k_0 is a non-vanishing continuous function in a *nbd.* of x_0. Hence, for some $\delta > 0$,

$$|K(x_0 + iy)| \geq \delta \left| \left(1 - \frac{x_0 + iy}{x_0}\right) \right|^{2/p} = \frac{\delta y^{2/p}}{|x_0|^{2/p}},$$

for all positive small y. This along with Lemma 1.2.8 imply that

$$Const.\ y^{2/p} \leq |K(x_0 + iy)| \leq \ Const.\ e^{-d/y},$$

for all positive small y, where $d = \frac{\mu\{x_0\}}{2\pi}$; but this is impossible. Therefore $supp(\mu)$ is empty.

To see the uniqueness of the parameters; first note (from the proof of the if part above) that $\{\beta_n\} \bigcup \{\gamma_n\}$ in (1.12) are the zeros of the entire function R (the extension of $\mathbf{K}_T \mathbf{F}_T$) and $2N$ in (1.12) is the multiplicity of the zero of R at 0. Therefore, these parameters are uniquely determined. Now fixing an analytic branch $z^{2N/p}$ on $\mathcal{H}$, the uniqueness of A, λ and a easily follows. ∎

Proof of Proposition 1.2.4. The continuity of $T \longmapsto \mathbf{P}_T(\varphi)$ is clear from Lemma 1.2.9 and the continuity of $T \longmapsto e_{-T}\varphi$ on $(0, \infty)$. Using this and (1.4) one immediately gets the continuity of $T \longmapsto \Psi_T(\varphi)$ on $(0, \infty)$. Further, since $e_{-T}\varphi \longrightarrow \varphi$ in L^p as $T \downarrow 0$, we have, by Lemma 1.2.9, that $\lim_{T\downarrow 0}\mathbf{P}_T(\varphi) = \varphi$. This and (1.4) also gives $\lim_{T\downarrow 0}\Psi_T(\varphi) = 0$.

Now we proceed to prove that $\lim_{T\uparrow\infty} \Psi_T(\varphi) = \varphi$. First we consider the case $1 < p \leq 2$. Let $\{T_n\}$ be a sequence such that $T_n \uparrow \infty$, we must show that the L^p-limit of $\{\Psi_{T_n}\}$ exists and is equal to φ as $n \to \infty$. It is equivalent to prove that every subsequence of $\{\Psi_{T_n}\}$ has in turn a subsequence which converges to φ in L^p. For simplicity of notation, denote an arbitrary subsequence also by $\{\Psi_{T_n}\}$. By the definition of Ψ_{T_n}'s, $\|\Psi_{T_n}\|_p \leq \|\varphi\|_p$, for all n. Thus, by the weak sequential compactness of the unit ball of L^p, we can find a subsequence $\{n(j)\}$ of positive integers such that $\{\psi_j \equiv \Psi_{T_{n(j)}}\}$ converges weakly to a function ψ belonging to L^p. Therefore, by the continuity of the Fourier transform from L^p into L^q, $\{\hat{\psi}_j\}$ converges weakly to $\hat{\psi}$. Using this and the fact that $\hat{\psi}_j = 0$ a.e. on $(-\infty, 0)$ and $\hat{\psi}_j = \hat{\varphi}$ a.e. on $[0, T_{n(j)}]$, we have that $\hat{\psi} = 0$ a.e. on $(-\infty, 0)$ and $\hat{\psi} = \hat{\varphi}$ a.e. on $[0, \infty)$. Thus, $\psi = \varphi$; and, hence, using $\|\Psi_{T_n}\|_p \leq \|\varphi\|_p$ and the weak convergence of $\{\psi_j\}$ to $\psi (= \varphi)$, we have

$$\|\varphi\|_p \leq \underline{\lim}_j \|\psi_j\|_p \leq \overline{\lim}_j \|\Psi_{T_{n(j)}}\|_p \leq \|\varphi\|_p.$$

Thus $\lim_j \|\psi_j\|_p = \|\varphi\|_p$. This along with the weak convergence of $\{\Psi_{T_{n(j)}}\}$ to φ and uniform convexity of L^p now implies $\Psi_{T_{n(j)}} \to \varphi$ in L^p, as $n \to \infty$.

Now we proceed to prove that the L_p-limit of $\Psi_T(\varphi)$ is φ as $T \uparrow \infty$, when $p = 1$. Let $T_n \uparrow \infty$; and let $\mathcal{C}_0$ be the Banach space (in the sup norm) of continuous functions on $\mathbf{R}$ vanishing at ∞ and $\mathcal{A}_0 = \mathcal{C}_0 \bigcap \mathbf{H}^\infty$; then the dual of $\mathcal{C}_0/\mathcal{A}_0 = \mathbf{H}^1$ (see [14, p. 193]). Since the unit ball of $\mathbf{H}^1$ is weak* compact and metrizable, there exists a subsequence $\{\psi_j \equiv$

$\Psi_{T_{n(j)}}\}$ of $\{\Psi_{T_n}\}$ which converges in weak* topology to a $\psi \in \mathbf{H}^1$. This along with the facts that $\|\psi_j\|_1 \leq \|\varphi\|_1$ that, for every $z \in \mathcal{H}$, the Poisson kernel $Q_z(t) = \frac{y}{(x-t)^2+y^2} \in \mathcal{C}_0$ and that, for any $\zeta \in \mathbf{H}^p, p \geq 1, \zeta(z) = \int_{\mathbf{R}} \zeta(t)Q_z(t)dt, z \in \mathcal{H}$, [14, p. 153], implies that

$$\|\psi\|_1 \leq \liminf_{j\to\infty}\|\psi_j\|_1 \leq \limsup_{j\to\infty}\|\psi_j\|_1 \leq \|\varphi\|_1 \text{ and } \lim_j \psi_j(z) = \psi(z). \tag{1.31}$$

On the other hand, since $|\hat{\psi}_j(t)| \leq \frac{1}{\sqrt{2\pi}} \|\psi_j\|_1 \leq \frac{1}{\sqrt{2\pi}}\|\varphi\|_1, \hat{\psi}_j = \hat{\varphi}$ a.e. on $[0, T_{n(j)}]$ and $T_{n(j)} \uparrow \infty$, it follows, by the dominated convergence theorem and BF-6, that

$$\lim_j|\psi_j(z) - \varphi(z)| = \lim_j \frac{1}{\sqrt{2\pi}} \mid \int_{T_{n(j)}}^{\infty} e^{itz}(\hat{\psi}_j(t) - \hat{\varphi}(t))dt \mid = 0;$$

for all $z \in \mathcal{H}$. Therefore, by (1.31), $\psi = \varphi$ and $\lim_j\|\psi_j\|_1 = \|\varphi\|_1$.

Now let $f_j = \psi_j/\|\psi_j\|_1$ and $f = \varphi/\|\varphi\|_1$ (note $\varphi \not\equiv 0$); then, form above, $\{f_j\}$ converges in weak* topology to f and $\|f_j\|_1 = \|f\|_1 = 1$. Using these facts, we will show that $\|f_j - f\|_1 \to 0$, as $j \to \infty$; this will complete the proof of $\|\psi_j - \varphi\|_1 \to 0$, as $j \to \infty$, since $\lim_j\|\psi_j\|_1 = \|\varphi\|_1$. Once again, it is sufficient to show that every subsequence of $\{fj\}$ has in turn a subsequence which converges to f in L^1 and, for simplicity, we denote an arbitrary subsequence of $\{f_j\}$ by $\{f_j\}$ itself.

We write $f_j = g_j h_j$, where $g_j, h_j \in \mathbf{H}^2$ and $\|g_j\|_2 = \|h_j\|_2 = 1$. (To see this, write $f_j = f'_j B_j$, where f'_j is a non-vanishing $\mathbf{H}^1$-function and B_j is the Blaschke product of f_j, we then take $g_j = (f'_j)^{1/2}B_j$ and $h_j = (f'_j)^{1/2}$, where $(f'_j)^{1/2}$ is any analytic square root of f'_j). We can find a subsequence $\{j(k)\}$ of positive integers so that $\{g_{j(k)}\}$ and $\{h_{j(k)}\}$ converge weakly in L^2, respectively, to functions g and h in $\mathbf{H}^2$. Therefore, since the Poisson kernel $Q_z(.)$ belongs to $\mathcal{C}_0 \bigcap \mathbf{H}^2$, for all $z \in \mathcal{H}$, it follows, using the above noted property of Q_z, that

$$\lim_k h_{j(k)}(z) = \lim_k \int_{\mathbf{R}} h_{j(k)}(t)Q_z(t)dt = \int_{\mathbf{R}} h(t)Q_z(t)dt = h(z),$$

for all $z \in \mathcal{H}$; similarly, $\lim_k g_{j(k)}(z) = g(z)$, for all $z \in \mathcal{H}$. However, since $f_{j(k)} = g_{j(k)}h_{j(k)}$, we have $f = gh$; therefore $1 = \|f\|_1 = \|gh\|_1 \leq \|g\|_2\|h\|_2$. On the other hand, by weak convergence, $\|g\|_2 \leq \liminf_k \|g_{j(k)}\|_2 \leq 1$ and $\|h\|_2 \leq \limsup_k \|h_{j(k)}\|_2 \leq 1$. It follows $\|g\|_2 = \|h\|_2 = 1$. Hence, since $\|g_{j(k)}\|_2 = \|h_{j(k)}\|_2 = 1$, by the weak convergence again, we have $\|g_{j(k)} - g\|_2 \to 0$ and $\|h_{j(k)} - h\|_2 \to 0$, as $k \to \infty$. Hence, since

$$\|f_{j(k)} - f\|_1 = \|g_{j(k)}h_{j(k)} - gh\|_1 \leq \|g_{j(k)}(h_{j(k)} - h)\|_1 + \|(g_{j(k)} - g)h\|_1$$

$$\leq \|g_{j(k)}\|_2\|h_{j(k)} - h\|_2 + \|g_{j(k)} - g\|_2\|h\|_2,$$

it follows that $\|f_{j(k)} - f\|_1 \to 0$, as $k \to \infty$. This completes the proof of $\|f_j - f\|_1 \to 0$, as $j \to \infty$.

The proof of $\lim_{T\uparrow\infty} \mathbf{P}_T(\varphi) = 0$ now follows from above and (1.4). ∎

Proof of Proposition 1.2.5. Let F be the right continuous version of $\hat{\psi}$ and let μ be the finite Borel measure on $[0, \infty)$ such that

$$\mu((0,t]) = F(0) - F(t), \quad t > 0 .$$

Let $z \epsilon \mathcal{H}$. If $\psi(z) = 0$, then, by BF-6, we have

$$\begin{aligned} 0 = \psi(z) &= \frac{1}{\sqrt{2\pi}} \int_0^\infty \hat{\psi}(t)e^{itz}\, dt \\ &= \frac{1}{\sqrt{2\pi}} \int_0^\infty F(t)e^{itz}\, dt \\ &= \frac{1}{\sqrt{2\pi}} \int_0^\infty (F(0) - \mu(0,t])e^{itz}\, dt \\ &= \frac{1}{\sqrt{2\pi}} \frac{i}{z} F(0) - \frac{1}{\sqrt{2\pi}} \frac{i}{z} \int_0^\infty e^{isz}\, d\mu(s) ; \end{aligned}$$

so that $F(0) = \int_0^\infty e^{isz} d\mu(s)$. But this is a contradiction, since

$$\left| \int_0^\infty e^{isz} d\mu(s) \right| \leq \int_0^\infty e^{-sy} d\mu(s) < \mu((0,\infty)) \leq F(0)$$

(recall $\psi \not\equiv 0$ and hence $F \not\equiv 0$). ∎

Proof of Lemma 1.2.6. We begin by noting the simple fact that if $h \in \mathbf{H}^p \cap \mathbf{H}^\infty$ and h is non-vanishing on $\mathcal{H}$, then, for any $r > p/2, h^r \in \mathbf{H}^2 \cap \mathbf{H}^\infty$, where h^r is any analytic branch of the r-th root of h. This follows, since for any $y > 0$,

$$\int_{\mathbf{R}} |h^r(x+iy)|^2 \; dx = \int_{\mathbf{R}} |h(x+iy)|^{2r-p} \; |h(x+iy)|^p \; dx$$
$$\leq C \int_{\mathbf{R}} |h(x+iy)|^p \; dx \; ,$$

where $C = \sup_{z \epsilon \mathcal{H}} \; |h(z)|^{2r-p}$. This fact will be used at several places in the following proof without any specific reference.

Let $y_0 > 0$ and consider the functions $f_{y_0}(z) \equiv f(z+iy_0)$ and $g_{y_0}(z) \equiv g(z+iy_0)$, $z \in \overline{\mathcal{H}}$. From [14, p. 149], we have

$$|f_{y_0}(z)| \leq \frac{Const.}{(y+y_0)^{1/2}} \; , \tag{1.32}$$

for all $z \in \overline{\mathcal{H}}$. Then (1.32) clearly implies that $f_{y_0} \in \mathbf{H}^2 \cap \mathbf{H}^\infty$. Using BF-6, we have $f_{y_0}(x) = f(x+iy_0) = \frac{1}{\sqrt{2\pi}} \int_0^\infty e^{it(x+iy_0)} \hat{f}(t)dt$; thus $\hat{f}_{y_0}(t) = \overline{e}^{ty_0} \hat{f}(t)$ a.e. Similarly, $g_{y_0} \in \mathbf{H}^2 \cap \mathbf{H}^\infty$ and $\hat{g}_{y_0}(t) = \overline{e}^{ty_0} \hat{g}(t)$ a.e. Hence, since $\hat{f} = \hat{g}$ a.e. on $[0, T]$, we have

$$\hat{f}_{y_0}(t) = \overline{e}^{ty_0} \hat{f}(t) = \overline{e}^{ty_0} \hat{g}(t) = \hat{g}_{y_0}(t) \quad \text{a.e. on} \quad [0, T] \; . \tag{1.33}$$

Now let $f^{2/p}$ be any analytic branch on $\mathcal{H}$; and set $f_{y_0}^{2/p} \equiv [f^{2/p}]_{y_0}$ and $g_{y_0}^{2/p} \equiv [g^{2/p}]_{y_0}$. Then, clearly, $f_{y_0}^{2/p}$ and $g_{y_0}^{2/p} \in \mathbf{H}^p \cap \mathbf{H}^\infty$; and, hence, these also belong to $\mathbf{H}^2$. Applying (1.33) with f and g replaced, respectively, by $f^{2/p}$ and $g^{2/p}$, we get

$$\widehat{f_{y_0}^{2/p}}(t) = e^{-y_0 t} \widehat{f^{2/p}}(t) \quad \text{a.e. and} \quad \widehat{g_{y_0}^{2/p}}(t) = e^{-y_0 t} \widehat{g^{2/p}}(t) \quad \text{a.e. on} \quad [0, T] \; . \tag{1.34}$$

Thus, if using (1.33), we can prove the analogs of (1.14) and (1.15) for $f_{y_0}^{2/p}$ and $g_{y_0}^{2/p}$, the proof of (1.14) and (1.15) will follow from (1.34). In view of this, without loss of generality, we can and will assume that

$$f, g \quad \text{and, hence, also} \quad f^{2/p}, g^{2/p} \in \mathbf{H}^2 \cap \mathbf{H}^\infty . \tag{1.35}$$

(Note that $g^{2/p}$ is the given analytic branch and $f^{2/p}$, so far, is any analytic branch).

We shall now prove that

$$\widehat{f^{2/p}} = \omega(p)\widehat{g^{2/p}} \quad \text{a.e. on} \quad [0, T] , \tag{1.36}$$

where $\omega(p) \in \mathbf{C}$, with $|\omega(p)| = 1$. First, we prove (1.36), in the case when $2/p$ is rational, say, $2/p = n/m$. It follows, from (1.35), that $f^k, g^k \in \mathbf{H}^2 \cap \mathbf{H}^\infty$, for $1 \leq k \leq n$. Hence, using Theorem 7.8 of [21, p. 172] and standard limiting arguments, it follows that $\widehat{f^k} = \widehat{f^{k-1}} * \hat{f}$ a.e., for $2 \leq k \leq n$; similarly, $\widehat{g^k} = \widehat{g^{k-1}} * \hat{g}$ a.e. But as $\hat{f} = \hat{g} = 0$ a.e. on $(-\infty, 0]$, we have $\widehat{f^k}(t) = \int_0^t \widehat{f^{k-1}}(t-s)\hat{f}(s)ds$ and $\widehat{g^k}(t) = \int_0^t \widehat{g^{k-1}}(t-s)\hat{g}(s)ds$ a.e. Thus, since $\hat{f} = \hat{g}$ a.e. on $[0, T]$, we have, by induction, $\widehat{f^n} = \widehat{g^n}$ a.e. on $[0, T]$. For simplicity of notation, we set $f_1 = f^{n/m}$ and $g_1 = g^{n/m}$. Since $0 \in supp(\hat{f}) \cap supp(\hat{g})$, it follows, from BF-4, that f, g and, hence, f_1 and g_1 do not have e_τ factors, $\tau > 0$ in their canonical factorizations. Therefore, using BF-4 again, $0 \in supp(\hat{f}_1) \cap supp(\hat{g}_1)$; and, as noted above, $\widehat{f_1^m} = \widehat{g_1^m}$ a.e. on $[0, T]$. Now write

$$(f_1^m - g_1^m) = [(f_1 - g_1)(f_2 - \omega_m g_1) \cdots (f_1 - \omega_m^{m-1} g_1)] , \tag{1.37}$$

where ω_m is a primitive m-th root of unity. Observe that 0 is in the support of the Fourier transforms of all but at most one of the factors $f_1 - \omega_m^j g_1$. (For if not, then for two distinct $0 \leq j_1, j_2 < m$, $\hat{f}_1 = \omega_m^{j_1} \hat{g}_1$ a.e. and $\hat{f}_1 = \omega_m^{j_2} \hat{g}_1$ a.e. in a *nbd.* of 0; and, hence, $\hat{g}_1 = 0$ a.e. in a *nbd.* of 0. But this contradicts the fact that $0 \in supp(\hat{g}_1)$). Further, since $[f_1^m - g_1^m\hat{]} = 0$ a.e. on $[0, T]$, by BF-4, $f_1^m - g_1^m$ must have a factor $e_\tau, \tau \geq T$, in its canonical representation. Thus, using these two facts, (1.37) and the canonical representation result, it follows that $f_1 - \omega_m^j g_1$ must have a factor $e_\tau, \tau \geq T$, in its representation, for some j. Hence, by BF-4, $\hat{f}_1 = \omega_m^j \hat{g}_1$ a.e. on $[0,T]$, proving (1.36).

Now we prove (1.36), for arbitrary $2/p$. Let $\{r_k\}$ be a sequence of rational numbers such that $2 > r_k \downarrow 2/p$, as $k \to \infty$; then $f^{r_k}(z) \to \beta_1 f^{2/p}(z)$ and $g^{r_k}(z) \to \beta_2 g^{2/p}(z)$, $z \epsilon \mathcal{H}$, as $k \to \infty$, where $|\beta_j| = 1, j = 1,2$. Then, since $f^{r_k} \in \mathbf{H}^2 \cap \mathbf{H}^\infty$ and since

$$\begin{aligned}
\|f^{r_k} - \beta_1 f^{2/p}\|_2^2 &= \int_{\mathbf{R}} |f^{r_k}(x) - \beta_1 f^{2/p}(x)|^2 \, dx \\
&= \int_{\mathbf{R}} |f^{r_k}(x) - \beta_1 f^{2/p}(x)|^p \, |f^{r_k}(x) - \beta_1 f^{2/p}(x)|^{2-p} \, dx \\
&= \int_{\mathbf{R}} |f(x)|^2 \, |f^{r_k - 2/p}(x) - \beta_1|^p \, |f^{r_k}(x) - \beta_1 f^{2/p}(x)|^{2-p} \, dx \ ,
\end{aligned}$$

and the second and the third term in the last integral are bounded a.e., we have, by the dominated convergence theorem, that $\|f^{r_k} - \beta_1 f^{2/p}\|_2 \to 0$, as $k \to \infty$. Similarly, $\|g^{r_k} - \beta_2 g^{2/p}\|_2 \to 0$, as $k \to \infty$. Therefore, $\|\widehat{f^{r_k}} - \beta_1 \widehat{f^{2/p}}\|_2$ and $\|\widehat{g^{r_k}} - \beta_2 \widehat{g^{2/p}}\|_2 \to 0$, as $k \to \infty$; hence some subsequence of $\{\widehat{f^{r_k}}\}$ (resp. of $\{\widehat{g^{r_k}}\}$) converges a.e. to $\beta_1 \widehat{f^{2/p}}$ (resp. to $\beta_2 \widehat{g^{2/p}}$). But, as shown above, $\widehat{f^{r_k}} = \omega'(k)\widehat{g^{r_k}}$ a.e. on $[0,T]$, for all k, with $|\omega'(k)| = 1$; it follows, passing to further subsequences if necessary, that $\widehat{f^{2/p}} = \beta_0 \widehat{g^{2/p}}$ a.e. on $[0,T]$, where $\beta_0 = \frac{\beta_3 \beta_2}{\beta_1}$ and β_3 is the limit of a suitable subsequence of $\{\omega'(k)\}$. Thus (1.36) is proved for any $1 \leq p \leq 2$.

Now we proceed to prove (1.14) and (1.15). A first step here is to prove that

$$\lim_{y \uparrow \infty} \frac{f(iy)}{g(iy)} = 1 \ . \tag{1.38}$$

Using BF-6, write, for any $y > 0$,

$$\frac{f(iy)}{g(iy)} = \frac{e^{yT} f(iy)}{e^{yT} g(iy)} = \frac{\int_0^T e^{y(T-t)} \hat{f}(t)dt + \int_T^\infty e^{y(T-t)} \hat{f}(t)dt}{\int_0^T e^{y(T-t)} \hat{g}(t)dt + \int_T^\infty e^{y(T-t)} \hat{g}(t)dt} \ . \tag{1.39}$$

Since $\hat{f} = \hat{g}$ a.e. on $[0, T]$, it follows, from (1.39), that the proof of (1.38) will be complete, if we can show that both $\int_T^\infty e^{y(T-t)} \hat{f}(t)dt$ and $\int_T^\infty e^{y(T-t)} \hat{g}(t)dt$ converge to zero as $y \uparrow \infty$ and that $\liminf_{y\uparrow\infty} \mid \int_0^T e^{y(T-t)} \hat{f}(t)dt \mid > 0$. The proof of the first two assertions follows easily using Hölder's inequality and the fact that $\int_T^\infty e^{-2y(t-T)}dt \to 0$, as $y \uparrow \infty$. In view of $\lim_{y\uparrow\infty} \int_T^\infty e^{y(T-t)} \hat{f}(t)dt = 0$ and the decomposition of $e^{yT} f(iy)$ used in (1.39), the third assertion will be implied by $\lim_{y\uparrow\infty} |e^{yT} f(iy)| = \infty$, which can be proved as follows: Since $0 \in supp(\hat{f})$, f does not have a factor $e_\tau, \tau > 0$, in its representation (see BF-4). Therefore, using (0-4) and (0-5),

$$\log |e^{yT} f(iy)| = y[T + \frac{1}{\pi} \int_{\mathbf{R}} \frac{\log |f(t)|}{y^2 + t^2} dt - \frac{1}{\pi} \int_{\mathbf{R}} \frac{1}{y^2 + t^2} \, d\mu(t)] ,$$

for all $y > 0$; but since

$$\int_{\mathbf{R}} \frac{\log |f(t)|}{y^2 + t^2} dt \leq \int_{\mathbf{R}} \frac{|\log f(t)|}{1 + t^2} dt < \infty$$

and

$$\int_{\mathbf{R}} \frac{1}{y^2 + t^2} \, d\mu(t) \leq \int_{\mathbf{R}} \frac{1}{1 + t^2} \, d\mu(t) < \infty ,$$

for all $y \geq 1$, we have $\lim_{y\uparrow\infty} |e^{yT} f(iy)| = \infty$.

Now let $f = e^{u_1 + iv_1}, f^{2/p} = e^{\frac{2}{p}(u_1 + iv_1)}$ and $g = e^{u_2 + iv_2}, g^{2/p} = e^{\frac{2}{p}(u_2 + iv_2)}$, where $u_j, v_j, j = 1, 2$ are real functions. Then, using (1.38) and the continuity of $(v_1 - v_2)$, we have that $(u_1 - u_2)(iy) \to 0$ and $(v_1 - v_2)(iy) \to 2k_0\pi$, as $y \uparrow \infty$, for some integer $k_0 \geq 0$. Define the new analytic branch of the $2/p$-th root of f on $\mathcal{H}$ by $f^{2/p} = e^{\frac{2}{p}(u_1 + i(v_1 - 2k_0\pi))}$. Then, clearly

$$\frac{f^{2/p}(iy)}{g^{2/p}(iy)} = \exp\left[\frac{2}{p}\{(u_1 - u_2)(iy) - i(v_1 - v_2)(iy) - 2k_0\pi\}\right] \to 1 , \tag{1.40}$$

as $y \uparrow \infty$. On the other hand, (1.38) applied to $f^{2/p}$ (new branch) and $\omega(p)g^{2/p}$ (note that $f^{2/p}, \omega(p)g^{2/p} \in \mathbf{H}^2 \cap \mathbf{H}^\infty$ and $\widehat{f^{2/p}} = [\omega(p)g^{2/p}\hat{]}$, see (1.36)), gives $\frac{f^{2/p}(iy)}{g^{2/p}(iy)} \to \omega(p)$, as $y \uparrow \infty$. Thus $\omega(p) = 1$, and $\widehat{f^{2/p}} = \widehat{g^{2/p}}$ a.e. on $[0, T]$. This and (1.40) completes the proof. ∎

Proof of Lemma 1.2.7. Let $g(z) = f(z)\bar{f}(-\bar{z}), z \in \mathbf{C}$. Since $L^2 \cap L^\infty \subseteq L^4$ and $f \in L^2 \cap L^\infty$, we have $g \in L^2$. Let $f_1(t) = \overline{f(-t)},\ t \in \mathbf{R}$. Then, observing that $\hat{f}_1 = \bar{\hat{f}}$ and using Theorem 7.8 of [21, p. 172] and standard limiting arguments, we have $\hat{g} = \hat{f} * \bar{\hat{f}}$; and hence $supp(\hat{g}) = cl(supp(\hat{f}) + supp(\hat{f}))$. Thus, recalling that $0 \in supp(f) \subseteq [0, T]$, it follows that

$$\inf(supp(\hat{g})) = 0, \ \sup(supp(\hat{g})) = 2\tau \quad \text{and} \quad \tau \leq T\ , \tag{1.41}$$

where $\tau = \sup(supp(\hat{f}))$. Now, we observe that

$$\begin{aligned} g(z) &= (-1)^N z^{2N} e^{i2bz} \prod_n \left(1 - \frac{z}{\bar{\zeta}_n}\right)\left(1 + \frac{z}{\zeta_n}\right) \\ &= (-1)^N e^{i2bz} h(z)\ , \end{aligned}$$

where $h(z) = z^{2N} \prod_n \left(1 - \frac{z}{\bar{\zeta}_n}\right)\left(1 + \frac{z}{\zeta_n}\right)$. Since $g \in L^2$, we have $h \in L^2$; and, clearly, $\hat{g}(t) = (-1)^N \hat{h}(t - 2b)$; therefore, $supp(\hat{h}) = supp(\hat{g}) - 2b$. Consequently, from (1.41), $\sup(supp(\hat{h})) = 2(\tau - b)$ and $\inf(supp(\hat{h})) = -2b$. In view of these, the proof of the lemma will be complete if we can show

$$\sup(supp(\hat{h})) = -\inf(supp(\hat{h}))\ . \tag{1.42}$$

For then we would have $2(\tau - b) = 2b$ or $b = \frac{\tau}{2} \leq T/2$ (see (1.41)). The proof of (1.42) is given in the following.

Observing that $\hat{h}$ has compact support, the function h is entire of order 1 (see BF-6). Thus, using a version of Paley-Wiener Theorem [14, p. 179] and a related remark [14, p. 181], we have

$$\limsup_{y \to \infty} \frac{log|h(iy)|}{y} = \sup(supp(\hat{h})),\ \limsup_{y \to -\infty} \frac{log|h(iy)|}{|y|} = -\inf(supp(\hat{h}))\ . \tag{1.43}$$

Further, observing that $\int_{\mathbf{R}} \frac{\log^+ |h(x)|}{1+x^2}\, dx \le \|h\|_2 \times \left(\int_{\mathbf{R}} \frac{1}{(1+x^2)^2}\, dx\right)^{1/2} < \infty$ and applying the result stated in Problem 5(a) of [15, p.77] for $h(z)$ and $\overline{h}(\overline{z})$, we have

$$\limsup_{y\to\infty} \frac{\log|h(iy)|}{y} = \lim_{R\to\infty} \frac{1}{2R}\int_0^\pi \log|h(Re^{i\theta})| d\theta$$

and

$$\limsup_{y\to-\infty} \frac{\log|h(iy)|}{|y|} = \lim_{R\to\infty} \frac{1}{2R}\int_0^\pi \log|h(Re^{-i\theta})|\, d\theta\ .$$

Thus, in view of (1.43), we must show that

$$\lim_{R\to\infty} \frac{1}{2R}\int_0^\pi \log|h(Re^{i\theta})|\, d\theta = \lim_{R\to\infty} \frac{1}{2R}\int_0^\pi \log|h(Re^{-i\theta})|\, d\theta$$

or, equivalently, that $\lim_{R\to\infty} \frac{1}{2R}\int_0^\pi \log\left|\frac{h(Re^{-i\theta})}{h(Re^{\theta})}\right| d\theta = 0$. But this follows from the result stated in Problem 5(b) of [15, p. 77], since $\left|\frac{\overline{h}(\overline{z})}{h(z)}\right|$ is equal to the absolute value of the Blaschke product $\prod_n \left(\frac{1-z/\lambda_n}{1-z/\overline{\lambda}_n}\right)$, where $\{\lambda_n\} = \{\zeta_m\} \cup \{-\zeta_m\}$. ∎

Proof of Lemma 1.2.8. From BF-2, for some $\tau \ge 0$ we have

$$\begin{aligned} |f(x_0+iy)| &= e^{-\tau y}\left| \exp\left\{\frac{i}{\pi}\int_{\mathbf{R}}\left(\frac{1}{x_0+iy-t}+\frac{t}{1+t^2}\right)\log|f(t)|\, dt \right.\right. \\ &\qquad \left.\left. -\frac{i}{\pi}\int_{\mathbf{R}}\left(\frac{1}{x_0+iy-t}+\frac{t}{1+t^2}\right) d\mu(t)\right\}\right| \\ &= e^{-\tau y}\exp\left\{\frac{1}{\pi}\int_{\mathbf{R}}\left(\frac{y}{(x_0-t)^2+y^2}\right)\log|f(t)|\, dt \right. \\ &\qquad \left. -\frac{1}{\pi}\int_{\mathbf{R}}\left(\frac{y}{(x_0-t)^2+y^2}\right) d\mu(t)\right\}, \end{aligned} \tag{1.44}$$

for all $y > 0$. Using the fact $\int_{\mathbf{R}} \frac{|\log|f(t)||}{1+t^2}\, dt < \infty$ (see (0.6)) and the dominated convergence theorem, we have

$$\lim_{y\downarrow 0}\int_{\mathbf{R}} \frac{y^2}{(x_0-t)^2+y^2}\log|f(t)|\, dt = 0\ .$$

Therefore,

$$\int_{\mathbf{R}} \frac{y}{(x_0-t)^2+y^2}\log|f(t)|dt \le \frac{\mu\{x_0\}}{2\pi y}\ , \tag{1.45}$$

for all small $y > 0$. Further, since $\frac{y}{(x_0-t)^2+y^2} \leq \frac{1}{(x_0-t)^2}$, for $0 < y \leq 1$, and $\int_{\mathbf{R}\setminus\{x_0\}} \frac{\mu(dt)}{(x_0-t)^2} < \infty$ (note that $\frac{1}{(x_0-t)^2} \leq$ *Const.* $\frac{1}{1+t^2}$ on $supp(\mu)\setminus\{x_0\}$ and use (0-5)), we have, by the dominated convergence theorem, that

$$\lim_{y\downarrow 0} \int_{\mathbf{R}\setminus\{x_0\}} \frac{y}{(x_0-t)^2+y^2} d\mu(t) = 0 .$$

Therefore, for all small $y > 0$, we have

$$\exp -\frac{1}{\pi}\left\{\int_{\mathbf{R}} \left(\frac{y}{(x_0-t)^2+y^2}\right) d\mu(t)\right\} \leq Const. \exp\left(-\frac{\mu\{x_0\}}{\pi y}\right) .$$

This along with (1.44) and (1.45) yields (1.18). ∎

Proof of Lemma 1.2.9. As we noted prior to the statement of Proposition 1.2.4, this result is well known for $1 < p < \infty$ (see, [24, pp. 368 and 388]). We shall prove the lemma for $p = 1$. In view of the isomorphism between the H^p-spaces of the unit disc and the upper half plane [14, p. 159], it is sufficient (in fact, is equivalent) to prove the lemma for the case of the unit disc. Throughout the following proof, $\mathbf{H}^1$ and L^1 will denote the corresponding spaces of the unit disc; and m will denote the normalized Lebesgue measure on $(0, 2\pi]$.

Now let $f_n, f \in L^1, f_n \to f$ in L^1, and let $g_n \equiv \mathbf{P}(f_n), g \equiv \mathbf{P}(f)$. Since, clearly, $\|f-g\|_1 \leq \|f-g_n\|_1$, for all n, and

$$\begin{aligned}\|f-g_n\|_1 &\leq \|f_n-g_n\|_1 + \|f-f_n\|_1 \\ &\leq \|f_n-g\|_1 + \|f-f_n\|_1 \\ &\leq \|f_n-f\|_1 + \|f-g\|_1 + \|f-f_n\|_1 \\ &= \|f-g\|_1 + 2\|f-f_n\|_1 ,\end{aligned}$$

for all n, it follows, using $\|f-f_n\|_1 \to 0$, that

$$\lim_n \|f-g_n\|_1 = \|f-g\|_1 . \tag{1.46}$$

Let $\tilde{f} = \frac{f-g}{\|f-g\|_1}$ (note that if $\|f-g\|_1 = 0$, then $f = g$ and, by (1.46), $\lim_n \|g_n - g\|_1 = 0$ and we are done); and let $\tilde{g}_n = \frac{g_n - g}{\|f-g\|_1}$. Then $\tilde{f} \in L^1, \|\tilde{f}\|_1 = 1$, $\tilde{g} \equiv \mathbf{P}(\tilde{f}) = 0$, $\tilde{g}_n \in \mathbf{H}^1$, and, by (1.46), $\lim_n \|\tilde{f} - \tilde{g}_n\|_1 = 1 = \|\tilde{f} - \tilde{g}\|_1$. To complete the proof we need to prove $\|\tilde{g}_n\|_1 \to 0$. We prove this in the following.

By Theorem 1.2 of [9, p. 134], there exists a unique $K \in \mathbf{H}_0^\infty$ (see [9, p. 134], for this notation) such that $\|K\|_\infty = 1$ and $\int \tilde{f} K dm = 1$; these imply $\tilde{f}K \geq 0$ and $|K| = 1$ a.e. $[m]$. Further, since $\int \tilde{g}_n K dm = 0$, we also have $\int (\tilde{f} - \tilde{g}_n) K dm = 1$; and, from above, $\lim_n \int |(\tilde{f} - \tilde{g}_n)| dm = 1$. Now we prove that

$$\lim_n \int |\Im((\tilde{f} - \tilde{g}_n)K)| dm = 0 \,. \tag{1.47}$$

To see this let $h_n = (\tilde{f} - \tilde{g}_n)K = u_n + iv_n$ for all n, where u_n and v_n are real functions. Then we have

$$\int u_n dm = 1, \quad \text{for all } n \quad \text{and} \quad \lim_n \int |h_n| dm = 1 \,, \tag{1.48}$$

(recall that $\int h_n dm = 1$, for all n). The proof of (1.47) now follows from (1.48) and the following inequalities

$$\begin{aligned}\int |v_n| dm &= \int [v_n^2 + u_n^2 - u_n^2]^{1/2} \, dm \\ &= \int (|h_n| - u_n)^{1/2} \, (|h_n| + u_n)^{1/2} \, dm \\ &\leq \left(\int [|h_n| - u_n] dm \right)^{1/2} \left(\int [|h_n| + u_n] dm \right)^{1/2} .\end{aligned}$$

Since $\Im(\tilde{f}K) = 0$ a.e. (recall $\tilde{f}K \geq 0$ a.e.), we have, by (1.47),

$$\lim_n \int |\Im(\tilde{g}_n K)| dm = 0 \,. \tag{1.49}$$

Let $A_n = \{|\tilde{g}_n K| \leq 1\}$ and $B_n = \{|\tilde{g}_n K| > 1\}$; the next step in the proof is to show that $\int_{A_n} |\tilde{g}_n K| dm$ and $\int_{B_n} |\tilde{g}_n K| dm$ converge to 0, as $n \to \infty$. This will complete the proof of the fact that $\|\tilde{g}_n\|_1 \to 0$, as $n \to \infty$, since $|K| = 1$ a.e. Denote by w_n and w'_n the real and imaginary parts of $K\tilde{g}_n$, respectively; and let $\widetilde{w'_n}$ be the harmonic conjugate of w'_n, then $\widetilde{w'_n} = -w_n$ (recall that $w_n(0) = w'_n(0) = 0$, since $K \in \mathbf{H}_0^\infty$). Since by the Kolmogorov's and the Chybeshev's inequalities, $m\{|w_n| > \lambda\} = m\{|\widetilde{w'_n}| > \lambda\} \leq \frac{Const.}{\lambda}\|w'_n\|$, $m\{|w'_n| > \lambda\} \leq \frac{\|w'_n\|_1}{\lambda}$, for any $\lambda > 0$, it follows, by (1.49) and the inclusion $B_n \subseteq \{|w_n| > \frac{1}{\sqrt{2}}\} \cup \{|w'_n| > \frac{1}{\sqrt{2}}\}$, that $m(B_n) \to 0$, as $n \to \infty$. Further, since for any $0 < \lambda \leq 1$, $m\{\lambda < |w_n| \leq 1\} \leq \min\{\frac{Const.\ \|w'_n\|_1}{\lambda}, 1\}$ (Kolmogorov's inequality used again here), it follows, by (1.49) and the dominated convergence theorem, that $\int_0^1 m\{\lambda < |w_n| \leq 1\}d\lambda \to 0$, as $n \to \infty$. Hence, since $\int_{A_n} |w_n|dm = \int_0^\infty m(A_n \cap \{|w_n| > \lambda\})d\lambda \leq \int_0^1 m(\lambda < |w_n| \leq 1)d\lambda$, we have $\int_{A_n} |w_n|dm \to 0$, as $n \to \infty$. This along with (1.49) implies

$$\lim_n \int_{A_n} |\tilde{g}_n K|dm = 0 \ . \tag{1.50}$$

Since $\tilde{f}K \in L^1$ and $m(B_n) \to 0$, as $n \to \infty$, we have that $\int_{B_n} |\tilde{f}K|dm \to 0$, as $n \to \infty$. Using this, (1.50) and the following inequality (which uses the fact that $\int \tilde{f}Kdm = 1$ and $\tilde{f}K \geq 0$ a.e.)

$$\begin{aligned}\int_{A_n} |\tilde{f}K - \tilde{g}_n K|dm &\geq \int_{A_n} |\tilde{f}K|dm - \int_{A_n} |\tilde{g}_n K|\ dm \\ &= 1 - \int_{B_n} |\tilde{f}K|dm - \int_{A_n} |\tilde{g}_n K|\ dm\ ,\end{aligned}$$

we have $\liminf_n \int_{A_n} |\tilde{f}K - \tilde{g}_n K|dm \geq 1$. This together with the above noted fact that $\lim_n \int |(\tilde{f} - \tilde{g}_n)K|dm = 1$ shows that $\lim_n \int_{B_n} |(\tilde{f} - \tilde{g}_n)K|dm = 0$. This along with the already proved fact that $\lim_n \int_{B_n} |\tilde{f}K|dm = 0$, yields $\lim_n \int_{B_n} |\tilde{g}_n K|dm = 0$. ∎

1.3. Examples

Here we present two examples. In these examples, we compute, more explicitly, the kernels $\mathbf{P}_T$ and Ψ_T for two concrete functions $\varphi \in \mathbf{H}^p$.

Example 1.3.1. Let $\varphi(z) = \frac{i}{z+i}$, $z \neq -i$; then, since φ is analytic on $\mathcal{H}$ and $\int_{\mathbf{R}} |\frac{i}{x+i(1+y)}|^p \, dx \leq \int_{\mathbf{R}} \frac{dx}{(1+x^2)^{p/2}} < \infty$, for all $y > 0$ and $1 < p < \infty$, we have that $\varphi \in \mathbf{H}^p$. Clearly φ has no zeros in $\mathcal{H}$.

Fix $1 < p < 2$. We shall compute $\widehat{\varphi^{p/2}}$, where $\varphi^{p/2} = e^{\frac{p}{2}\log\varphi}$ and $\log\varphi$ is that analytic branch of log of φ on $\mathcal{H}$ for which $\log\varphi(i)$ is real. (Note that φ maps $\mathbf{C}\backslash\{z : \Im(z) \leq -1\}$ onto $\mathbf{C}\backslash\{z : \Re(z) \leq 0\}$ and $\log\varphi(z) = \text{Log } w$, where $w = \varphi(z)$ and Log w is the principle branch of log of $w \in \mathbf{C}\backslash\{z : \Re(z) \leq 0\}$. This observation will be useful in the following computations). Let $t > 0$; then, considering the following contour

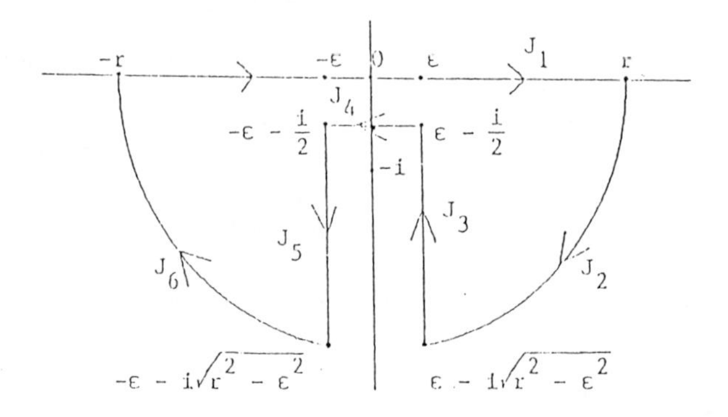

and using Cauchy's theorem, it follows that, for any fixed $r > 1$,

$$\int_{-r}^{r} e^{-itx} \varphi^{p/2}(x) dx = -\sum_{\ell=2}^{6} \int_{J_\ell} e^{-itz} \varphi^{p/2}(z) dz \ , \tag{1.51}$$

for any $\varepsilon < 1$. Clearly, $\int_{J_4} e^{-itz}\varphi^{p/2}(z)dz \to 0$, as $\varepsilon \downarrow 0$; further, it is easy to see that

$$
\begin{aligned}
\left| \int_{J_6} e^{-itz}\varphi^{p/2}(z)dz \right| &\leq \frac{r}{(r-1)^{p/2}} \int_0^{\pi/2} e^{-tr\sin\theta} d\theta \\
&\leq \frac{r}{(r-1)^{p/2}} \int_0^{\pi/2} e^{-\frac{2tr\theta}{\pi}} d\theta \\
&= \frac{r}{(r-1)^{p/2}} \frac{\pi}{2rt} \left(1 - e^{-tr}\right) \\
&\leq \frac{\pi}{2t(r-1)^{p/2}} ,
\end{aligned}
$$

uniformly in $\varepsilon > 0$, since $t > 0$. Thus $\int_{J_6} e^{-itz}\varphi^{p/2}(z)dz$ (and, similarly, $\int_{J_2} e^{-itz}\varphi^{p/2}(z)dz) \to$ 0, uniformly in ε, as $r \to \infty$. Hence, from (1.51),

$$
\lim_{r\to\infty} \int_{-r}^{r} e^{-itx}\varphi^{p/2}(x)dx = -\lim_{r\to\infty}\lim_{\varepsilon\downarrow 0}\left[\int_{J_3\cup J_5} e^{-itz}\varphi^{p/2}(z)dz\right] . \tag{1.52}
$$

Next, using the dominated convergence theorem, we observe that (note that $p < 2$ is important here)

$$
\begin{aligned}
\lim_{\varepsilon\downarrow 0} \int_{J_5} e^{-itz}\varphi^{p/2}(z)dz &= \lim_{\varepsilon\downarrow 0} - ie^{it\varepsilon} \int_{1/2}^{\sqrt{r^2-\varepsilon^2}} e^{-ty} \left\{\frac{i}{-\varepsilon + i(1-y)}\right\}^{p/2} dy \\
&= -i \int_{1/2}^{r} e^{-ty} \lim_{\varepsilon\downarrow 0} \left\{\frac{i}{-\varepsilon + i(1-y)}\right\}^{p/2} dy ;
\end{aligned} \tag{1.53}
$$

and, similarly,

$$
\lim_{\varepsilon\downarrow 0} \int_{J_3} e^{-itz}\varphi^{p/2}(z)dz = i \int_{1/2}^{r} e^{-ty} \lim_{\varepsilon\downarrow 0} \left\{\frac{i}{\varepsilon + i(1-y)}\right\}^{p/2} dy . \tag{1.54}
$$

Now, recalling the definition of $\log\varphi$, it follows that

$$
\lim_{\varepsilon\downarrow 0} \left\{\frac{i}{\varepsilon + i(1-y)}\right\}^{p/2} = \lim_{\varepsilon\downarrow 0} \left\{\frac{i}{-\varepsilon + i(1-y)}\right\}^{p/2} = \left(\frac{1}{1-y}\right)^{p/2} ,
$$

if $0 < y < 1$; and that

$$
\lim_{r\to\infty} \left\{\frac{i}{-\varepsilon + i(1-y)}\right\}^{p/2} = \frac{1}{(y-1)^{p/2}} e^{-i\pi p/2} , \lim_{\varepsilon\downarrow 0} \left\{\frac{i}{\varepsilon + i(1-y)}\right\}^{p/2} = e^{i\pi p/2} \frac{1}{(y-1)^{p/2}} ,
$$

if $y > 1$. Hence, using (1.52)-(1.54),

$$\lim_{\epsilon \downarrow 0} \int_{-r}^{r} e^{-itx} \varphi^{p/2}(x) dx = 2 \sin(\pi p/2) \int_{1}^{\infty} \frac{e^{-ty}}{(y-1)^{p/2}} \, dy \, . \tag{1.55}$$

Note that if $p = 2$, the integral on the right side of (1.55) is not finite. For this case, we use the contour

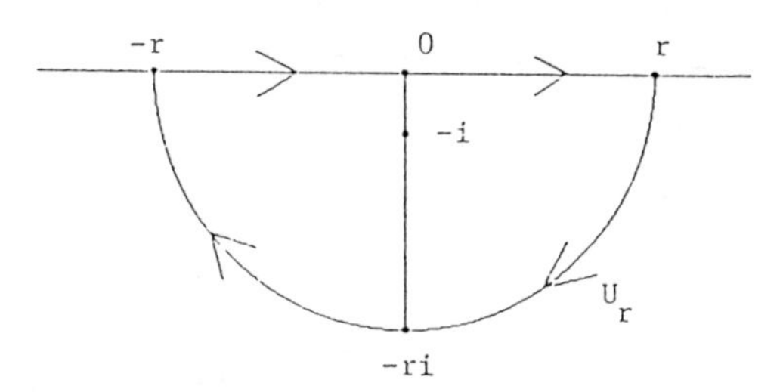

and use the residue theorem to get

$$\int_{-r}^{r} e^{-itx} \left(\frac{i}{i+x} \right) \, dx = 2\pi e^{-t} - \int_{U_r} e^{-itz} \varphi(z) dz$$

for any $r > 1$. Using standard calculations, one sees that $\lim_{r \to \infty} \int_{U_r} e^{-itz} \varphi(z) dz = 0$, if $t > 0$. Hence, for any $t > 0$,

$$\lim_{r \to \infty} \int_{-r}^{r} e^{-itx} \left(\frac{i}{i+x} \right) \, dx = 2\pi e^{-t} \, . \tag{1.56}$$

From (1.55) and (1.56), we finally have that, if $t > 0$, then, for $1 < p < 2$,

$$\begin{aligned} \widehat{\varphi^{p/2}}(t) &= \frac{2 \sin(\pi p/2)}{\sqrt{2\pi}} \int_{1}^{\infty} \frac{e^{-ty}}{(y-1)^{p/2}} \, dy \\ &= \sqrt{\frac{2}{\pi}} \, t^{p/2-1} e^{-t} \sin(\pi p/2) \int_{0}^{\infty} e^{-x} x^{-p/2} \, dx \\ &= \left(\sqrt{2\pi} \, / \, \Gamma(p/2) \right) \, t^{p/2-1} e^{-t} \, , \end{aligned} \tag{1.57}$$

(since $\sin(\pi p/2) = \frac{\pi}{\Gamma(p/2)\Gamma(1-p/2)}$); and, for $p = 2$, $\widehat{\varphi^{p/2}}(t) = \sqrt{2\pi} \, e^{-t}$. Note that the formula (1.57) does indeed reduce to $\sqrt{2\pi} \, e^{-t}$ when $p = 2$. Thus $\widehat{\varphi^{p/2}}(t)$ is given by (1.57) for all $1 < p \leq 2$ and $t > 0$. We also note that, since $\varphi^{p/2} \in \mathbf{H}^2$, $\widehat{\varphi^{p/2}} = 0$ a.e. on $(-\infty, 0]$. This fact can easily be directly deduced using contour integration along the semicircle in the upper half plane with center at 0. Thus, for any $1 < p \leq 2$, we have

$$\widehat{\varphi^{p/2}}(t) = \frac{\sqrt{2\pi}}{\Gamma(p/2)}\, t^{p/2-1} e^{-t} I_{[0,\infty)}(t) \;, \tag{1.58}$$

a.e. on $\mathbf{R}$. Thus, in particular, $0 \in supp(\hat{\varphi})$. Further, since, by (1.58) $\widehat{\varphi^{p/2}}$ is non-negative and $\downarrow$ on $[0, \infty)$, it follows, from Proposition 1.2.5, that

$$\mathcal{P}_T(\varphi^{p/2})(z) = \frac{1}{\sqrt{2\pi}} \int_0^T e^{itz} \widehat{\varphi^{p/2}}(t)dt, \quad z \in \mathcal{H} \;, \tag{1.59}$$

has no zeros in $\mathcal{H}$ (see (1.6)). Thus, $\varphi, \hat{\varphi}$ and $\mathcal{P}_T(\varphi^{p/2})$ satisfy all the hypotheses of Theorem 1.2.1.

Using (1.57) and (1.59), simple calculations now show that

$$\mathcal{P}_T(\varphi^{p/2})(z) = \frac{\gamma(p/2, T(1-iz))}{\Gamma(p/2)(1-iz)^{p/2}} \;,$$

for all $z \in \mathcal{H}$, where

$$\gamma(p/2, T(1-iz)) = \int_0^{T(1-iz)} e^{-w} w^{p/2-1}\, dw \;,$$

(see [1, p. 260] for the properties of the function $\gamma(\cdot,\cdot)$). Hence, from Theorem 1.2.1, we have

$$\mathbf{P}(e_{-T}\varphi) = e^{-iTz} \frac{i}{z+i} - e^{-iTz} \left[\left(\frac{1}{\Gamma(p/2)} \right) \left(\frac{i}{z+i} \right)^{p/2} \gamma\left(p/2, T(1-iz)\right) \right]^{2/p}$$

and

$$\Psi_T = \left[\left(\frac{1}{\Gamma(p/2)} \right) \left(\frac{i}{z+i} \right)^{p/2} \gamma\left(p/2, T(1-iz)\right) \right]^{2/p} \;,$$

where $2/p$-root is chosen according to the condition (1.8) (note that, in view of Remark 1.2.2(d), this root coincides with one which is positive at iy, $y > 0$). The error δ_T can also be evaluated explicitly; in fact, by (1.9) and (1.58), we have

$$\delta_T^p = \frac{2\pi}{(\Gamma(p/2))^2} \int_0^T e^{-2t} t^{p-2} dt = \frac{2^{2-p}\pi}{(\Gamma(p/2))^2} \gamma(p-1, 2T) \, .$$

Note that, as $T \to \infty$, δ_T^p converges to $\frac{2^{2-p}\pi\Gamma(p-1)}{(\Gamma(p/2))^2}$, which is equal to π, if $p = 2$.

Example 1.3.2. One can provide a simple example for the case $p = 1$ by using the calculations of Example 1.3.1. Let φ be as in Example 1.3.1. Let $\varphi_1(z) = \left(\frac{i}{z+i}\right)^2 = \varphi^2(z), z \in \mathcal{H}$, and $\varphi_1^{1/2} = \varphi$. Then, it follows, from calculations in Example 1.3.1, that $\varphi_1 \in \mathbf{H}^1, \widehat{\varphi_1^{1/2}} = \hat{\varphi} = \sqrt{2\pi}\; e^{-t} I_{[0,\infty)}$. Thus, it follows that φ_1 satisfies all the hypotheses of Theorem 1.2.1 with $p = 1$. Now

$$\begin{aligned} \mathcal{P}_T\left(\varphi_1^{1/2}\right)(z) = \mathcal{P}_T(\varphi)(z) &= \int_0^T e^{itz} e^{-t} dt \\ &= \frac{e^{T(iz-1)} - 1}{iz - 1} \, . \end{aligned}$$

Hence, from Theorem 1.2.1,

$$\begin{aligned} \mathbf{P}(e_{-T}\varphi_1) &= e^{-iTz}\left(\frac{i}{z+i}\right)^2 - e^{-iTz}\left\{\frac{e^{T(iz-1)} - 1}{(iz-1)}\right\}^2 , \\ \Psi_T(\varphi_1) &= \left(\frac{e^{T(iz-1)} - 1}{(iz-1)}\right)^2 \end{aligned}$$

and

$$\delta = \pi[1 - e^{-2T}] \, .$$

1.4. Appendix

As noted in the Introduction, we collect here some known facts from the theory of complex variables and H^p-spaces. For most facts included here, we have supplied references.

A particular fact is either available directly in the quoted reference(s) or can be proved using idea(s) and/or result(s) from the references. A few facts, for which no references are given, are standard. Throughout, unless stated otherwise, we shall take, $1 \leq p < \infty$, even though some of the quoted facts hold for $0 < p \leq \infty$.

BF-1. A sequence $\{z_n\} \subseteq \mathcal{H}$ is the zero set of an $\mathbf{H}^p$-function $\Longleftrightarrow$

$$\sum_{n \geq 1} \frac{\Im(z_n)}{1 + |z_n|^2} < \infty . \tag{0-1}$$

If this condition is met, then

$$B(z) = \prod_n \left(\frac{1 + |z_n|^2}{1 + z_n^2} \right) \left(\frac{z - z_n}{z - \overline{z}_n} \right), \quad z \in \mathcal{H},$$

with the convention that the converging factor $\frac{1+|z_n|^2}{1+z_n^2} = 1$, if some $z_n = i$, is called the *Blaschke* product with zeros $\{z_n\}$; it is an $\mathbf{H}^\infty$-function [9, pp. 53–56].

We shall consider sequences $\{z_n\} \subseteq \mathcal{H}$ which are bounded away from zero; in this case (0-1) will be replaced by

$$\sum_n \frac{\Im(z_n)}{|z_n|^2} < \infty ; \tag{0-2}$$

and, in this case, we define the Blaschke product by

$$B(z) = \prod_n \left(\frac{\overline{z}_n}{z_n} \right) \left(\frac{z - z_n}{z - \overline{z}_n} \right).$$

BF-2. [14, p. 163]. (Canonical representation). An $\mathbf{H}^p$-function f has a unique factorization:

$$f(z) = C\ e_\lambda(z) S_\mu(z) B(z) \varphi(z), z \in \mathcal{H}, \tag{0-3}$$

where $C \in \mathbf{C}$, $|C| = 1$, $e_\lambda(z)$ (as before) $\equiv e^{i\lambda z}$, $\lambda \geq 0$,

$$S_\mu(z) = \exp \left\{ -\frac{i}{\pi} \int_{\mathbf{R}} \left(\frac{1}{z - t} + \frac{t}{1 + t^2} \right) d\mu(t) \right\} , \tag{0-4}$$

where μ is a non-negative singular measure on $\mathbf{R}$ satisfying

$$\int_{\mathbf{R}} \frac{d\mu(t)}{1+t^2} < \infty \ , \tag{0-5}$$

$$\varphi(z) = \exp\left\{\frac{i}{\pi}\int_{\mathbf{R}}\left(\frac{1}{z-t}+\frac{t}{1+t^2}\right)\log|f(t)|\,dt\right\} \ , \tag{0-6}$$

and B is the Blaschke product. The function $|f|$ satisfies

$$|f| = |\varphi| \text{ a.e. on } \mathbf{R} \text{ and } \int_{\mathbf{R}} \frac{|(\log|f(t)|)|}{1+t^2}dt < \infty \ . \tag{0-7}$$

The functions S_μ and φ are, respectively, called the singular (inner) and outer factor of f. We also have

$$|e_\lambda| = |S_\mu| = |B| = 1, \text{ a.e. on } \mathbf{R} \ . \tag{0-8}$$

Further, if a point $x \in \mathbf{R}$ is either in the support of μ or in the closure of zeros of B, then f cannot be continued analytically across any open interval containing x (for this point, see [9, pp. 75–76]).

BF-3. [9, p. 88]. Let $1 \le p \le 2$. A function $f \in L^p$ is in $\mathbf{H}^p$ (i.e., it is the boundary value of a function in $\mathbf{H}^p$) if and only if $supp(\hat{f}) \subseteq [0,\infty)$.

BF-4. Let $1 \le p \le 2$ and let $f \in \mathbf{H}^p$. Using BF-2 and BF-3 and the fact $\widehat{e_\eta f}(t) = \hat{f}(t-\eta)$, it follows that the number λ appearing in (0-3) is given by

$$\lambda = \max\{\eta : e_{-\eta}f \in \mathbf{H}^p\} = \min\left\{\eta : \eta \in supp(\hat{f})\right\}.$$

BF-5. [14, p. 109]. Let $1 \le p_j < \infty$, $j = 1,2,3$, and let $f \in \mathbf{H}^{p_1}$, $g \in \mathbf{H}^{p_2}$. Suppose g has no factor of the form e_λ and S_μ in (0-3), f/g is analytic on $\mathcal{H}$ and the boundary values of $f/g \in L^{p_3}$, then $f/g \in \mathbf{H}^{p_3}$.

BF-6. **(a).** [14, p. 177]. Let $1 \le p \le 2$ and $f \in \mathbf{H}^p$, then

$$f(z) = \frac{1}{\sqrt{2\pi}}\int_0^\infty e^{itz}\hat{f}(t)dt, z \in \mathcal{H}.$$

(b). If $g \in L^p[a,b]$, $-\infty < a < b < \infty$ and $1 \leq p \leq \infty$, then

$$G(z) = \int_a^b e^{itz} g(t)dt$$

is an entire function of order 1.

BF-7. [15, p. 16]. (Hadamard Factorization Theorem). Let f be an entire function of order 1 with N zeros at $z = 0$ and let $z_1, z_2, \ldots,$ be its other zeros. Then $\sum_n |z_n|^{-2} < \infty$, and

$$f(z) = Cz^N e^{az} e^{ibz} \prod_n \left(1 - \frac{z}{z_n}\right) e^{z/z_n}, \; z \in \mathbf{C},$$

where $C \in \mathbf{C}$, $a, b \in \mathbf{R}$. The product is uniformly convergent on compact subsets of $\mathbf{C}$.

BF-8. Let Ω be a simply connected open subset of $\mathbf{C}$ containing $\mathcal{H}$ and a *nbd.* of 0. Let f_n's be non-vanishing analytic functions on Ω satisfying $f_n(0) = 1$, and let $\log f_n$ be the (unique) analytic branch of the logarithm of f_n satisfying $\log f_n(0) = 0$, $n = 1, 2, \ldots$. If $\sum_n |1 - f_n|$ converges uniformly on compact subsets of Ω, then the product $\prod_n f_n$ and the series $\sum_n |\log f_n|$ converge uniformly on compact subsets of Ω and $\sum_n \log f_n$ defines the unique analytic branch of log of $f \equiv \prod_n f_n$ satisfying $\log f(0) = 0$.

BF-9. Let $\{\alpha_n\} \subseteq \mathbf{C}$ satisfying $\sum_n |\alpha_n|^{-2} (1 + |\Im(\alpha_n)|) < \infty$. Let f_n be any of the three functions $\left(1 - \frac{z}{\alpha_n}\right)^2 e^{z\left(\frac{1}{\alpha_n} + \frac{1}{\overline{\alpha_n}}\right)}$, $\left(1 - \frac{z}{\alpha_n}\right)\left(1 - \frac{z}{\overline{\alpha_n}}\right) e^{z\left(\frac{1}{\alpha_n} + \frac{1}{\overline{\alpha_n}}\right)}$ or $\left(1 - \frac{z}{\overline{\alpha_n}}\right) e^{\frac{z}{2}\left(\frac{1}{\alpha_n} + \frac{1}{\overline{\alpha_n}}\right)}$ on $\mathbf{C}$, $n = 1, 2, \ldots$. Then $\sum_n |1 - f_n|$ converges uniformly on compact subsets of $\mathbf{C}$; and, hence, by BF-8, the product $\prod_n f_n$ converges uniformly on compact subsets of $\mathbf{C}$ and defines an entire function.

BF-10. [9, p. 134]. Let $1 \leq p < \infty$ and let q be the conjugate exponent of p. Let $k \in L^p$, and let K and F be complex functions on $\mathbf{R}$. Then K and F are, respectively, equal to the extremal kernel $\mathbf{K}$ and the dual extremal function $\mathbf{F}$ of k if and only if they satisfy the following conditions:

(*i*) $K \in k + \mathbf{H}^p$, $F \in \mathbf{H}^q$, $FK \geq 0$ a.e.,

(*ii*) $|F|^q = C|K|^p$ a.e., for some $C > 0$, and $\|F\|_q = 1$, if $1 < p$; $|F| = 1$ a.e., if $p = 1$.

II. Linear Prediction of L^p-representable Processes

2.1. Preliminaries

We begin by introducing some notations and conventions. If ν is a σ-finite measure on a measurable space $(\Omega, \mathcal{F})$, then, throughout, $L^p(\Omega, \nu) \equiv L^p(\nu)$ will denote the usual Banach space (resp. the usual linear metric space) of complex functions, if $1 \leq p < \infty$ (resp. $0 < p < 1$). In this part of the paper, we shall be considering L^p-spaces relative to more than one measure. Therefore, in order to keep the notations straight, we shall denote by $\|\cdot\|_{L^p(\nu)}$ the usual norm or the quasinorm on $L^p(\nu)$ according as $1 \leq p < \infty$ or $0 < p < 1$. If $\nu = P$, a probability measure, then the standard non-homogeneous norm which metrizes the convergence in probability in $L^0(P)$ will be denoted by $\|\cdot\|_{L^0(P)}$.

Now we recall some definitions and certain facts about random measures and integrals relative to such measures. We also fix some additional notations and conventions. Let $(\Omega, \mathcal{F}, P)$ be a probability space, and let M be a real or complex finitely additive random measure on the ring $\mathcal{B}_0(\mathbf{R})$ of bounded Borel sets of $\mathbf{R}$; i.e., M is a map: $\mathcal{B}_0(\mathbf{R}) \longmapsto L^0(\Omega)$ satisfying $M\left(\bigcup_{j=1}^{n} A_j\right) = \sum_{j=1}^{n} M(A_j)$, for every finite sequence of disjoint Borel sets A_j's. Let μ be a finite measure on $\mathcal{B}(\mathbf{R})$, the Borel σ-algebra of $\mathbf{R}$, and let $\mathcal{I}(M)$ be the set of properly defined M-integrable complex functions (identified a.e. $[\mu]$). We will be interested here in pairs (M, μ) of measures for which the following requirements are met:

A-1. $\mathcal{I}(M)$ is a linear space, and the stochastic integral map $\mathbf{I}_M(f) = \int_{\mathbf{R}} f dM$ is well defined, 1-1 and linear from $\mathcal{I}(M)$ into $L^0(P)$.

A-2. For some $1 \leq p < \infty$, $L^p(\mu) \subseteq \mathcal{I}(M)$.

There are a host of examples of pairs of measures (M, μ) for which A-1 and A-2 are satisfied:

(i) Let M be a symmetric (or, more generally, a centered) real or complex α-stable or (r, α)-semistable random measure on $\mathcal{B}(\mathbf{R})$ with $1 \leq \alpha < 2$ and $0 < r < 1$ (α is assumed not equal to 1, if M is non-symmetric), and let μ be the (properly defined) finite control measure of M. Then, for any such pair (M, μ), A-1 and A-2 are satisfied with $p = \alpha$; and, moreover, $\mathcal{I}(M) = L^\alpha(\mu)$ and the following inequalities hold: For every $0 < q < \alpha$,

$$c(q,\alpha)\|f\|_{L^\alpha(\mu)} \leq \left(\int_\Omega |\mathbf{I}_M(f)|^q \, dP\right)^{\frac{1}{q}} \leq C(q,\alpha)\|f\|_{L^\alpha(\mu)}, \tag{2.1}$$

for all $f \in L^\alpha(\mu)$; and

$$c(0,\alpha)\|f\|_{L^\alpha(\mu)} \leq \|\mathbf{I}_M(f)\|_{L^0(P)} \leq C(0,\alpha)\|f\|_{L^\alpha(\mu)}, \tag{2.2}$$

for all $f \in L^\alpha(\mu)$, where $c(q, \alpha)$ and $C(q, \alpha)$ are finite positive constants (independent of M, μ and f) (for these facts see [2, 16-18]). Further, if M is complex rotationally invariant α-stable, $1 \leq \alpha < 2$, then we have, for any $0 < q < \alpha$,

$$\left(\int_\Omega |\mathbf{I}_M(f)|^q \, dP\right)^{\frac{1}{q}} = C_0(q,\alpha)\|f\|_{L^\alpha_{(\mu)}}, \tag{2.3}$$

for every $f \in L^\alpha(\mu)$, where $C_0(q, \alpha)$ is a constant independent of M, μ, f; see [5; 2, p. 77]. (Analog of (2.3), in fact, holds for $0 < \alpha < 1$ also, but this fact is not pertinent here).

(ii) Let M be a real or complex orthogonally scattered (o.s.) random measure on $\mathcal{B}(\mathbf{R})$ and let μ be the associated structure measure [8, pp. 231–232]. Then $\mathcal{I}(M) = L^2(\mu)$; and A-1 and A-2 hold with $p = 2$. Furthermore, one has

$$\|I_M(f)\|_{L^2(P)} = \|f\|_{L^2(\mu)}, \tag{2.4}$$

for all $f \in L^2(\mu)$.

Now we formulate the concept of *L^p-representable processes*; this is motivated from the remarks noted in the Introduction concerning the linear space of certain second order and α-stable processes. Let (M, μ) be a pair of measures satisfying A-1 and A-2, for some $1 \le p < \infty$. Set

$$\mathcal{L}(M) = \left\{ \mathbf{I}_M(f) \equiv \int_{\mathbf{R}} f dM : \quad f \in L^p(\mu) \right\} ;$$

and define the norm on $\mathcal{L}(M)$ by

$$\|\mathbf{I}_M(\cdot)\|_p = \|\cdot\|_{L^p(\mu)} \,. \tag{2.5}$$

Then, clearly, the stochastic integral map $\mathbf{I}_M$

$$L^p(\mu) \ni f \longmapsto \int_{\mathbf{R}} f dM \in \mathcal{L}(M)$$

establishes an isometric isomorphism between complex Banach spaces $L^p(\mu)$ and $(\mathcal{L}(M), \|\cdot\|_p)$. Let now $X \equiv \{X(t) : t \in \mathbf{R}\}$ be a real or complex stochastic process on (Ω, F, P). If there exist functions $f_t \in L^p(\mu)$ such that $X(t) = \mathbf{I}_M(f_t), t \in \mathbf{R}$, then we say that X is *stochastic integral representable* or $\mathbf{I}_M$*-representable* in $L^p(\mu)$ (or simply that X is an $\mathbf{I}_M$*-representable process* in $L^p(\mu)$). The functions f_t's will be called the *representing functions* of X. If $f_t(\lambda) = e^{it\lambda} \equiv e_t(\lambda)$, then we say that X is *a harmonizable* $\mathbf{I}_M$*-representable process* in $L^p(\mu)$. The linear space $\mathcal{L}(X)$ of X is, by definition, the $\|\cdot\|_p$-closure of $sp(X)$, where $sp(X)$ denotes the linear span of X over $\mathbf{C}$. We note that if $sp\{f_t : t \in \mathbf{R}\}$ is dense in $L^p(\mu)$; then $\mathcal{L}(X) = \mathcal{L}(M)$ and $\mathbf{I}_M$ establishes an isometric isomorphism between $L^p(\mu)$ and $(\mathcal{L}(X), \|\cdot\|_p)$. We now define various concepts needed for the purpose of prediction.

Let $X = \{X(t) : t \in \mathbf{R}\}$ be an $\mathbf{I}_M$-representable process in $L^p(\mu)$; and let $\mathcal{L}(X, \infty) \equiv \mathcal{L}(X)$. For $t \in \mathbf{R}$, let $\mathcal{L}(X, t)$ be the $\|\cdot\|_p$-closure of $sp\{X(s) \le t\}$ and $\mathcal{L}(X, -\infty) =$

$\bigcap_{t\in\mathbf{R}} \mathcal{L}(X,t)$. We recall that X is called *regular* (resp. *singular*) if $\mathcal{L}(X,-\infty) = \{0\}$ (resp. $\mathcal{L}(X,-\infty) = \mathcal{L}(X,\infty)$). Let $T > 0$, if there exists a unique element $\tilde{X}(t,T)$ in $\mathcal{L}(X,t)$ satisfying

$$\inf_{Y\in\mathcal{L}(X,t)} \|X(t+T) - Y\|_p = \|X(t+T) - \tilde{X}(t,T)\|_p \ ,$$

then we say that $\tilde{X}(t,T)$ is the *optimal linear predictor* (or just the *linear predictor*) of $X(t+T)$ based on $\{X(s) : s \leq t\}$ and the *prediction error* is defined by

$$\delta(T) \equiv \delta_p(t,T) \equiv \|X(t+T) - \tilde{X}(t,T)\|_p \ .$$

All the notions defined above are in terms of the norm $\|\cdot\|_p$; thus, in order for the results obtained here to be useful, it is necessary that $\|\cdot\|_p$ be (up to a constant) equal to or at least equivalent to $\|\cdot\|_{L^q(P)}$ on $sp(X)$, for some $q \geq 0$. If X is an $\mathbf{I}_M$-representable process in $L^p(\mu)$, where (M,μ) is any of the pairs of measures discussed in (i) and (ii) above, then, clearly from (2.1)-(2.4), the corresponding norm $\|\cdot\|_p$ meets this requirement.

2.2. Statements and Proofs of Results

Now we are ready to formulate the results. First we shall state a result which gives a characterization of regularity and singularity and provides a W-C type decomposition for processes which are $\mathbf{I}_M$-representable in $L^p(\mu)$ $1 \leq p < \infty$ and are harmonizable. The analog of this result, as pointed out earlier, was formulated and proved by Cambanis and Soltani [3] for processes which are Fourier transforms of symmetric α-stable random measures, $1 < \alpha < 2$. They also noted, in their concluding remarks, that their proof indeed yields the analog of their result for processes which (in our terminology) are $\mathbf{I}_M$-representable in $L^p(\mu)$, $1 < p < \infty$ and are harmonizable. We show here that, by making some modifications in their proof, one also gets the last noted result for $p = 1$; in addition, we also prove the uniqueness in the W-C decomposition. For completeness, the following result is stated for all $1 \leq p < \infty$.

We shall need some more notations and a couple of definitions. Let μ be a finite measure on $\mathcal{B}(R)$, then μ_a and μ_s will, respectively, denote the absolutely continuous and singular components of μ. Further, S_a and S_s will, respectively, denote disjoint Borel sets on which μ_a and μ_s are concentrated; and ρ will denote the density $d\mu_a/dLeb$. Let ξ and η be two processes $\mathbf{I}_M$-representable in $L^p(\mu)$, we say ξ is *subordinate* to η if $\mathcal{L}(\xi, t) \subseteq \mathcal{L}(\eta, t)$, for every t; or equivalently, $\xi(t) \in \mathcal{L}(\eta, t)$, for all t. This condition signifies that $\xi(t)$ should be determined linearly by the development of η up to (and including) the time t. We say that ξ_t and η_s are $\mathbf{I}_M$-*independent* if

$$f_t g_s = 0 \quad \text{a.e. } [\mu] \,, \tag{2.6}$$

where f_t and g_s are, respectively, the representing functions of ξ_t and η_s. Note that if M is infinitely divisible (i.d.) independently scattered (i.s.) random measure [20, 25], then condition (2.6) implies that ξ_t and ξ_s are independent (this follows from Lemma 2.1 of [27] and the definition of $\mathbf{I}_M(\cdot)$); conversely, if the measure M is *purely Poissonian* (i.e., $M(A)$ is neither Gaussian nor is a non-zero constant for any $A \in \mathcal{B}(\mathbf{R})$), then independence of ξ_t and ξ_s implies (2.6). This is a corollary of an important result of Urbanik (Theorem 2.1 of [25, p. 74]). Thus, in purely Poissonian case, (2.6) is equivalent to the independence of $\mathbf{I}_M(\xi_t)$ and $\mathbf{I}_M(\eta_s)$. This result in the case of special i.d. random measures (e.g., stable, semistable) is obtained using different methods by several authors (e.g., [2, 18, 22]). In Theorems 2.2.1 and 2.2.3, the process X can be real or complex; however, when X is real one should use these results with caution (see Remark 2.3.3). We also mention here the works of Cambanis, Hardin and Weron [4], Hosoya [10], Miamee and Pourahmadi [17],

Urbanik [26, 27], and Weron [28] which are related to the results presented here.

Theorem 2.2.1. *Let $1 \leq p < \infty$ and let $X = \{\int_{\mathbf{R}} e^{itu} dM(u) : t \in \mathbf{R}\}$ be a harmonizable $\mathbf{I}_M$-representable process in $L^p(\mu)$. Then we have:*

(a) The process X is regular $\iff \mu \ll Leb$ and

$$\int_{\mathbf{R}} \frac{\log \rho(x)}{1+x^2}\, dx > -\infty \;;$$

and X is singular $\Leftrightarrow$

$$\int_{\mathbf{R}} \frac{\log \rho(x)}{1+x^2}\, dx = -\infty \;.$$

(b) (W-C decomposition) The process X admits a unique representation

$$X(t) = X_1(t) + X_2(t) \;,$$

where X_1 and X_2 are $\mathbf{I}_M$-representable processes in $L^p(\mu)$ such that X_1 and X_2 are subordinate to X, X_1 is regular, X_2 is singular and $X_1(s)$ and $X_2(t)$ are $\mathbf{I}_M$-independent for all $s, t \in \mathbf{R}$.

In fact, if $X_1(t) = \mathbf{I}_M(f_t)$ and $X_2(t) = \mathbf{I}_M(g_t)$, then $f_t = 0$ and $g_t = e_t$ a.e. $[\mu]$, in the case when X is singular; whereas, in all other cases, $f_t = e_t I_{S_a}$ and $g_t = e_t I_{S_s}$ a.e. $[\mu]$. Furthermore, $X_2(t)$ is the (unique) nearest element to $X(t)$ in $\mathcal{L}(X, -\infty)$.

Remark 2.2.2. Let X be as in Theorem 2.2.1 (b) with M i.d. and i.s. Then, in view of the remarks about the independence of stochastic integrals, the existence part of the theorem asserts that X has a decomposition in terms of X-subordinate $\mathbf{I}_M$-representable components X_1 and X_2 such that X_1 is regular, X_2 is singular and $X_1(t)$ and $X_2(s)$ are independent, for all $t, s, \in \mathbf{R}$ (in fact, the processes X_1 and X_2 are independent). Further, if, additionally, M is purely Poissonian, then the uniqueness part of the theorem and the above noted remarks imply that this decomposition in terms of X_1 and X_2 is unique.

In view of the W-C decomposition of harmonizable $\mathbf{I}_M$-representable processes X and in view of the fact that the singular component is perfectly predictable, one can obtain information about the linear predictor of X provided one can pinpoint the linear predictor of the regular component of the process X.

Let now X be a regular harmonizable $\mathbf{I}_M$-representable process in $L^p(\mu)$. Thus, according to Theorem 2.2.1 (a), $\mu \ll Leb$ and $\int_{\mathbf{R}} \frac{\log \rho(u)}{1+u^2}\, du > -\infty$, where recall $\rho = d\mu/dLeb$. Then, from [9, p. 66], the function

$$\varphi(z) = \exp\left\{\frac{2i}{\pi p}\int_{\mathbf{R}}\left(\frac{1}{z-u}+\frac{u}{1+u^2}\right)\log\ \rho(u)du\right\} \tag{2.7}$$

is an $\mathbf{H}^p$-function whose boundary function satisfies the condition

$$|\varphi(u)|^p = \rho(u) \qquad \text{a.e.}[Leb]\ . \tag{2.8}$$

Theorem 2.2.3. *Let $X = \{\int_{\mathbf{R}} e^{itu} dM(u) : t \in \mathbf{R}\}$ be a regular harmonizable $\mathbf{I}_M$-representable process in $L^p(\mu), 1 \leq p \leq 2$; if $p < 2$, assume that $\mathcal{P}_T(\varphi^{2/p})$ has no zeros in $\mathcal{H}$. Let $T > 0$; then the (optimal) linear predictor $\tilde{X}(t,T)$ of $X(t+T)$ based on $\{X(s) : s \leq t\}$ exists and is given by the formula*

$$\tilde{X}(t,T) = \int_{\mathbf{R}} e^{i(t+T)u}\left\{1 - \frac{(\mathcal{P}_T(\varphi^{p/2}))^{2/p}(u)}{\varphi(u)}\right\}dM(u)\ , \tag{2.9}$$

where $\varphi^{p/2}$ is the natural $p/2$-the root of φ (see 2.7) and the root $[\mathcal{P}_T(\varphi^{p/2})]^{2/p}$ is determined by the condition (1.8); the prediction error is given by

$$\delta(T)^p = \int_{\mathbf{R}} |\mathcal{P}_T(\varphi^{p/2}|^2(u)du = \int_0^T |\widehat{\varphi^{p/2}}(v)|^2\ dv\ . \tag{2.10}$$

Further, the map

$$[0,\infty) \ni T \longmapsto \tilde{X}(t,T) \in \mathcal{L}(M) \tag{2.11}$$

is continuous.

Remark 2.2.4(a). If $X \equiv \{X(t) : t \in \mathbf{R}\}$ is a complex process which is the Fourier transform of any one of the α-stable or (r, α)-semistable random measures noted in (i) above, then X is a harmonizable $\mathbf{I}_M$- representable process in $L^p(\mu)$, where μ is the control measure of M and $p = \alpha$. Thus, if X is regular, then, under the noted conditions, the best linear predictor of $X(t+T)$ will be given by (2.9) with p replaced by α. This form of the linear predictor for the case when X is the Fourier transform of a complex symmetric α-stable random measure, $1 < \alpha < 2$, as noted earlier, was conjectured by Cambanis and Soltani [3], who obtained a similar formula in the discrete parameter α-stable case, $1 < \alpha < 2$. We note that, since the analog of Theorem 1.2.1 for the H^1-space of the unit disc is known [6, p. 140], the analog of their Theorem 5.1 [3] for $\alpha = 1$ can be proved using the formulation and the line of proof presented here. In this context, we must emphasize that the norm $\|\cdot\|_p$ introduced on the space $\mathcal{L}(M)$ here coincides with the norm introduced in [3] using the co-variation function in the case when M is complex α-stable and $1 < \alpha = p < 2$. For $\alpha = 1$, even though there is no co-variation function, there is the natural choice of the norm which coincides with $\|\cdot\|_1$.

(b). Let $X = \{X(t) : t \in \mathbf{R}\}$ be a real or complex Gaussian or, more generally, a second order mean zero continuous co-variance stationary process. Then, as is well known [8, p. 246], X is $\mathbf{I}_M$-representable in $L^p(\mu)$ with $p = 2$, and is harmonizable where M is an o.s. random measure and μ is the associated structure measure of M (see (ii)). Therefore if X is regular, then the formula for the linear predictor $\tilde{X}(t, T)$ will be given by (2.9) with $p = 2$; thus we recover the famous Kolmogorov [11-13] Krein [16] Wiener [29] formula for the linear predictor.

(c). We will see in the proof of Theorem 2.2.3 that, if X is a regular harmonizable $\mathbf{I}_M$-representable process in $L^p(\mu), 1 \leq p < \infty$, then, without any other hypotheses, the linear predictor $\tilde{X}(t,T)$ of $X(t+T)$ exists and is given by

$$\tilde{X}(t,T) = \mathbf{I}_M\left(\frac{e_t}{\bar{\varphi}}\overline{\mathbf{P}(e_{-T}\varphi)}\right) \equiv \int_{\mathbf{R}} \frac{e^{itu}}{\bar{\varphi}}\left(\overline{\mathbf{P}(e_{-T}\varphi)(u)}\right) dM(u) , \tag{2.12}$$

and that the map (2.11) is continuous. The first statement follows from two applications of the Beurling's Theorem; and the second follows from Proposition 1.2.4. Thus, as noted in the Introduction, the central point in obtaining the formula (2.9) for the linear predictor is to find a suitable formula for $\mathbf{P}(e_{-T}\varphi)$. For this, we use Theorem 1.2.1; and, it is for this reason, we need the hypotheses that $1 \leq p \leq 2$ and that $\mathcal{P}_T(\varphi^{p/2})$ be non-vanishing on $\mathcal{H}$.

Now we are ready to prove the results. First we introduce a few more notations; these will be needed in the proofs. Let $g \equiv \{g_t\} \subseteq L^p(\nu)$; then we shall denote, by $H_t(g)$, the sub-space $L^p(\nu)$ $cl(sp\{g_s \leq t\})$, and, by $H_{-\infty}(g)$, the sub-space $\bigcap_{t\in\mathbf{R}} H_t(g)$, where $L^p(\nu)$-$cl(\{\cdot\})$ denotes the closure of $\{\cdot\}$ in $L^p(\nu)$.

Outline of the Proof of Theorem 2.2.1. The proof of Part (a) and the proof of the existence part of the decomposition in Part (b) for $1 < p < \infty$, as noted above, is as in [3]. The proofs of these, for the case $p = 1$, need only minor modifications except for one improtant point: The proof of the necessity part of Lemma 4.3 in [3] breaks down towards the end in the case when $p = 1$. We outline below how this can be fixed. After this, an outline of the proof of the uniqueness part is provided.

Let f be as in the last paragraph of the proof of Lemma 4.1 of [3]. Since $f \in L^\infty(\mu)$, we can find a set E of μ-measure zero such that $|f| \leq k$ (constant) off E. But then $\mu_a(E) = \int_E \rho(x)dx = 0$; therefore $\rho = 0$ a.e. [Leb] on E (note our ρ is the same φ in [3]).

Thus, since $\rho \neq 0$ a.e. on $\mathbf{R}$ (recall $f\rho \in \mathbf{H}^1$), we have Leb $(E) = 0$. Therefore $|f| \leq k$ a.e. [Leb]. Now using that $\int_{\mathbf{R}} \frac{\log \rho(x)|f(x)|}{1+x^2}\, dx > -\infty$, we conclude that $\int_{\mathbf{R}} \frac{\log \rho(x)}{1+x^2}\, dx > -\infty$ as follows:

$$\int_{\mathbf{R}} \frac{\log \rho(x)}{1+x^2}\, dx = \int_{\mathbf{R}} \frac{\log \rho(x)|f(x)|}{1+x^2}\, dx - \int_{\mathbf{R}} \frac{\log |f(x)|}{1+x^2}\, dx$$
$$\geq \int_{\mathbf{R}} \frac{\log \rho(x)|f(x)|}{1+x^2}\, dx - k \int_{\mathbf{R}} \frac{dx}{1+x^2} > -\infty\ .$$

By replacing the last two lines in the proof of Lemma 4.3 in [3] by the above, one completes the proof of the analog of this lemma for p=1.

Now we outline the proof of the uniqueness of the decomposition in Part (b). Let $X_1(t) = \mathbf{I}_M(f_t)$ and $X_2(t) = \mathbf{I}_M(g_t)$, where X_1, X_2 satisfy the given requirements. Thus, in particular, using the decomposition and $\mathbf{I}_M$-independence, we have

$$e_t = f_t + g_t \quad \text{a.e.}[\mu] \quad \text{and} \quad f_t g_s = 0 \quad \text{a.e.}[\mu]\ , \tag{2.13}$$

for all $t, s \in \mathbf{R}$. We need to prove that if X is singular, then, for every t, $f_t = 0$ and $g_t = e_t$ a.e. $[\mu]$; and, in all other cases, $f_t = e_t I_{S_a}$ and $g_t = e_t I_{S_s}$ a.e. $[\mu]$. Let t_0, t, be fixed $t_0 \neq t_1$. Let $F_{t_j} = \{f_{t_j} \neq 0\}$, $j = 0, 1$. Then, from (2.13),

$$f_{t_j} = e_{t_j} I_{F_{t_j}} \qquad \text{and} \qquad g_{t_j} = e_{t_j} I_{F^c_{t_j}}\ , \quad \text{a.e. } [\mu]\ ,$$

for each $j = 0, 1$. Using second equation in (2.13), it is easy to show that $\mu(F_{t_0} \Delta F_{t_1}) = 0$. Therefore, setting $F_{t_0} = F$, we conclude that, for every fixed $t \in \mathbf{R}$,

$$f_t = e_t I_F \qquad \text{and} \qquad g_t = e_t I_{F^c} \quad \text{a.e.}[\mu]\ . \tag{2.14}$$

Then, using the subordinate property and the fact $e_t = f_t + g_t$, from (2.13), we get

$$H_t(e) = H_t(f) \oplus H_t(g)\ .$$

Therefore, since X_2 is singular, we get $H_t(e) = H_t(f) \oplus L^p(\mu_{F^c})$, where for a Borel set A, $\mu_A \equiv \mu/\mathcal{B}(A)$. This implies $H_{-\infty}(e) = H_{-\infty}(f) \oplus L^p(\mu_{F^c})$. Therefore, since X_1 is regular, we have $H_{-\infty}(e) = L^p(\mu_{F^c})$. Thus, if X is itself singular, then $H_{-\infty}(e) = L^p(\mu) = L^p(\mu_{F^c})$; which, along with (2.14), implies that $f_t = 0$ and $g_t = e_t$ a.e. $[\mu]$. If X is not singular, then, by Lemma 4.1 of [3] and above remarks, $H_{-\infty}(e) = L^p(\mu_s)$; thus $L^p(\mu_s) = L^p(\mu_{F^c})$. This shows that $I_{FC} = I_{S_s}$ a.e. $[\mu]$; therefore, using (2.14) again, we get that, for every $t \in \mathbf{R}$, $f_t = e_t I_{Sa}$ and $g_t = e_t I_{S_s}$ a.e. $[\mu]$. ∎

Proof of Theorem 2.2.3. We shall first prove formula (2.12); this formula holds true for any $1 \leq p < \infty$. We first show that

$$Y_0 \equiv \mathbf{I}_M \left(\frac{e_t}{\varphi} \overline{\mathbf{P}(e_{-T}\varphi)} \right) \in \mathcal{L}(X, t) . \tag{2.15}$$

Recall that the Beurling's Theorem asserts that $L^p(du)\text{-}cl(sp\{e_s\varphi : s \geq 0\}) = \mathbf{H}^p$ [14]. Therefore, by the definition of $\mathbf{P}(e_{-T}\varphi), \mathbf{P}(e_{-t}\varphi) \in L^p(du)\text{-}cl(sp\{e_s\varphi : s \geq 0\})$; which implies that $f_0 \equiv \frac{e_t}{\bar{\varphi}} \overline{\mathbf{P}(e_{-T}\varphi)} \in L^p(\mu)\text{-}cl(\{e_s : s \leq t\})$, since $|\bar{\varphi}|^p du = d\mu$ (see (2.8)). Hence $Y_0 = \mathbf{I}_M(f_0) \in \mathcal{L}(X, t)$; thus (2.15) holds.

Next we show that $Y_0 = \tilde{X}(t, T)$; which is clearly equivalent to showing that

$$\|e_{t+T} - f_0\|_{L^p(\mu)} = \inf \left\{ \|e_{t+T} - f\|_{L^p(\mu)} : \quad f \in L^p(\mu)\text{-}cl(sp\{e_s : s \leq t\}) \right\} . \tag{2.16}$$

Using again $|\varphi(u)|^p du = d\mu$, we have $\|e_{t+T} - f\|^p_{L^p(\mu)} = \|e_{-T}\varphi - e_t\bar{f}\varphi\|_{L^p(du)}$, for any $f \in L^p(\mu)$. Therefore the left side of (2.16)

$$\|e_{t+T} - f_0\|_{L^p(\mu)} = \|e_{-T}\varphi - \mathbf{P}(e_{-T}\varphi)\|_{L^p(du)} ; \tag{2.17}$$

and the right side of (2.16) is

$$= \inf\{\|e_{-T}\varphi - e_t\overline{f}\varphi\|_{L^p(du)} : f \in L^p(\mu)\text{-}cl(sp\{e_s : s \leq t\})\}$$

$$= \inf\{\|e_{-T}\varphi - g\|_{L^p(du)} : g \in L^p(du)\text{-}cl(sp\{e_s\varphi : s \geq 0\})\}$$

$$= \inf\{\|e_{-T}\varphi - g\|_{L^p(du)} : g \in \mathbf{H}^p\}, \text{ by Beurling's Theorem,}$$

$$= \|e_{-T}\varphi - \mathbf{P}(e_{-T}\varphi)\|_{L^p(du)} \ , \text{ by the definition of } \mathbf{P}(e_{-T}\varphi) \ .$$

This along with (2.17) completes the proof of (2.16). Thus the proof of (2.12) is complete.

The continuity of the map in (2.11) follows now from (2.12) and Proposition 1.2.4; once again this fact is true for any $1 \leq p < \infty$. Finally, the formulae (2.9) and (2.10) follow now easily from (2.12) and Theorem 1.2.1 (note that clearly φ is non-vanishing on $\mathcal{H}$; and, by BF-1, $0 \in supp(\hat{\varphi})$). ∎

2.3. Examples and Remarks

For two $L^p(\mu)$-representable processes for which the density $\rho = d\mu/dLeb$ is concretely given, we compute here the formulae for the linear predictors more explicitly. In these, we make use of the formulae obtained for $\mathbf{P}_T$ in Examples 1.3.1 and 1.3.2. After the examples, we make some concluding remarks.

Example 2.3.1. Let $1 < p \leq 2$, and let μ be the finite measure on $\mathcal{B}(\mathbf{R})$ with $\rho(u) = d\mu/dLeb = (1+u^2)^{-p/2}$. It is easy to verify that $\int_{\mathbf{R}} \frac{\log \rho(u)}{1+u^2}\, du > -\infty$. Now let $X = \{X(t) = \mathbf{I}_M(e_t) : t \in \mathbf{R}\}$ be a harmonizable $\mathbf{I}_M$-representable process in $L^p(\mu)$, where the pair (M, μ), of course, is such that A-1 and A-2 are satisfied. From Theorem 2.2.1, X is regular. Let $\varphi_0(z) = \frac{i}{z+i}, z \in \mathcal{H}$. From Example 1.3.1, we know that $\varphi_0 \in \mathbf{H}^p, 0 \in supp(\hat{\varphi}_0)$ and that φ_0 has no zeros on $\mathcal{H}$. Further, since φ_0 is analytic on $\{z : \Im(z) > -1\}$, it follows, using $\mid \frac{i}{z+i} \mid = (1+u^2)^{-1/2}, z \in \mathbf{R}$, and BF-2, that

$$\begin{aligned}\varphi_0(z) &= C_0 \ \exp \left\{ \frac{i}{\pi} \int_{\mathbf{R}} \left(\frac{1}{z-u} + \frac{u}{1+u^2} \right) \ \log(1+u^2)^{-1/2} du \right\} \\ &= C_0 \ \exp \left\{ \frac{i}{\pi p} \int_{\mathbf{R}} \left(\frac{1}{z-u} + \frac{u}{1+u^2} \right) \ \log(1+u^2)^{-p/2} du \right\} \\ &= C_0 \varphi(z) \ ,\end{aligned}$$

for some $C_0 \in \mathbf{C}$ with $|C_0| = 1$, where φ is as in (2.7) with $\rho(u)$ replaced by $(1+u^2)^{-1/2}$. Therefore $\varphi(z) = C_0^{-1} \varphi_0(z), z \in \mathcal{H}$. Now, using formula (2.9) and Example 1.3.1, we get

$$\begin{aligned}\tilde{X}(t,T) &= \int_{\mathbf{R}} e^{i(t+T)u} \left\{ 1 - \frac{[\mathcal{P}_T(\varphi^{p/2})]^{2/p}(u)}{\varphi(u)} \right\} \ dM(u) \\ &= \int_{\mathbf{R}} e^{i(t+T)u} \left\{ 1 - \frac{[\mathcal{P}_T(\varphi_0^{p/2})]^{2/p}(u)}{\varphi_0(u)} \right\} \ dM(u) \\ &= \int_{\mathbf{R}} e^{i(t+T)u} \left\{ 1 - \left[\frac{[\gamma(\frac{p}{2}, T(1-iz))]}{\Gamma(p/2)} \right]^{p/2} \right\} \ dM(u)\end{aligned}$$

and

$$\delta_T^p = \frac{2^{2-p}}{\Gamma(p/2)^2} \ \gamma(p-1, 2T) \ .$$

Example 2.3.2. Let $\rho(u) = (1-u^2)^{-1}, u \in \mathbf{R}$ and $\varphi_1 = \left(\frac{i}{i+z}\right)^2 = (\varphi_0(z))^2, z \in \mathcal{H}$. Then, with a similar argument as in the above example, it follows that $\varphi_1 = C_1 \varphi$, where φ is defined as in the above example with $(1+u^2)^{-1/2}$ replaced by $(1+u^2)^{-1}$. Thus, if $d\mu/dLeb = (1+u^2)^{-1}$ and if $X = \{\mathbf{I}_M(e_t) : t \in \mathbf{R}\}$ is an $\mathbf{I}_M$-representable process in $L^p(\mu)$, then it follows, from Example 1.3.2 and formula (2.9), that

$$\begin{aligned}\tilde{X}(t,\tau) &= \mathbf{I}_M \left(e_{t+T} \left\{ 1 - \frac{(z+i)^2}{i^2} \ \left(\frac{e^{T(iz-1)} - 1}{iz-1} \right)^2 \right\} \right) \\ &= \mathbf{I}_M \left(e_{t+T} \left\{ 1 - \left(e^{T(iz-1)} - 1 \right)^2 \right\} \right) \\ &= \int_{\mathbf{R}} \left\{ 2e^{-T} e^{itu} - e^{-2T} e^{i(t-T)u}) \right\} dM(u) \\ &= 2e^{-T} X(t) - e^{-2T} X(t-T) \ ;\end{aligned}$$

and the prediction error is given by

$$\delta_T = \pi[1 - e^{-2T}] .$$

Remark 2.3.3 (a). The problem of expressing the linear predicter $\tilde{X}(t,T)$(see Theorem 2.2.3) in terms of a "nice" function (e.g., a suitably defined integral) of the past $\{X(s) : s \le t\}$ is not pursued here. We believe that this is an interesting problem and should be pursued in the future research.

(b). It is well known that every *real* mean zero second order mean square continuous stationary process Y admits the representation $Y(t) = \int_{\mathbf{R}} e^{itu} dZ(u)$, where Z is an o.s. random measure; this fact plays a crucial role in the linear prediction theory of such processes. On the other hand, if a process X admits a representation $X(t) = \int e^{itu} dM(u)$ with M i.d. and i.s. random measure, then X is stationary if and only if M is rotationally invariant [25, p. 80]. Thus the only *real* stationary process which can be expressed as a Fourier transform of an i.d. i.s. random measure is the zero process. It appears that a similar situation seems to prevail even when X is non-stationary. We provide one example which supports this assertion; but we do not investigate this here in the general case. Specifically we prove that the following assertion holds:

Proposition. *Let $0 < \alpha < 2$; and let M be an α-stable random measure on $\mathcal{B}(\mathbf{R})$ such that the ch. function of $M(A), A \in \mathcal{B}(\mathbf{R})$, is given by*

$$E\left[e^{i\Re(M(A)\bar{z})}\right] = \exp - \iint_{A\times(-\pi,\pi]} |\Re(\bar{z}e^{i\theta})|^\alpha d\nu(u,\theta) ,$$

$z \in \mathbf{C}$, where ν is a finite measure on $\mathcal{B}(\mathbf{R} \times (-\pi,\pi])$. Let $X_n = \int_{\mathbf{R}} e^{inu} dM(u), n \in Z \equiv \{0, \pm 1, \pm 2, \dots\}$; and let X_n's be real. Then $X_{2n} = X_n$ a.s.[P] and $X_{2n-1} = X_1$ a.s.[P], for all n; further, if $\mu(\cdot) \equiv \nu(\cdot \times (-\pi,\pi])$ is continuous, then $X_n = 0$ a.s.[P], for all n.

Proof. Using Theorem 5.2 of [2], we have

$$1 = E(e^{i\Re(iX_n)}) = \exp - \iint_{\mathbf{R}\times(-\pi,\pi]} |\Re(ie^{inu+i\theta})|^\alpha \, d\nu(u,\theta) ;$$

this yields $\iint_{\mathbf{R}\times(-\pi,\pi]} |\sin(nu+\theta)|^\alpha d\nu(u,\theta) = 0$. Thus ν must be supported by $\{(u,\theta) : \frac{nu+\theta}{\pi} \in \mathbf{Z}, \theta \in (-\pi,\pi], u \in \mathbf{R}\}$. Since this is true for all $n \in \mathbf{Z}$, it follows that ν is supported by $\{(u,\theta) : \frac{u}{\pi} \in \mathbf{Z},\ \theta = 0 \text{ or } \pi, u \in \mathbf{R}\}$. Thus, again by Theorem 5.2 of [2], we have

$$E\left(e^{i\Re\int_{\mathbf{R}} f dM}\right) = \exp\left\{-\sum_m [\nu\{(m\pi,0)\} + \nu\{(m\pi,\pi)\}]\,|\Re f(m\pi)|^\alpha\right\}, \tag{2.18}$$

for every complex function $f \in L^\alpha(\mu)$. Taking $f(u) = a(e^{2niu}-1), a \in \mathbf{R}$; and using (2.18), we get $E[e^{ia(X_{2n}-X_0)}] = 1$, for all $n \in \mathbf{Z}$ and $a \in \mathbf{R}$. Thus $X_{2n} = X_0$ a.s. [P]; similarly, taking $f(u) = a(e^{i(2n-1)u} - e^{iu})$, we see that $X_{2n-1} = X_1$ a.s. [P], for all n. Finally, if μ is continuous, then $\nu\{(m\pi,\theta\} \le \mu(\{m\pi\}) = 0$, for all $m \in Z$ and $\theta = 0, \pi$; and, hence, from (2.18), $X_n = 0$ a.s. [P]. ∎

The point of these comments is that if X is a real $\mathbf{I}_M$-representable process in $L^p(\mu)$ with M i.s. and i.d., then X may be a degenerate process; and, hence, in such a case Theorems 2.2.1 and 2.2.3 should be used with caution. Despite the above negative results, we note here that for (complex) i.d., i.s. and rotationally invariant random measurs M, the processes $\xi_t = \Re\left(\int e^{itu} dM(u)\right), t \in \mathbf{R}$, provide a wealth of examples of real stationary processes with integral representions. Several problems related to special classes of such processes (e.g. α-stable processes) have been studied recently by several authors.

(c). In Shiryayev [23, p.404] it is implied, contrary to the above mentioned fact about real stationary processes, that if the process Y (introduced in (b) above) is Gaussian then the o.s. random measure Z in its representation is i.s. (see the first paragraph on p. 405 in [23]). In view of the above remarks, this statement is not true.

Remark 2.3.4. A careful look at the proofs of Theorems 2.2.1 and 2.2.3 reveals that the 1-1 linear map $\mathbf{I}_M : L^p(\mu) \longmapsto \mathcal{L}(M) \subseteq L^0(P)$ is a stochastic integral is nowhere used. (The only place where this fact is needed is in Remark 2.2.2, where we pointed out the relationship of condition (1.6) to the independence of random variables). In fact, the whole analysis can be carried out by replacing $\mathbf{I}_M$ be any 1-1 linear map $\mathbf{\Phi} : L^p(\mu) \longmapsto L^0(P)$ (note that there is no random measure involved here). In this setting, the norm $\|\cdot\|_p$ on $\mathcal{L}(\mathbf{\Phi}) \equiv \{\mathbf{\Phi}(f) : f \in L^p(\mu)\}$ will be defined as in (2.5) by replacing $\mathbf{I}_M$ by $\mathbf{\Phi}$. The definitions of $\mathbf{\Phi}$-representable processes in $L^p(\mu)$, representing functions, $\mathbf{\Phi}$-independence etc. will be given in a natural way. However, since there is no interesting nontrivial example of a $\mathbf{\Phi}$-representable process which is not $\mathbf{I}_M$-representable for a suitable pair (M, μ) of measures satisfying A-1 and A-2, we confined our analysis to $\mathbf{I}_M$-representable processes.

Note. After the results of this paper were presented at the Workshop at Cornell, Professor V. Mandrekar informed us about his joint forthcoming paper with A. Makagon entitled "The Spectral Representation of Stable Processes: Harmonizability and Regularity." In this work, done independently of ours, they obtained an analog of Theorem 2.2.1 for processes which are Fourier transforms of α-stable random measures $0 < \alpha < 2$. Thus, Theorem 2.2.1 has an overlap with their work; however, the methods of proof used in the two papers for this result are different.

References

1. M. Abramowitz and I.A. Stegun, "Handbook of Mathematical Functions," Dover Publications, Inc., New York, 1965.
2. S. Cambanis, *Complex symmetric stable variables and processes*, P.K. Sen, Ed., in "Contributions to Statistics: Essays in Honour of Norman L. Johnson," North Holland, New York, 1982, pp. 63-79.
3. S. Cambanis and A.R. Soltani, *Prediction of stable processes: Spectral and moving average representations*, Z. Wahrsch. verw. Gebiete **66** (1984), 593-612.
4. S. Cambanis, C.D. Hardin Jr. and A. Weron, *Innovations and Wold decompositions of stable sequences*, Probab. Th. Rel. Fields **79** (1988), 1-27.

5. S. Cambanis and A.G. Miamee, *On prediction of harmonizable stable processes.* Sankhya, (in press) (1989).
6. P.L. Duren, "Theory of H^p Spaces," Academic Press, New York, 1970.
7. H. Dym and H.P. McKean, "Gaussian Processes, Function Theory, and the Inverse Spectral Problem," Academic Press, New York, 1976.
8. I.I. Gihman and A.V. Skorohod, "The Theory of Stochastic Processes I," Springer Verlag, New York, 1974.
9. J.B. Garnett, "Bounded Analytic Functions," Academic Press, New York, 1981.
10. Y. Hosoya, *Harmonizable stable processes*, Z. Wahrsch. verw. Gebiete **60** (1982), 517-533.
11. A.N. Kolmogorov, *Sur l'interpolation et extrapolation des suites stationnaires*, C.R. Acad. Sci. Paris **208** (1939), 2043-2045.
12. A.N. Kolmogorov, *Stationary sequences in Hilbert space*, Bull. Math. Univ. Moscow **2** (1941), 1-40.
13. A.N. Kolmogorov, *Interpolation und Extrapolation von stationären zufälligen Folgen*, Izv. Akad. Nauk SSSR **5** (1941), 3-14.
14. P. Koosis, "Introduction to H^p Spaces," Cambridge University Press, New York, 1980.
15. P. Koosis, "The Logarithmic Integral," Cambridge University Press, New York, 1988.
16. M.G. Krein, *On a problem of extrapolation of A.N. Kolmogorov*, Dokl. Akad. Nauk SSSR **46** (1945), 306-309.
17. A.G. Miamee and M. Pourahmadi, *Wold decomposition, prediction and parameterization of stable processes with infinite variance*, Probab. Th. Rel. Fields **79** (1988), 145-164.
18. B.S. Rajput and K. Rama-Murthy, *Spectral representations of semi-stable processes, and semistable laws on Banach spaces*, J. Multivariate Anal. **21** (1987), 139-157.
19. B.S. Rajput and K. Rama-Murthy, *Spectral representations of complex semi-stable and other infinitely divisible stochastic processes*, Stochastic Processes Appl. **26** (1987), 141-159.
20. B.S. Rajput and J. Rosinski, *Spectral representations of infinitely divisible processes*, Probab. Th. Rel. Fields **82** (1989), 451-487.
21. W. Rudin, "Real and Complex Analysis," McGraw-Hill, New York, 1966.
22. M. Schilder, *Some structure theorems for the symmetric stable laws*, Ann. Math. Stat. **41** (1970), 412-421.
23. A.N. Shiryayev, "Probability," Springer-Verlag, New York, 1984.
24. I. Singer, "Best Approximation in Normed Linear Space by Elements of Linear Subspaces," Springer-Verlag, New York, 1970.
25. K. Urbanik, *Random measures and harmonizable sequences*, Studia Math. **31** (1968), 61-88.
26. K. Urbanik, *Prediction of strictly stationary sequences*, Coll. Math. **12** (1964), 115-129.
27. K. Urbanik, *Some prediction problems for strictly stationary processes*, Part I, Proc. Fifth Berkeley Symp. Math. Stat. Probab. **2** (1967), 235-258. Univ. California Press.
28. A. Weron, *Harmonizable stable processes on groups: spectral, ergodic and interpolation properties*, Z. Wahrsch. verw. Gebiete **681** (1985), 473-491.
29. N. Wiener, "Extrapolation, Interpolation and Smoothing of Stationary Time Series," Wiley, New York, 1949.

Balram S. Rajput and Carl Sundberg
Mathematics Department
University of Tennessee
Knoxville, TN 37996, USA

Kavi Rama-Murthy
Indian Statistical Institute
Bangalore, India
and
Mathematics Department
University of Tennessee
Knoxville, TN 37996, USA

On Multiple Markov $S\alpha S$ Processes[1]

by

V. Mandrekar[2] and B. Thelen

Michigan State University University of Michigan

0. Introduction. In case $\{X_t;\ t \in \mathbb{R}\}$ is a Gaussian process, Mandrekar [11] and Pitt [12] have studied the structure of n–ple Markov processes. It is shown in ([11], [12]) that a Gaussian process is n–ple Markov if and only if it has the so called Goursat representation introduced in [11]. In [11], a conjecture of Lévy [9], [10] is solved (Theorem 3.12). This says that the solution of the nth order stochastic differential equation with white noise input is n–ple Markov if and only if the differential operator satisfies the so called Polya condition; namely, the Wronksians of all orders are different from zero [7] for the solutions. Recently, Adler, Cambanis, Samorodnitsky [1] have studied a representation for symmetric stable left Markov processes. Our purpose is to study n–ple Markov processes and their representations for the symmetric stable case and also to study the Markov properties of the solutions of the nth order ordinary stochastic differential equations with stationary symmetric α–stable input. The latter generalizes earlier work of Hida [6] and Lévy [10] and the former allows us to show that a process of type $X_t = \varphi(t)L(\tau(t))$, L Lévy process and $\tau(t)$ measuring time change in [1] is first order Markov.

1. Goursat Representation. Let $\{X_t;\ t \in \mathbb{R}\}$ be a symmetric α–stable (for short, $S\alpha S$) stochastic process, i.e., a stochastic process such that all finite linear combinations of $\{X_t;\ t \in \mathbb{R}\}$ are $S\alpha S$. We denote the characteristic function of a $S\alpha S$ random variable X by $e^{-\gamma|t|^{\alpha}}$, $t \in \mathbb{R}$, and we define

$$\|X\|_{\alpha} = \begin{cases} \gamma^{\frac{1}{\alpha}} & 1 \le \alpha \le 2 \\ \gamma & 0 < \alpha < 1 \end{cases}$$

following Schilder [13]. We note that for $1 \le p < \alpha$, we have [2]

(1.1) $$\|X\|_{\alpha} = C(p, \alpha)\|X\|_{p}.$$

Here $\|\cdot\|_p$ denotes the $L_p(\Omega, \mathcal{F}, P)$ norm. Hence for $\alpha > 1$, the completion of

---

[1]Supported ONR N00014–85–K–0150

[2]On leave at the University of Michigan

$sp\{X_t;\ t \in \mathbb{R}\}$ in $\|\cdot\|_\alpha$ (or $\|\cdot\|_p$, $1 < p < \alpha$) is a uniformly convex space. In view of (1.1), the $\|\cdot\|_p$–completion is the same. We define the following subspaces of

$$H(X) = \overline{sp}^\alpha\{X_t;\ t \in \mathbb{R}\},$$

the $\|\cdot\|_\alpha$ closure of the span,

$$H(X:t) = \overline{sp}^\alpha\{X_s;\ -\infty < s \leq t\},$$

and

$$H(X:-\infty) = \cap_t \overline{sp}^\alpha\{X_s;\ -\infty < s \leq t\}.$$

We denote by $\mathcal{E}_t$ the metric projection onto $H(X:t)$. The metric projection exists in view of (1.1) and the fact that $L_p(\Omega, \mathcal{F}, P)$, $p > 1$ is reflexive (Singer [14], Corollary 2.4). We note that in general $\mathcal{E}_s\mathcal{E}_t \neq \mathcal{E}_s$ on $H(X)$ for $s \leq t$ ([15]) and in general $\mathcal{E}_t$ is not linear. However for $\alpha = 2$, $\mathcal{E}_t$ being the orthogonal projection $\mathcal{E}_s\mathcal{E}_t = \mathcal{E}_s$ for $s \leq t$ and $\mathcal{E}_t$ is linear. In the Gaussian case, the concept of n–ple Markov processes have been studied in ([6], [10]). We now define it in the stable case.

1.2 *Definition.* Let $\{X_t;\ t \in \mathbb{R}\}$ be a SαS process and $Y(t,s) = \mathcal{E}_s X(t)$. It is said to be n–ple Markov if for $s \leq t_1 < t_2 \cdots < t_n < t_{n+1}$, $\{Y(t_1, s), \ldots, Y(t_n, s)\}$ are linearly independent and $\{Y(t_1, s), \ldots, Y(t_{n+1}, s)\}$ are linearly dependent.

1.3. *Definition.* An n–dimensional SαS process $\{\underline{U}(t);\ t \in \mathbb{R}\}$ is said to be a (wide–sense) n–variate martingale if $\mathcal{E}_s U_i(t) = U_i(s)$, $s \leq t$, $i = 1, \ldots, n$.

Note that for $\alpha = 2$, this coincides with the definition of Doob [4]. We now introduce a generalization of Goursat representation in [11].

1.4. *Definition.* A process $\{X_t;\ t \in \mathbb{R}\}$ has Goursat representation if

$$X(t) = \sum_{i=1}^{n} \phi_i(t) U_i(t), \tag{1.5}$$

where

(i) for any $t_1 < t_2 < \cdots < t_n$, the matrix $(\phi_i(t_j))$ is <u>non–singular</u>,

(ii) for $s \leq t_1 < t_2 \ldots < t_n$ $(U_1(s), \ldots, U_n(s))^* = ((\phi_i(t_j))^{-1}(\mathcal{E}_s X(t_1), \ldots, \mathcal{E}_s X(t_n))^*$

with $*$ being the transpose and $\{(U_1(t), U_2(t), \ldots, U_n(t)\}$ is linearly independent.

We note that if $\{X_t;\ t \in \mathbb{R}\}$ is Gaussian, then $\mathcal{E}_s X(t) = E(X(t)|\mathcal{F}_s)$ where $\mathcal{F}_s = \sigma\{X_u;\ u \leq s\}$. This gives $\underline{U}(s) = (U_1(s), \ldots, U_n(s))^*$ is a $\mathcal{F}_s$–martingale for $s < t$. In addition, $\sup_{s \leq t} E|U_i(s)|^2 < \infty$ gives, using martingale convergence theorem, that $\{\underline{U}(s), \mathcal{F}_s, s \leq t\}$ is a martingale. Hence in Gaussian case, $X(t) = \Sigma_{i=1}^n \phi_i(t) U_i(t)$ with

(i) matrix $\phi_i(t_j))$ being non–singular;

(ii) $(U_1(t),\ldots,U_n(t))^*$ being a $\mathcal{F}_t$ martingale and $(U_1(t),\ldots,U_n(t))$ being linearly independent.

Thus the above definition reduces to that in [11] in the Gaussian case. In general, if X has a Goursat representation, then

$$H(X: t) \subset H(\underline{U}: t) = \overline{sp}^{\alpha}\{U_i(s);\ s \le t,\ i = 1,\ldots,n\}\ \forall t.$$

If $H(\underline{U} : t) = H(X : t)\ \forall\, t$, we call the Goursat representation proper. We now state our main representation result.

1.6. **Lemma.** *Let* $\{X_t;\, t \in \mathbb{R}\}$ *be a* SαS *process for* $\alpha > 1$ *with* $H(X: t) = \overline{sp}^{\alpha}\{\cup_m H(X: t - \frac{1}{m})\} = H(X: t-)$. *Then it is n–ple Markov if and only if* X(t) *has a proper Goursat representation. In case* $\alpha = 2$, $(U_1(t),\ldots,U_n(t))$ *is a martingale.*

Proof. Assume X is n–ple Markov. Let $s < \tau$. Then for $s \le t_1 < t_2 < \ldots < t_n < \tau$,

$$Y(\tau, s) = \sum_{i=1}^{n} a_i(\tau, t_1,\ldots,t_n)Y(t_i, s) \tag{1.7}$$

Also for $s \le s_1 < s_2 < \ldots < s_n < t_1 < \ldots < t_n < \tau$, we have

$$Y(\tau, s) = \sum_{i=1}^{n}\sum_{k=1}^{n} a_k(\tau, t_1,\ldots,t_n)a_j(t_k,s_1,\ldots,s_n)Y(s_j, s). \tag{1.8}$$

For (1.7) and (1.8), we get

$$\text{(1.9)}\quad \begin{aligned} &\text{(i)} \quad \sum_{k=1}^{n} a_k(\tau, t_1,\ldots,t_n)a_j(t_k,s_1,\ldots,s_n) = a_j(\tau,s_1,\ldots,s_n),\\ &\text{(ii)} \quad \det\,(a_j(t_k,s_1,\ldots,s_n)) \neq 0. \end{aligned}$$

By following arguments in Hida ([6], pg 129–130), (1.9) implies that $\exists\, \underset{\sim}{\phi} = (\phi_1,\ldots,\phi_n)$: $\mathbb{R} \to \mathbb{R}^n$ such that $\forall\, \tau > t_n > \ldots. > t_1$ have,

$$a(\tau, \underset{\sim}{t}) = (a_1(\tau, \underset{\sim}{t}),\ldots,a_n(\tau, \underset{\sim}{t})) = \underset{\sim}{\phi}(\tau)\, B(\underset{\sim}{t}).$$

where $B(\underset{\sim}{t})$ is a non–singular n by n matrix. In view of (1.7) (ii), $\det\,(\phi_i(t_j)) \neq 0$.

Rewrite (1.7) with obvious notation as

$$\begin{aligned} Y(\tau, s) &= a(\tau, \underset{\sim}{t})Y(\underset{\sim}{t}, s)\\ &= \phi(\tau)\, B(\underset{\sim}{t})Y(\underset{\sim}{t}, s) \qquad (1.10)\\ &= \underset{\sim}{\phi}(\tau)\, U^*(s, \underset{\sim}{t}), \qquad s < \tau, \end{aligned}$$

with $U^*(s, \underset{\sim}{t})$ denoting $B(\underset{\sim}{t})Y(\underset{\sim}{t}, s)$.

Also for $\tau > t_n' > t_{n-1}' \ldots. > t_1'$,

$$Y(\tau, s) = \underset{\sim}{\phi}(\tau)\, U^*(s, \underset{\sim}{t}'), \quad \tau > \max\,(t_n' > t_n). \tag{1.11}$$

Now (1.10), (1.11) and (1.9) (ii) imply that $U(s, \underset{\sim}{t}) = U(s, \underset{\sim}{t}')$, i.e., $U(s) = U(s, \underset{\sim}{t})$ is independent of $\underset{\sim}{t}$ for $s < \tau$. Thus we get that for $s < t$,

$$\mathcal{E}_s X(t) = \sum_{i=1}^{n} \varphi_i(t) U_i(s). \tag{1.12}$$

We prove in Lemma 1.14 that $\mathcal{E}_s X(t) \to \mathcal{E}_{t-} X(t)$ as $s \uparrow t$ where $\mathcal{E}_t$ is the metric projection on $H(X : t -)$. Using this we define

$$(U_1(t), U_2(t),...,U_n(t))^* = \lim_{s \uparrow t} ((\varphi_i(t_j)))^{-1} (\mathcal{E}_s X(t_1),...,\mathcal{E}_s X(t_n))^*. \tag{1.13}$$

From (1.12) and (1.3) we get from $H(X : t) = H(X : t-)$

$$X(t) = \mathcal{E}_{t-} X(t) = \sum_{i=1}^{n} \varphi_i(t) U_i(t).$$

Clearly the representation is proper. We observe that the limit in (1.13) is independent of $(t_1,...,t_n)$ in view of (1.9).

Suppose, conversely, that $X(t)$ has proper Goursat representation. Then for $s \leq t_1 < t_2 < ... < t_n$,

$$((\varphi_i(t_j))(U_1(s),...,U_n(s))^* = (\mathcal{E}_s X(t_1), ..., \mathcal{E}_s X(t_n))^*$$

giving $\{\mathcal{E}_s X(t_1), ..., \mathcal{E}_s X(t_n)\}$ are linearly independent and for $s \leq t_1 < t_2 .... < t_{n+1}$ we have $(\mathcal{E}_s X(t_1), ..., \mathcal{E}_s X(t_{n+1})\}$ lies in the subspace generated by $(U_1(s),...,U_n(s))$ giving linear dependence. Thus $\{X(t), t \in \mathbb{R}\}$ is n–ple Markov.

Note that in the Gaussian case we get that $\{\underline{U}(s): s \leq t\}$ is a martingale.

To prove the following lemma we need some notation. We denote for x, y in $L_p(\mathcal{X}, \Sigma, \mu)$ $(p > 1)$, $[x, y] = \int x\, y^{<p-1>} d\mu$ and as in [3], for $S \alpha S$ random variables X, Y, $[X, Y]_\alpha$ is defined through their Kuelbs representation. We need some facts about projections on L_p. Let $M \subseteq L_p$ be a subspace and $x \in L_p$. Then the metric projection on M of x is denoted by $P_M x$ and is characterized by $[y, x - P_M x] = 0$ for all $y \in M$. From [3] and [15] we have the facts, $x \in L_p$ has unique representation $x = x_1 + x_2$ where $x_1 \in M$ and $x_2 \perp M$.

1.14 Lemma. *Let* H_s *be subspaces of* $H(X)$ *and* P_s *be the metric projection on* H_s. *Then* $\lim_{s \uparrow t} P_s$ *is the projection on* $\overline{sp}(\cup_{s<t} H_s) = H_{t-}$.

Proof. Let P_{t-} be projection on $\overline{sp}(\cup_{s<t} H_s) = H_{t-}$. Then $x \in H(X)$ can be written as $x_1 + x_2$ where $[y, x_2] = 0 \; \forall \; y \in H_{t-}$. By ([3])

$$P_{t-}(x_1 + x_2) = P_{t-}x_1 + P_{t-}x_2 = x_1.$$

By remark 3.22 pg. 56 of [15]. $P_{t-}x_2 = 0$. It therefore suffices to prove for $x \in H_{t-}$ that $P_s x \to x$. Let $x \in \cup_{s<t} H_s$ say, $x \in H_{s_0}$. Then $P_s x = x$ for $s > s_0$. Let $x \in H_{t-}$ and $\epsilon > 0$. Then $\exists\, s_0 < t$ and $x_0 \in H_{s_0}$ such that $\|x - x_0\| < \epsilon$. Now for $s \geq s_0$, $P_s x_0 = x_0$ and $P_s x = P_s x_0 + P_s(x - x_0)$ as $x_0 \in H_s$. Hence $\|P_s x - P_s x_0\| \leq \|x - x_0\| \leq \epsilon$ since P_s is idempotent and bounded. Hence $\lim_{s\uparrow t} P_s x$ exists and is x for $x \in H_{t-}$.

In case that $\mathcal{E}_s$ is linear, we get the following corollary.

1.15 Corollary. *If* $\mathcal{E}_s$ *is linear,* $(U_1,\ldots,U_n)$ *is a wide–sense* SαS *martingale, i.e,* X *has a proper non–singular Goursat representation in the sense of*([11]).

Proof. $\mathcal{E}_s X(t) = \Sigma_{i=1}^n \phi_i(t)\mathcal{E}_s U_i(t)$, $s < t$. Also $\mathcal{E}_s X(t) = \Sigma_{i=1}^n \phi_i(t) U_i(s)$. Using properties of ϕ_i, we get the result.

In case that $n = 1$, we get that $X(t) = \phi(t)U(t)$ and $U(t)$ is a wide–sense SαS martingale and we get the covariation of $\phi(t)K(s)^{<\alpha-1>}$. We also get that a process of the form

$$X(t) = H(t)L(\tau(t))$$

introduced in [1] is 1–ple Markov in our sense.

An interesting class of examples of processes with proper Goursat representation is provided by solutions of ordinary stochastic differential equations.

2. Multiple Markov Processes and Polya Condition.

Let $L_t = \Sigma_{k=0}^n a_k(t) \dfrac{d^{n-k}}{dt}$ be a differential operator. We consider the solution in M(X) of the linear stochastic differential equation

$$L_t Y dt = dX(t)$$

where X is a Sα S Lévy process, i.e.,

$$E(e^{i\gamma(X(t)-X(s))}) = e^{-|\gamma|^{\alpha}|t-s|}$$

and the initial condition $Y^{(k)}(0) = 0$, $k = 0, 1,\ldots, n-1$. Here $M(X) = \overline{sp}^{\alpha}\{X(b)-X(a): 0 < a < b < \infty\}$. The Green's function of L_t can be written in the form $\Sigma_{i=1}^n f_i(t) g_i(u)$ and hence repeating exactly the proofs in [5], we get

$$Y(t) = \int_0^t \sum_{i=1}^n f_i(t) g_i(u) dX(u). \tag{2.1}$$

Here $\{f_i\}_{n-1}^n$ is a fundamental system of solutions of $L_t f = 0$ and $\{g_i\}_{n-1}^n$ is a fundamental system of solutions of $L_u^* g = 0$ with L_u^* being the formal adjoint of L_u.

Here we assume that the Wronskian $W_n(f_1,\dots,f_n) \neq 0$. This gives that $\{f_1,\dots,f_n\}$ are linearly independent for each u. Note that the integral in (2.1) is in the sense of Schilder [13]. Observe that

$$Y(t) = \sum_{i=1}^{n} f_i(t)U_i(t) \tag{2.2}$$

where $U_i(t) = \int_0^t g_i(u)dX(u)$. Let us assume

(2.3) for each interval (a, b], $\exists\, i \in \{1,\dots,n\}$ such that $g_i \neq 0$ on (a, b].

Under (2.3), we get

$$\overline{sp}^{\alpha}\{U_j(s):\ s \leq t,\ j = 1, 2,\dots,n\} = \overline{sp}^{\alpha}\{X(b) - X(a);\ 0 \leq a \leq b \leq t\},$$

ie., $H(\underline{U}: t) = H(\Delta X : t)$. So we get by the continuity in probability of X, $H(U: t - 0) = H(\underline{U}: t)$.

Let us now consider (2.2) under (2.3). With $\mathcal{E}_s^U$ being the projection on $H(\underline{U} : s)$ we get

$$\mathcal{E}_s^U Y(t) = \mathcal{E}_s^U \left[\sum_{j=1}^{n} f_j(t)(U_j(t) - U_j(s))\right] + \sum_{j=1}^{n} f_j(t)(U_j(s).$$

We note that the first term is independent of $H(\underline{U}: t)$ and hence the first term is zero, i.e.,

$$\mathcal{E}_s^U Y(t) = \sum_{j=1}^{n} f_j(t)\, U_j(s).$$

Now since $f_j(t)$ is (n–1) times continuously differentiable we get that the RHS is (n–1) times differentiable in L_p and hence in $\|\cdot\|_\alpha$ norm. Thus we have

$$\mathcal{E}_s^U Y^{(k)}(t) = \sum_{j=1} f_j^{(k)}(t)U_j(s),\ k = 1, 2,\dots, n,$$

for each $s < t$. But

$$\mathcal{E}_s^U Y^{(k)}(t) = \sum_{j=1}^{n} f_j^{(k)}(t)\mathcal{E}_s^U U_j(t).$$

Let $s \uparrow t$ and use Lemma 1.14 to get

$$Y^{(k)}(t) = \mathcal{E}_t^U\, Y^{(k)}(t) = \sum_{j=1}^{n} f_j^{(k)}(t)U_j(t)$$

as $H(\underline{U}: t) = H(\underline{U} : t-)$. Since $W_n(f_1,\dots,f_n) \neq 0$, we have $H(\underline{U} : t) = H(Y : t)$. We thus have the following result.

2.4 Proposition. *Let* Y_t *be the solution of the equation* $L_t Ydt = dX(t)$, $Y^{(k)}(0) = 0$ *for* $k = 0, 1,2,\dots,n-1$ *with* X *a* SαS Lévy *process* $(\alpha > 1)$, $W_n(f_1,\dots,f_n) \neq 0$ *and* $\{g_i\}$ *satisfying* (2.3). *Then* X *has a proper representation,* $X(t) = \Sigma_{i=1}^{n} f_i(t)U_i(t)$ *where* $\{U_1(t),\dots,U_n(t)\}$ *is a martingale.*

2.5 Theorem. *Let* $\{Y(t)\}$ *be the solution of the stochastic differential equation satisfying conditions of Proposition* 2.4. *Then* $\{Y(t)\}$ *is n–ple Markov if and only if* $\det\,(f_i(t_j)) \neq 0$

Proof. We know that

$$Y(t) = \sum_{i=1}^{n} f_i(t)U_i(t),$$

$H(\underline{U}: t) = H(\underline{Y} : t)$ and $(U_1(t),...,U_n(t))$ are linearly independent for each t.

$$\mathcal{E}_s^Y Y(t_j) = \sum_{j=1}^{n} f_i(t_j)U_i(s)$$

are linearly independent if and only if $\det\,(f_i(t_j)) \neq 0$.

We say that the set $\{f_1,...,f_n\}$ of $(n-1)$ times continuously differentiable functions satisfies Polya property (property (W)) if $W_k(f_1,...,f_k) \neq 0\ \forall\ k = 1, 2,...,n$. Here $W_k(f_1,...,f_k) = \det\,([f_j^{(i)}]_{1\le j\le k}^{0\le i\le k-1})$. Now following arguments in ([11], Lemma 3.10 and Remark 3.11) we have the following result.

2.3 Theorem. *Let* $\{Y(t)\}$ *be the solution of the stochastic differential equation with* $Y^{(k)}(0) = 0$, $k = 0,...,n-1$, *and suppose the fundamental solutions of* $L_t f = 0$ *satisfy* $W_n(f_1,...,f_n) \neq 0$ *and the adjoint system satisfies* (2.3). *Then the following are equivalent.*

(*i*) Y *is n–ple Markov.*

(*ii*) L_t *is factorable in the sense of Ince* ([7], p. 120).

The result for $\alpha = 2$, settled a conjecture of Lévy ([10, 2, p. 159). We note that since condition (ii) is independent of the stochastic process, the above result seems more natural. We note that in case L_t satisfies (ii), then in view of Lemma 3.13, [11], we get the condition (2.3) is satisfied for $\alpha > 1$. Thus, we get a generalization of the earlier results of Hida and Lévy (Theorem II.7, [10 p. 159).

REFERENCES

1. Adler, R. J., Cambanis, S., and Samorodnitsky, G. (1990). On stable Markov processes. Stochastic Proc. Appl., 34, 1–17.

2. Cambanis, S. and Miller, G. (1981). Linear problems in pth order and stable processes, SIAM J. App. Math., **41**, 43–69.

3. Cambanis, S., Hardin, C.D. Jr., and Weron, A. (1988). Innovations and Wold decomposition of stable sequences, *Probability Theory and Related Fields*, 79, 1–27.

4. Doob, J. L. (1953). *Stochastic Processes*, John Wiley, New York.

5. Dolph, C. L., and Woodbury, M.A. (1952). On the relation between Green's functions and covariances of certain stochastic processes and its application to unbiased linear predication, *Trans. Amer. Math. Soc.*, **72**, 519–550.

6. Hida, T. (1960). Canonical representtions of Gaussian processes and their applications, *Mem. Coll. Sci. Kyoto*, **A33**, 1090–155.

7. Ince, E. L. (1926). *Ordinary Differential Equations,* Dover, New York.

8. Karlin, S. (1968). *Total Positivity,* Stanford University Press, Stanford.

9. Lévy, P. (1956). A special problem of Brownian motion and general theory of Gaussian random functions, *Proc. Third Berk. Symp. Stat. and Prob. II*, 133–175.

10. Lévy, P. (1957). Fonctions linéarement Markovieness d'ordre n, *Math Japonicae*, 4, 113–121.

11. Mandrekar, V. (1974). On the multiple Markov property of Lévy–Hida for Gaussian processes, *Nagoya Math.* J., **4**, 113–121.

12. Pitt, L. D. (1975). Hida–Cramér multiplicity theory for multiple Markov processes and Goursat representations, *Nagoya Math.* J., **57**, 199–228.

13. Schilder, M. (1970). Some structure theorems for symmetric stable laws, *Ann. Math. Stat.*, 41, 412–421.

14. Singer, I. (1970). *Best Approximation in Normed Linear Spaces by Elements of Linear Subspaces*, Springer–Verlag, New York.

15. Taraporevala, A. P. (1988). Series representation for processes with infinite energy and their prediction, Thesis, Michigan State University.

Department of Statistics and Probability
Michigan State University
E. Lansing, MI 48824

Department of Statistics
University of Michigan
Ann Arbor, MI 48109–1027

On shot noise processes attracted to fractional Lévy motion

L. GIRAITIS AND D. SURGAILIS

Convergence in distribution of an integrated shot noise process to α-stable fractional Lévy motion ($1 < \alpha < 2$) is discussed. We show also that the class of limiting processes contains some non-stable self-similar processes.

1. Introduction

We consider shot noise processes $X_t,\ t \in \mathbf{R}$ of the form:

$$X_t = \sum_j U_j G((t - \tau_j)V_j), \tag{1.1}$$

where $G(x),\ x \in \mathbf{R}$ is a real function, $G(x) = 0\ (x < 0),\ \{\tau_j\}_{j \in \mathbf{Z}}$ is Poisson point process on $\mathbf{R}$ with intensity $\lambda > 0$, and $(U_j, V_j)_{j \in \mathbf{Z}}$ is i.i.d. sequence of $\mathbf{R} \times \mathbf{R}_+$-valued vectors with $P(U_0 \in du,\ V_0 \in dv) := F(du, dv)$, independent of $\{\tau_j\}$. The X_t process can be interpreted as the sum of "G-shaped signals" triggered at the moments τ_j's in the past, with random amplitudes U_j's and time-scales V_j's. Shot noise processes were found useful in many areas of physical sciences (see e.g. [6] and the references therein). The X_t processes represent an important special case of the more general definition of shot noise [9].

Here we discuss convergence of finite dimensional distributions of the integrated process

$$S_T := \int_0^T X_t dt \tag{1.2}$$

to an α-stable fractional Lévy motion (FLM), $1 < \alpha < 2$. (The convergence to fractional Brownian motion is discussed in a parallel paper [5].) Our basic condition on X_t for such a convergence requires the conditional distribution

$$F(du|v) := P(U_0 \in du | V_0 \in dv) \tag{1.3}$$

to belong to the domain of attraction of the α-stable law, for each $v \in \mathbf{R}_+ = (0, \infty)$.

The characteristic function of FLM $Z_t,\ t \in \mathbf{R}_+$ is given by

$$E\exp\{i\sum_{k=1}^{n}\theta_k Z_{t_k}\} =$$

$$= \exp\{-\int_{\mathbf{R}} \Xi(\sum_{k=1}^{n}\theta_k c((t_k - s)_+^{H-1/\alpha} - (-s)_+^{H-1/\alpha}))ds\} \tag{1.4}$$

where $\theta_1,\ldots,\theta_n \in \mathbf{R},\ t_1,\ldots,t_n \in \mathbf{R}_+,\ (x)_+ := \max(x,0),\ c > 0$ is some multiplicative constant, $\Xi(x) := |x|^\alpha(1 - i\beta\ sgn(x)\tan(\pi\alpha/2)),\ x \in \mathbf{R}$, $H \in (0,1),\ H \neq 1/\alpha$ is the index of self-similarity and $\beta \in [-1,+1]$ is the skewness parameter; see e.g. [10]. Write $\stackrel{\Delta}{=}, \stackrel{\Delta}{\Rightarrow}$ for equality and weak convergence of finite dimensional distributions, respectively. We recall that a real stochastic process $Z_t,\ t \in \mathbf{R}_+$ with stationary increments is self-similar with index $H \in (0,1)$ (H-sssi) if for each $a \in \mathbf{R}_+$

$$Z_{at} \stackrel{\Delta}{=} a^H Z_t. \tag{1.5}$$

Let us formulate conditions about the existence of X_t (1.1) and S_T (1.2). Set $F_+(u|v) := F((u,\infty)|v),\ F_-(u|v) := F((-\infty,-u)|v),\ F(u|v) := F_+(u|v) + F_-(u|v),\ u,v \in \mathbf{R}_+$.

Theorem 1. *Let $G \in L^\alpha(\mathbf{R})$ $(1 < \alpha < 2)$, and let $q(v) > 0$ be a measurable function such that* *

$$\sup_{u,v>0} u^\alpha F(u|v) q^{-1}(v) \leq 1 \tag{1.6}$$

and

$$\int_{\mathbf{R}_+} q(v) v^{-1} F(dv) < \infty, \tag{1.7}$$

where $F(dv) = P(V_0 \in dv)$. Assume also

$$\int_{\mathbf{R}} uF(du|v) = 0 \quad \forall v \in \mathbf{R}_+. \tag{1.8}$$

Then:

(i) The series X_t (1.1) converges a.s. and in $L^1(\Omega)$ for each $t \in \mathbf{R}$, and defines a strictly stationary process, which is integrable a.s. on each finite interval.

---

* *Editor's Note:* If "≤ 1" is replaced by "$< \infty$", subsequent formulas may require a multiplicative constant.

(ii) The characteristic function of S_T (1.2) equals

$$E\exp\{i\sum_{k=1}^{n}\theta_k S_{t_k}\} = \exp\{\int_{\mathbf{R}\times\mathbf{R}\times\mathbf{R}_+}(e^{i\phi}-1-i\phi)d\mu\}, \tag{1.9}$$

where $\theta_1,\ldots,\theta_n \in \mathbf{R}$, $t_1,\ldots,t_n \in \mathbf{R}_+$, $\mu(ds,du,dv) := \lambda ds F(du,dv)$ *and*

$$\phi = \phi(s,u,v) := \sum_{k=1}^{n}\theta_k\int_0^{t_k} uG((\tau-s)v)d\tau. \tag{1.10}$$

Our main result is Theorem 2 below about convergence to FLM. In Theorem 2, we assume that (1.6) - (1.8) are satisfied and that $G(x)$ is bounded. Moreover, we put $\lambda = 1$ for simplicity.

Let $\chi_\pm \geq 0$ be the parameters in the Lévy-Khintchine representation of the α-stable law with parameters $(\alpha, 1, \beta, 0)$, i.e. $\chi_+ + \chi_- =$

$$= -\alpha(\alpha-1)/((\cos(\pi\alpha/2)\Gamma(2-\alpha)),\ (\chi_+ - \chi_-)/(\chi_+ + \chi_-) = \beta.$$

Write $a_\gamma \sim b_\gamma$ if $\lim a_\gamma/b_\gamma = 1$ (a_γ, b_γ are real numbers).

Theorem 2. *Assume that for some* $H \in (0,1)$, $H \neq 1/\alpha$ $(1 < \alpha < 2)$

$$(H-1/\alpha)^{-1}G(x) \sim x^{H-1-1/\alpha} \quad (x \to \infty) \tag{1.11}$$

and for any $v \in \mathbf{R}_+$ *

$$F_\pm(u|v) \sim \chi_\pm q(v)u^{-\alpha} \quad (u \to \infty), \tag{1.12}$$

where $\chi_\pm, q(v)$ *are the same as above and, moreover,*

$$\int_{\mathbf{R}_+} q(v)v^{\alpha H-1-\alpha}F(dv) < \infty. \tag{1.13}$$

In the case $H \in (0, 1/\alpha)$ *we assume also*

$$\int_{\mathbf{R}} G(x)dx = 0. \tag{1.14}$$

Then

$$T^{-H}S_{Tt} \stackrel{\Delta}{=} Z_t, \tag{1.15}$$

* *Editor's Note:* Since $\chi_+ + \chi_- < 1$, (1.12) is not inconsistent with (1.6).

where Z_t is the FLM with the characteristic function (1.4).

Theorem 2 is related to the results of Astrauskas [1], Maejima [8], Avram and Taqqu [2], who discussed convergence of adequately normalized partial sums of moving averages to FLM. (See also Davis and Resnick [3].) In particular, the shot noise process X_t (1.1) corresponding to $V_j = const$ is a direct continuous time analogue of discrete time moving averages. It is easy to understand, however, that random scaling with small V_j's in X_t (1.1) may influence dramatically the limiting behavior of S_T. This does not happen in Theorem 2 because of the condition (1.13), which restricts concentration of V_j's near $v = 0$. As we shall see in Section 4, if (1.13) is violated, then $T^{-H} S_{Tt}$ may converge in distribution to a H-sssi process which is different from FLM and whose one-dimensional distributions are not even stable.

Let us remark, finally, that general criteria for convergence in distribution of shot noise processes were obtained by Lane [7]. It seems, however, that shot noise processes with long range dependence were not studied previously.

2. Proof of Theorem 1

We shall prove the statement (i) only, as (ii) seems to be well-known ([9], [7]). It suffices to discuss the convergence of the series (1.1) for $t = 0$.

Integrating by parts, we have for any $a, v \in \mathbf{R}_+$ that

$$\int_{\{|u|>a^{-1}\}} |u| F(du|v) \le \frac{\alpha}{\alpha-1} q(v) a^{\alpha-1}, \tag{2.1}$$

$$\int_{\{|u|\le a^{-1}\}} u^2 F(du|v) \le \frac{2}{2-\alpha} q(v) a^{\alpha-2} \tag{2.2}$$

Set $Z_j := U_j G(\tau_j V_j)$, $Z_j^+ := Z_j \mathbf{1}(|Z_j| > 1)$,

$$Z_j^- := Z_j - Z_j^+,\ U_j^+ := U_j \mathbf{1}(|Z_j| > 1),\ U_j^- := U_j - U_j^+,$$

where $\mathbf{1}(\cdot)$ is the indicator function. Write

$$\sum_j Z_j = \sum_j Z_j^+ + \sum_j Z_{1,j}^- + \sum_j Z_{2,j}^-, \tag{2.3}$$

where

$$\begin{aligned} Z_{1,j}^- &:= (U_j^- - E[U_j^- | V_j]) G(\tau_j V_j), \\ Z_{2,j}^- &:= E[U_j^- | V_j] G(\tau_j V_j) \\ &= -E[U_j^+ | V_j] G(\tau_j V_j), \end{aligned}$$

see (1.8).

Let us show that

$$E\sum_j (|Z_j^+| + |Z_{2,j}^-|) < \infty,$$

which implies both $L^1(\Omega)$ and a.s. convergence of the series ΣZ_j^+ and $\Sigma Z_{2,j}^-$. In fact,

$$E[|Z_{2,j}^-|\ |\tau_j] \le E[|Z_j^+|\ |\tau_j]$$

$$= \int_{\mathbf{R}_+\{|u|>|G(\tau_j v)|^{-1}\}} |u|F(du|v)|G(\tau_j v)|F(dv)$$

$$\le \frac{\alpha}{\alpha-1}\int_{\mathbf{R}_+} q(v)|G(\tau_j v)|^\alpha F(dv) \tag{2.4}$$

according to (2.1). Hence

$$E\sum_j |Z_{2,j}^-| \le E\sum_j |Z_j^+| \le$$

$$\le \frac{3\alpha\lambda}{\alpha-1}\int_{\mathbf{R}_+} q(v)v^{-1}F(dv)\int_{\mathbf{R}} |G(x)|^\alpha dx < \infty. \tag{2.5}$$

Consider $\Sigma Z_{1,j}^-$. Since $Z_{1,j}^-$, $j \in \mathbf{Z}$ are orthogonal and independent with respect to the conditional probability $P[\cdot|\tau_j,\ j \in \mathbf{Z}]$, so

$$E(\sum_j Z_{1,j}^-)^2 = E\sum_j E[(Z_{1,j}^-)^2|\tau_j,\ j \in \mathbf{Z}]$$

$$\le E\sum_j \int_{\mathbf{R}_+}[\int_{\{|u|\le|G(\tau_j v)|^{-1}\}} u^2F(du|v)]|G(\tau_j v)|^2F(dv)$$

$$\le \frac{6}{2-\alpha}\int_{\mathbf{R}_+} q(v)v^{-1}F(dv)\int_{\mathbf{R}} |G(x)|^\alpha dx < \infty \tag{2.6}$$

according to (2.2). Hence $\Sigma_j Z_{1,j}^-$ converges in $L^2(\Omega)$ and in $L^1(\Omega)$, and thus converges a.s..

We have proved that (1.1) converges both in $L^1(\Omega)$ and a.s. Stationarity of X_t (1.1) follows from stationarity of the point process $\{\tau_j\}_{j\in\mathbf{Z}}$ and the conditions on $(U_j, V_j)_{j\in\mathbf{Z}}$ mentioned in the beginning of the paper. Finally, the last statement of (i) about a.s. integrability of X_t follows from $E|X_t| < const$ and the Fubini theorem (see e.g. [4, Th. 2.7]).

Q.E.D.

3. Proof of Theorem 2

According to (1.4) and (1.9), it suffices to show that for any $n \geq 1$, $t_1, \dots, t_n \in \mathbf{R}_+$, $\theta_1, \dots, \theta_n \in \mathbf{R}$

$$\int (e^{i\phi_T} - 1 - i\phi_T) d\mu \to \int (e^{i\phi} - 1 - i\phi) d\mu_0, \tag{3.1}$$

where the integrals are taken over $\mathbf{R} \times \mathbf{R} \times \mathbf{R}_+$, μ is the same as in (1.9),

$$\mu_0(ds, du, dv) := \lambda \chi(u) \alpha |u|^{-\alpha-1} q(v) ds du F(dv), \tag{3.2}$$

$$\chi(u) := \chi_+ \mathbf{1}(u > 0) + \chi_- \mathbf{1}(u < 0),$$

$$\phi_T := T^{-H} u \sum_{k=1}^{n} \theta_k \int_0^{Tt_k} G((\tau - s)v) d\tau \tag{3.3}$$

and

$$\phi := u v^{H-1-1/\alpha} \sum_{k=1}^{n} \theta_k [(t_k - s)_+^{H-1/\alpha} - (-s)_+^{H-1/\alpha}] \tag{3.4}$$

We shall prove (3.1) for $n = t_1 = \theta_1 = 1$ only, as the general case can be considered analogously. By changing the variables $s \to Ts$, $u \to T^{1/\alpha} u$, $\tau \to T\tau$ we have

$$\int (e^{i\phi_T} - 1 - i\phi_T) d\mu = \int (e^{iu\psi_T} - 1 - iu\psi_T) d\mu_t,$$

where

$$\psi_T := T^{1+1/\alpha-H} \int_0^1 G(Tv(\tau - s)) d\tau \tag{3.5}$$

and

$$\mu_T(ds, du, dv) := ds F(dv) F_T(du|v), \tag{3.6}$$

where $F_T(du|v) := TF(dT^{1/\alpha}u|v)$. Observe that $dsF(dv)$ - a.s.

$$\psi_T(s, v) \to \psi(s, v) := v^{H-1-1/\alpha}[(1 - s)_+^{H-1/\alpha} - (-s)_+^{H-1/\alpha}] \tag{3.7}$$

and

$$I_T := \int_{\mathbf{R}} (e^{iu\psi_T} - 1 - iu\psi_T) F_T(du|v) \to I, \tag{3.8}$$

where

$$I := q(v) \int_{\mathbf{R}} (e^{iu\psi} - 1 - iu\psi) \chi(u) \alpha |u|^{-1-\alpha} du. \tag{3.9}$$

Indeed, (3.7) follows from (1.11) and boundedness of G, as $G(Tx)T^{1+1/\alpha-H} \leq const\ x_+^{H-1-1/\alpha}$ for all $T \in \mathbf{R}_+$, $x \in \mathbf{R}$. The last estimate implies also

$$|\psi_T| \leq const \int_0^1 (v(\tau - s)_+)^{H-1-1/\alpha} d\tau \leq const\ |\psi| \tag{3.10}$$

provided either $H \in (1/\alpha, 1)$, or $H \in (0, 1/\alpha)$ and $s < 0$ hold. In view of (1.14), this is true also in the case $H \in (0, 1/\alpha)$, $s \in (0,1)$, as

$$|\psi_T| = T^{1+1/\alpha-H} \Big| \int_{1-s}^{\infty} G(Tv\tau) d\tau \Big|$$

$$\leq const \int_{1-s}^{\infty} (\tau v)^{H-1-1/\alpha} d\tau \leq const\ |\psi|$$

To prove (3.8), we use (3.10) and the Lebesgue dominated convergence theorem, as follows. Integration by parts yields

$$I_T = i\psi_T \int_{\mathbf{R}} F_T(u|v)(e^{iu\psi_T} - 1) du,$$

where $F_T(u|v) := F_T((u,\infty)|v)\mathbf{1}(u > 0) - F_T((-\infty, u)|v)\mathbf{1}(u < 0)$*, and

$$F_T(u|v) \to (\chi_+ \mathbf{1}(u > 0) - \chi_- \mathbf{1}(u < 0))|u|^{-\alpha} q(v) \tag{3.11}$$

for any $u, v \in \mathbf{R} \times \mathbf{R}_+$, $u \neq 0$, according to (1.12). Moreover, by (3.10),

$$|F_T(u|v)(e^{iu\psi_T} - 1)| \leq$$

$$\leq const\ q(v)|u|^{-\alpha}(|u||\psi|\mathbf{1}(|u\psi| < 1) + \mathbf{1}(|u\psi| > 1)),$$

where the right hand side is du-integrable and does not depend on T. By (3.7) and (3.11), this proves (3.8). Moreover, the argument above yields the estimate

$$|I_T| \leq const\ q(v)|\psi|^\alpha, \tag{3.12}$$

where

$$\int_{\mathbf{R} \times \mathbf{R}_+} q(v)|\psi(s,v)|^\alpha ds F(dv) < \infty \tag{3.13}$$

according to (1.13). The relation (3.1) now follows from (3.8),(3.9),(3.12) and (3.13).

---

* *Editor's Note:* Do not confuse this $F_T(u|v)$ with $F(u|v)$ defined before Theorem 1.

Q.E.D.

4. Convergence to non-stable processes

Below we assume $F(du, dv)$ to be absolutely continuous with density

$$f(u,v) = |u|^{-1} f_1(v|u|^{-1/(1-H)}) f_2(v), \tag{4.1}$$

where $H \in (1/\alpha, 1)$ $(1 < \alpha < 2)$ and $f_i \geq 0$, $i = 1,2$ satisfy the following conditions:

$$f_1(v) \sim v^{\alpha(1-H)} \quad (v \downarrow 0), \tag{4.2}$$

$$f_2(v) \to 1 \quad (v \downarrow 0), \tag{4.3}$$

$$\int_{\mathbf{R}_+} f_1(v) v^{-1} dv < \infty, \tag{4.4}$$

$$\int_{\mathbf{R}_+} f_2(v)(1 + v^{\alpha - \alpha H - 1}) dv < \infty. \tag{4.5}$$

Moreover, we assume that f_i, $i = 1,2$ are bounded, and $2(1-H)\int_{\mathbf{R}_+} f_1(v)v^{-1}dv = \int_{\mathbf{R}_+} f_2(v)dv = 1$.

Let us check that $F(du, dv)$ satisfies all conditions of Theorems 1 and 2 with the exception of (1.13). One has easily:

$$F(dv) = P(V_0 \in dv) = f_2(v)dv \tag{4.6}$$

and

$$F(u|v) = 2(1-H)\int_0^{vu^{-1/(1-H)}} f_1(x)x^{-1}dx, \tag{4.7}$$

$u, v \in \mathbf{R}_+$. Hence (1.12) follows from (4.2) and (4.7) with $\chi_+ = \chi_-$ and

$$q(v) = (\alpha\chi_+)^{-1} v^{\alpha(1-H)}. \tag{4.8}$$

Moreover, (4.2) and the boundedness of f_1 imply $f_1(v) \leq const\ v^{\alpha(1-H)}$, $v \in \mathbf{R}_+$, which in turn implies (1.6) by (4.7). Condition (1.8) holds in view of the symmetry: $f(u,v) = f(-u,v)$. Let us note, finally, that according to (4.3) and (4.5), the integral

$$\int_{\mathbf{R}_+} q(v)v^{\gamma-1}F(dv) = const \int_{\mathbf{R}_+} v^{\alpha(1-H)+\gamma-1} f_2(v)dv$$

converges for any $\alpha(H-1) < \gamma \geq 0$ and diverges for $\gamma = \alpha(H-1)$; i.e. $F(du, dv)$ does not satisfy (1.13) with H replaced by any $H' < H$.

Theorem 3. *Let $f(u,v)$, defined by (4.1), satisfy the conditions above and, moreover,*

$$|G(x)| \le const\ (1 \wedge x^{H'-1-1/\alpha}) \tag{4.9}$$

for some $1/\alpha < H' < H (x \in \mathbf{R}_+)$; $G(x) = 0\ (x \notin \mathbf{R}_+)$. Then

$$T^{-H} S_{Tt} \overset{\Delta}{\Rightarrow} Z_t, \tag{4.10}$$

where $Z_t,\ t \in \mathbf{R}_+$ is H-sssi process with the characteristic function:

$$E \exp\{i \sum_{k=1}^{n} \theta_k Z_{t_k}\} = \exp\{\int (e^{i\phi} - 1 - i\phi) d\mu_0\}, \tag{4.11}$$

where $t_1, \ldots, t_n,\ \theta_1, \ldots, \theta_n,\ \phi$ are the same as in (1.10) and

$$\mu_0(ds, du, dv) := \lambda |u|^{-1} f_1(v|u|^{-1/(1-H)}) ds du dv. \tag{4.12}$$

One-dimensional distributions of Z_t are not stable.

Proof. The convergence (4.10) will follow from

$$\int (e^{i\phi_T} - 1 - i\phi_T) d\mu \to \int (e^{i\phi} - 1 - i\phi) d\mu_0, \tag{4.13}$$

where ϕ_T is given by (3.3). Similarly as in Sect. 2, we shall restrict the proof to the case $t_1 = \theta_1 = n_1 = 1$.

By change of the variables: $s \to Ts,\ u \to T^{H-1}u,\ v \to T^{-1}v,\ \tau \to T\tau$, we obtain

$$\int (e^{i\phi_T} - 1 - i\phi_T) d\mu = \int (e^{iu\psi} - 1 - iu\psi) \rho_T d\mu_0, \tag{4.14}$$

where

$$\psi = \psi(s,v) := \int_0^1 G((\tau - s)v) d\tau, \tag{4.15}$$

and

$$\rho_T = \rho_T(v) := f_2(v/T) \to 1 \quad (T \to \infty) \tag{4.16}$$

according to (4.3). Write μ_0 (4.11) as

$$d\mu_0 = \lambda ds dv F(du|v), \tag{4.17}$$

where $F(du|v)$ is symmetric and $F(u|v) := F((u,\infty)|v) + F((-\infty,-u)|v)$ coincides with (4.7). Similarly as in the proof of Theorem 2,

$$|\int_{\mathbf{R}} (e^{iu\psi} - 1 - iu\psi) F(du|v)| \le$$

$$\leq const\ q(v)|\psi|^\alpha \leq const\ v^{\alpha(1-H)}|\psi|^\alpha. \tag{4.18}$$

Taking into account (4.14) - (4.18) and the boundedness of ρ_T (4.13) will now follow from

$$J := \int_{\mathbf{R}\times\mathbf{R}_+} |\psi(s,v)|^\alpha v^{\alpha(1-H)} ds dv < \infty. \tag{4.19}$$

Write

$$J = \int_{\mathbf{R}_+} dv\{v^{-1-\alpha H}\int_{\mathbf{R}} |\int_0^v G(\tau+s)d\tau|^\alpha ds\}$$
$$= \int_0^1 dv\{\ldots\} + \int_1^\infty dv\{\ldots\} \equiv J_1 + J_2$$

By Hölder's inequality

$$J_1 \leq \int_0^1 v^{\alpha-\alpha H-2}dv \int_{\mathbf{R}} ds \int_0^v |G(\tau+s)|^\alpha d\tau$$

$$\leq \int_{\mathbf{R}} |G(s)|^\alpha ds \int_0^1 v^{\alpha-\alpha H-1}dv < \infty$$

On the other hand, making use of (4.9) with $H' \in (1/\alpha, H)$, we have

$$J_2 \leq const \int_1^\infty v^{-1-\alpha H}dv \int_{\mathbf{R}} ds|\int_0^v (\tau+s)_+^{H'-1-1/\alpha}d\tau|^\alpha$$

$$\leq const \int_{\mathbf{R}} ds|\int_0^1 (\tau+s)_+^{H'-1-1/\alpha}d\tau|^\alpha \int_1^\infty v^{-1+\alpha(H'-H)}dv < \infty.$$

This proves (4.10).

Self-similarity of Z_t follows immediately from its characteristic function (4.11). It remains to show that one-dimensional distributions of Z_t are not stable. By self-similarity, it suffices to consider $t = 1$. By (4.11),

$$E\exp\{i\theta Z_1\} = \exp\{\int_{\mathbf{R}} (e^{i\theta x} - 1 - i\theta x)\Lambda(dx)\}, \tag{4.20}$$

where

$$\Lambda(A) := \int \mathbf{1}(u\psi(s,v) \in A)d\mu_0 \tag{4.21}$$

is the corresponding Lévy measure on the real line. Set $\Lambda(x) := \Lambda(u \in \mathbf{R}: |u| > x)$, $x \in \mathbf{R}_+$. By the symmetry of $\Lambda(dx)$ and the uniqueness of the Lévy-Khintchine representation (4.20), Z_1 is stable iff

$$\Lambda(x) = \sigma x^{-\alpha} \tag{4.22}$$

for some $\sigma \in \mathbf{R}_+$ and some $\alpha \in (0, 2)$. Clearly, (4.22) contradicts the two following relations:

$$\Lambda(x) \sim \sigma x^{-\alpha} \quad (x \to \infty) \tag{4.23}$$

and

$$\Lambda(x) \le \sigma' x^{-\alpha'} \tag{4.24}$$

for some $\sigma' \in \mathbf{R}_+$, $\alpha' < \alpha$, both of which will be verified below.

Starting with (4.23), according to the definition (4.21) of $\Lambda(dx)$ and elementary transformations, we obtain

$$x^\alpha \Lambda(x) = 2\lambda(1 - H) \int_{\mathbf{R}\times\mathbf{R}_+} dsdv \int_0^{v|\psi|^{1/(1-H)}} u^{\alpha-\alpha H-1} r_x(u) du, \tag{4.25}$$

where $r_x(u) := u^{\alpha(H-1)} x^\alpha f_1(ux^{-1/(1-H)}) \to 1$ as $x \to \infty$, see (4.2), and $r_x(u) \le const$ for all $u, x \in \mathbf{R}_+$. Therefore one can pass to the limit as $x \to \infty$ under the sign of the second integral on the right hand side of (4.25), which yields (4.23) with

$$\sigma = 2\lambda\alpha^{-1} \int_{\mathbf{R}\times\mathbf{R}_+} |\psi(s, v)|^\alpha v^{\alpha(1-H)} dsdv < \infty;$$

see (4.19).

Let us prove (4.24). With the help of the relations (4.2), (4.4) together with $f_1(v) \le const\ v^{\alpha(1-H)}$ and (4.17), (4.7), one has the estimate:

$$\Lambda(x) \le const\ (\Lambda_1(x) + \Lambda_2(x)),$$

where

$$\Lambda_1(x) := x^{-\alpha} \int_{\mathbf{R}\times\mathbf{R}_+} |\psi|^\alpha v^{\alpha(1-H)} \mathbf{1}(v^{1-H}|\psi| < x) dsdv$$

and

$$\Lambda_2(x) := \int_{\mathbf{R}\times\mathbf{R}_+} \mathbf{1}(v^{1-H}|\psi| > x) dsdv. \tag{4.26}$$

Clearly

$$\Lambda_1(x) \le x^{-\alpha'} \int_{\mathbf{R}\times\mathbf{R}_+} |\psi|^{\alpha'} v^{\alpha'(1-H)} dsdv \tag{4.27}$$

for any $0 < \alpha' < \alpha$. Moreover, one can choose $\alpha - \alpha' > 0$ sufficiently small so that the integral on the right hand side of (4.27) converges; see the proof of (4.19).

Consider $\Lambda_2(x)$. According to (4.9), one can replace $\psi(s,v)$ in (4.26) by

$$\psi'(s;v) := c\int_0^1((\tau-s)_+v)^{H'-1-1/\alpha}d\tau \geq \psi(s,v),$$

where $c \in \mathbf{R}_+$, $H' \in (1/\alpha, H)$. Therefore

$$\Lambda_2(x) \leq (c/x)^{\alpha''}\int_{\mathbf{R}} ds\left(\int_0^1 (\tau-s)_+^{H'-1-1/\alpha}d\tau\right)^{\alpha''},$$

where $\alpha'' = (H - H' + 1/\alpha)^{-1} < \alpha$. Since the last integral converges, this concludes the proof of (4.24).

Q.E.D.

Remark. The limiting H-sssi process Z_t can be represented as stochastic integral with respect to Poisson random measure on $\mathbf{R}\times\mathbf{R}\times\mathbf{R}_+$. A related class of self-similar processes was discussed by Surgailis [11] (the case of finite variance) and by Taqqu and Wolpert [12] (the case of infinite variance).

REFERENCES

[1] A. Astrauskas, Limit theorems for sums of linearly generated random variables, *Lithuanian Math. Journal* **23** (1983), 127-134.

[2] F. Avram, M.S. Taqqu, Weak convergence of moving averages with infinite variance, in *Dependence in Probability and Statistics*, pp. 319-415, Birkhäuser, 1986.

[3] R. Davis, S. Resnick, Limit theory for moving averages with regularly varying tail probabilities, *Ann. Prob.* **13** (1985), 179-195.

[4] J.L. Doob, *Stochastic Processes*, Wiley, 1953.

[5] L. Giraitis, D. Surgailis, On Shot Noise Processes with Long Range Dependence, Preprint.

[6] T. Hsing, J.T. Teugels, Extremal properties of shot noise processes, *Adv. Appl. Prob.* **21** (1989), 513-525.

[7] J.A. Lane, The central limit theorem for the Poisson shot-noise process, *J. Appl. Prob.* **21** (1984), 287-301.

[8] M. Maejima, On a class of self-similar processes, *Z. Wahrscheinlichkeitstheorie verw. Geb.* **62** (1983), 235-245.

[9] J. Rice, On generalized shot-noise, *Adv. Appl. Prob.* **9** (1977), 553-565.

[10] G. Samorodnitsky, M.S. Taqqu, The Various Linear Fractional Lévy Motions, in *Probability, Statistics and Mathematics: Papers in Honor of Samuel Karlin*, pp. 261-270, Academic Press, 1989.

[11] D. Surgailis, On infinitely divisible self-similar random fields, *Z. Wahrscheinlichkeitstheorie verw. Geb.* **58** (1981), 453-477.

[12] M.S. Taqqu, R. Wolpert, Infinite variance self-similar processes subordinate to a Poisson measure, *Z. Wahrscheinlichkeitstheorie verw. Geb.* **62** (1983), 53-72.

L. Giraitis and D. Surgailis
Institute of Mathematics and Cybernetics
232600 Vilnius
Lithuania, USSR

SELF-SIMILAR STABLE PROCESSES WITH STATIONARY INCREMENTS

NORIO KÔNO

AND

MAKOTO MAEJIMA

1. INTRODUCTION

Let T be $(-\infty,\infty)$, $[0,\infty)$ or $[0,1]$. A real- or complex-valued stochastic process $X=(X(t))_{t\in T}$ is said to be H-self-similar (H-ss) if all finite-dimensional distributions of $(X(ct))$ and $(c^H X(t))$ are the same for every $c>0$ and to have stationary increments (si) if the finite-dimensional distributions of $(X(t+b)-X(b))$ do not depend on $b\in T$. A real-valued process $X=(X(t))_{t\in T}$ is said to be symmetric α-stable (SαS), $0<\alpha\le 2$, if any linear combination $\sum_{k=1}^n a_k X(t_k)$ is SαS.

A complex-valued random variable $X=X'+iX''$ is SαS (or rotationally invariant α-stable), if X' and X'' are jointly SαS and have a rotationally invariant (radially symmetric) distribution, which is equivalent to requiring, for any $z\in\mathbf{C}$,

$$E\exp\{i\mathrm{Re}(\bar{z}X)\}=\exp\{-\sigma|z|^\alpha\}\qquad \text{for some } \sigma>0.$$

We call this X also SαS, or SαS(σ) if we want to emphasize the scaling parameter σ. A complex-valued stochastic process $(X(t))_{t\in T}$ is SαS if all linear combinations $\sum_{j=1}^n \bar{z}_j X(t_j)$, $z_j\in\mathbf{C}$, are complex-valued SαS random variables.

In section 2, we survey all H-ss si SαS processes we know, where $H>0$ and $0<\alpha\le 2$. To avoid trivialities, we assume throughout this paper that all ss si processes are non-degenerate in the sense that $P\{X(1)\neq 0\}>0$. In section 3, we discuss sample path properties of various processes, after their introduction in section 2. Section 4 deals with the LePage representation of SαS processes. It provides a characterization of ss si SαS processes and is applied in section 5 to show the Hölder continuity of sample paths of some ss si SαS processes. In section 6, more sample path properties (local time and nowhere differentiability) are discussed.

2. Examples

(1) $H > \max(1, 1/\alpha)$. Such H-ss si SαS processes do not exist. This is a consequence of the following two properties of ss si non-degenerate processes: (i) If $0 < \gamma < 1$ and $E[|X(1)|^\gamma] < \infty$, then $H < 1/\gamma$ ([Mae2]). (ii) If $E[|X(t)|] < \infty$, then $H \leq 1$. If we apply these to α-stable processes X, for which $E[|X(1)|^\gamma] < \infty$ for all $\gamma < \alpha$, we see that $H \leq \max(1, 1/\alpha)$.

(2) $H = 1/\alpha$.

(2-1) $\alpha = 2$. Brownian motion: $(B(t))_{t \geq 0}$.

(2-2) $0 < \alpha < 2$. α-stable Lévy motion: $(Z_\alpha(t))_{t \geq 0}$. Here, by α-stable Lévy motion, we mean the $1/\alpha$-ss si α-stable process with independent increments.

(2-3) $1 < \alpha < 2$. Log-fractional stable process ([KMV]):

$$X_1(t) = \int_{-\infty}^{\infty} \log \left| \frac{t-s}{s} \right| dZ_\alpha(s).$$

(2-4) $1 < \alpha < 2$. Sub-Gaussian process ([KMV]): $X_2(t) = Z^{1/2} B_{1/\alpha}(t)$, where Z is a positive strictly $\alpha/2$-stable random variable and B_H is an H-ss si 2-stable (Gaussian) process independent of Z. In fact, B_H is the fractional Brownian motion defined below.

(2-5) $1 < \alpha < 2$. (Complex-valued) harmonizable fractional stable process ([CM]):

$$X_3(t) = \int_{-\infty}^{\infty} \frac{e^{it\lambda} - 1}{i\lambda} |\lambda|^{1-2/\alpha} d\tilde{M}_\alpha(\lambda),$$

where $\tilde{M}_\alpha$ is a complex SαS motion. (For a definition, see section 4.)

(2-6) $\alpha = 1$. Linear function with random slope: $X_4(t) = tX(1)$.

(3) $0 < H < 1,\ 0 < \alpha \leq 2,\ H \neq 1/\alpha$.

(3-1) $\alpha = 2$. Fractional Brownian motion ([MV]):

$$B_H(t) = C \left[\int_{-\infty}^{0} \left\{ (t-s)^{H-1/2} - (-s)^{H-1/2} \right\} dB(s) + \int_0^t (t-s)^{H-1/2} dB(s) \right],$$

where B is a standard Brownian motion and C is a normalizing constant assuring $E|B_H(t)|^2 = 1$.

(3-2) $0 < \alpha < 2$. Linear fractional stable process ([TW], [Mae1], [KM], [CM], [ST1]):

$$X_5(t) = \int_{-\infty}^{\infty} \Big[a\Big\{(t-s)_+^{H-1/\alpha} - (-s)_+^{H-1/\alpha}\Big\} + b\Big\{(t-s)_-^{H-1/\alpha} - (-s)_-^{H-1/\alpha}\Big\}\Big]\, dZ_\alpha(s),$$

where $a, b \in \mathbf{R}$ with $ab \neq 0$, $x_+ = \max(x, 0)$, $x_- = \max(-x, 0)$ and Z_α is an α-stable motion.

(3-3) $0 < \alpha < 2$. (Complex-valued) harmonizable fractional stable process ([CM]):

$$X_6(t) = \int_{-\infty}^{\infty} \frac{e^{it\lambda} - 1}{i\lambda}\left(a\lambda_+^{1-H-1/\alpha} + b\lambda_-^{1-H-1/\alpha}\right) d\tilde{M}_\alpha(\lambda),$$

where $a, b \in \mathbf{R}$ and $\tilde{M}_\alpha$ is the same as in (2-5).

(4) $0 < H < 1/\alpha$, $0 < \alpha < 2$.

(4-1) Substable processes ([H], [KMV]):

$$X_7(t) = Z^{1/\beta} Y(t),$$

where $(Y(t))$ is an H-ss si symmetric β-stable process $(0 < \beta \leq 2)$ and Z is a positive strictly α/β-stable random variable (so $0 < \alpha < \beta$), independent of $(Y(t))$. As special cases we have

$$X_8(t) = Z^{1/2} B_H(t) \qquad \text{and} \qquad X_9(t) = Z^{1/\beta} Z_\beta(t),$$

where B_H is the fractional Brownian motion and Z_β is the β-stable motion.

(4-2) Recently, Takenaka [Take] has constructed a class of H-ss si α-stable processes, X_{10}, say, for any $0 < H < 1/\alpha$, $0 < \alpha < 2$, by means of integral geometry. His process can also be constructed by using the stable integral in the following way. Let $\mathbf{R}_+^2 = \{(x,y) | x > 0, y \in \mathbf{R}\}$, $\mathcal{F} = \{$ bounded Borel sets in $\mathbf{R}_+^2\}$ and $\{M(A) | A \in \mathcal{F}\}$ be an SαS random measures such that (i) $E[\exp\{i\theta M(A)\}] = \exp\{-m(A)|\theta|^\alpha\}$, where $m(dxdy) = x^{H\alpha-2}dxdy$, (ii) $M(A_j), j = 1, \cdots, n$, are mutually independent if $A_j \cap A_k = \phi, j \neq k$, and (iii) $M(\cup_j A_j) = \sum_j M(A_j)$ a.s. for any disjoint family $\{A_j, j = 1, 2, \cdots\}$. Let $c(t; x, y)$ be the indicator function of $\{(x,y) | \, |y| \leq x\} \,\triangle\, \{(x,y) | \, |y - t| \leq x\}$, where $\triangle$ means symmetric difference of two sets. Then

$$X_{10}(t) = \int_{\mathbf{R}_+^2} c(t; x, y)\, dM(x, y) \tag{2.1}$$

is an H-ss si SαS process. This class seems new, in any case, different from the other examples above. This fact has recently been proved by Sato [S], who determined the supports of their Lévy measures.

3. Sample path properties

If an H-ss si SαS process $(X(t))_{t\in T}$ satisfies $1/\alpha < H \leq 1$, then it has a sample continuous version, which can be shown by Kolmogorov's moment criterion.

The sample path properties examined in the literatures can be listed as follows:

Property I : There exists a sample continuous version.

Property II : Property I does not hold, but there is a version whose sample paths are right-continuous and have left limits.

Property III : Any version of the process is nowhere bounded, i.e., unbounded on every finite interval.

The examples in the previous section are classified as follows:

Property I : $B, X_2, X_3, X_4, B_H, X_5$ for $1/\alpha < H < 1$, X_6, X_8.

Property II : Z_α, X_9, X_{10}.

Property III : X_1, X_5 for $0 < H < 1/\alpha$.

Proofs are needed to justify the classifications of X_1, X_3, X_5 for $0 < H < 1/\alpha, X_6$ and X_{10}. They can be based on Theorem 3.1 below, due to Nolan [N].

Let an SαS process $X = (X(t))_{t\in T}$ be given by

$$X(t) = \int_U f(t,u) W_m(du),$$

where T is a finite interval, $(U, \mathcal{U}, m)$ is some σ-finite measure space, $f : T \times U \to \mathbf{R}$ is a function with the property that for each $t \in T$, $f(t,\cdot) \in L^\alpha(U, \mathcal{U}, m)$, and W_m is an SαS random measure with control measure m such that $E[\exp\{i\theta W_m(A)\}] \doteq \exp\{-m(A)|\theta|^\alpha\}, A \in \mathcal{U}$. We assume X is continuous in probability and take a separable version of X. A kernel $f_0(t,u)$ is a modification of $f(t,u)$ if for all $t \in T, f_0(t,\cdot) = f(t,\cdot)$ $m-a.e.$ on U. Then $X_0 = \left(X_0(t) \doteq \int_U f_0(t,u) W_m(du)\right)$ is a version of X.

When $1 \leq \alpha < 2$, define β by $1/\alpha + 1/\beta = 1$. For $\varepsilon > 0$ and d a metric or pseudo-metric on T, let

$$H_\beta(d;\varepsilon) = \begin{cases} (\log N(d;\varepsilon))^{1/\beta}, & 2 \leq \beta < \infty, \\ \log^+ \log N(d;\varepsilon), & \beta = \infty, \end{cases}$$

where $N(d;\varepsilon) = N(T,d;\varepsilon)$ is the minimum number of d-balls of radius ε with centers in T that cover T. Let

$$d_X(t,s) = (-\log[Ee^{i(X(t)-X(s))}])^{1/\alpha}.$$

We consider three conditions on the kernel $f(t,u)$ and one condition on $H_\beta(d_X;\varepsilon)$:
(C1) f has a modification f_0 such that for every $u \in U$, $f_0(t,u)$ is continuous.
(C2) $f^*(u) = \sup_{t\in T_0} |f(t,u)|$ is in $L^\alpha(U,\mathcal{U},m)$, where $T_0 \subset T$ is a countable separant for X that is dense in T.

$$\text{(C3)} \qquad \int \left(\sup_{s,t\in T} \frac{|f(t,u)-f(s,u)|}{d_X(t,s)} \right)^\alpha m(du) < \infty.$$

$$\text{(C4)} \qquad \int_0^\infty H_\beta(d_X;\varepsilon)d\varepsilon < \infty.$$

Theorem 3.1. ([N]) *Let* $0 < \alpha < 1$.
(i) X has Property I if and only if (C1) and (C2) hold.
(ii) X has a version with discontinuous, bounded sample paths if and only if (C1) fails to hold and (C2) holds.
(iii) X has Property III if and only if (C2) fails to hold.
Let $1 \le \alpha < 2$.
(iv) If (C1), (C2), (C3) and (C4) are fulfilled, then X has Property I.

We now give the proofs for our examples.

Proofs. (1) The fact that X_1 and X_5 for $0 < H < 1/\alpha$ have Property III is verified by Theorem 3.1 (iii) above, or by Theorem 4 of Rosinski [R2].

(2) Recall

$$X_6(t) = \int_{-\infty}^\infty f(t,\lambda)d\tilde{M}_\alpha(\lambda), \quad 0 \le t \le 1,$$

where

$$f(t,\lambda) = \frac{e^{it\lambda}-1}{i\lambda}(a\lambda_+^{1-H-1/\alpha} + b\lambda_-^{1-H-1/\alpha}).$$

(When $H = 1/\alpha$, $X_6 = X_3$.) Let $f(t,\lambda) = g(t,\lambda) + ih(t,\lambda)$ and $\tilde{M}_\alpha = M_\alpha^{(1)} + iM_\alpha^{(2)}$. Then

$$X_6(t) = \int (gdM_\alpha^{(1)} - hdM_\alpha^{(2)}) + i\int (hdM_\alpha^{(1)} + gdM_\alpha^{(2)}).$$

Obviously h and g satisfy (C1). Observe that g^* and h^* are in $L^\alpha(\mathbf{R}, \mathcal{B}, dx)$, satisfying (C2). Hence when $0 < \alpha < 1$, it follows from Theorem 3.1 (i) that X_6 has Property I.

If X is H-ss si SαS, then

$$d_X(t,s) = C|t-s|^H$$

for some positive constant C, and so when $T = [0,1]$,

$$N(d_X;\varepsilon) = \begin{cases} C([\varepsilon^{-1/H}]+1), & \text{if } \varepsilon < 1, \\ C, & \text{if } \varepsilon \geq 1. \end{cases}$$

Here and in what follows throughout the paper, C denotes a positive constant which may differ from one equation to another. Hence, when $1 \leq \alpha < 2$, we have (C4). Condition (C3) is also satisfied. Thus by Theorem 3.1 (iv), we conclude that X_6 has Property I.

(3) When $0 < \alpha < 1$, we can apply Theorem 3.1 (ii) to show that X_{10} has Property II. Recall

$$X_{10}(t) = \int_{\mathbf{R}^2_+} c(t;x,y)dM(x,y).$$

We then apply Theorem 3.1 (ii) with $U = \mathbf{R}^2_+, u = (x,y), m(du) = x^{H\alpha-2}dxdy, W_m = M$. We see that $c(\cdot;x,y)$ does not satisfy (C1). However, since

$$\sup_{t\in[0,1]} c(t;x,y) = c(1;x,y),$$

(C2) is fulfilled. Hence when $0 < \alpha < 1$, X_{10} has a version with discontinuous, bounded sample paths. By an observation due to Vervaat [V], H-ss si processes with such versions have Property II. In the case $1 \leq \alpha < 2$, we arrive at the same conclusion if we represent the stochastic integral $X_{10}(t)$ by the pathwise integral after the integration by parts.

Q.E.D.

Remark 3.1. For $1/\alpha < H < 1$, the processes X_5 and X_8 cannot be discriminated by Property I. More delicate path properties exhibit them as different. Takashima [Taka] has recently proved that for any $\varepsilon > 0$,

$$\limsup_{t\downarrow 0} \frac{|X_5(t)|}{t^H(\log 1/t)^{1/\alpha-\varepsilon}} = \infty \quad \text{a.s.}$$

However, X_8 has the form $X_8 = Z^{1/2}B_H$, where Z and B_H are independent. Hence, it follows from the law of the iterated logarithm for B_H (cf.[Mar]) that

$$\limsup_{t\downarrow 0} \frac{|X_8(t)|}{t^H(2\log\log 1/t)^{1/2}} < \infty \quad \text{a.s.}$$

We systematize the different contingencies of α, H, existence of processes and validity of path properties by the following picture of the (H, α) plane.

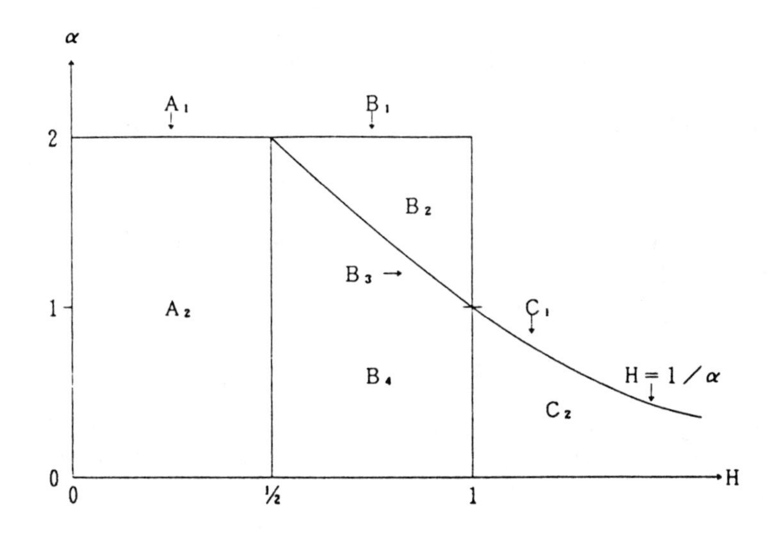

Let

$$A_1 = \{(H, \alpha) \mid \alpha = 2,\ 0 < H \leq 1/2\},$$
$$A_2 = \{(H, \alpha) \mid 0 < \alpha < 2,\ 0 < H \leq 1/2\},$$
$$B_1 = \{(H, \alpha) \mid \alpha = 2,\ 1/2 < H \leq 1\},$$
$$B_2 = \{(H, \alpha) \mid 1/H < \alpha < 2,\ 1/2 < H \leq 1\},$$
$$B_3 = \{(H, \alpha) \mid \alpha = 1/H,\ 1/2 < H \leq 1\},$$
$$B_4 = \{(H, \alpha) \mid 0 < \alpha < 1/H,\ 1/2 < H \leq 1\},$$
$$C_1 = \{(H, \alpha) \mid \alpha = 1/H,\ H > 1\}$$

and

$$C_2 = \{(H, \alpha) \mid 0 < \alpha < 1/H,\ H > 1\}.$$

In the region $B_1 \cup B_2$, H-ss si SαS processes have sample continuous versions as mentioned before.

The substable processes described in (4-1) in section 2 constitute a great variety of H-ss si SαS processes. Such processes with (H,α) in each of A_2, B_2, B_3 and B_4 are given by

$$X_8(t) = Z^{1/2}B_H(t)$$

with a positive strictly $\alpha/2$-stable random variable Z. They have Property I.

H-ss si SαS processes with (H,α) in either of B_4 and C_2 are given by

$$X_9(t) = Z^H Z_{1/H}(t)$$

with a positive strictly $\alpha H(<1)$-stable random variable Z. They have Property II.

An H-ss si SαS process with (H,α) in B_4 is given by

$$X_{11}(t) = Z^H X_1(t),$$

where X_1 is the log-fractional stable process in (2-3) in section 2 and Z is a positive strictly $\alpha H(<1)$-stable random variable. It has Property III.

An example of an H-ss si SαS processes with (H,α) in C_2 can be given in the following way. Kesten and Spitzer [KS] constructed a class of H-ss si processes $(\Delta_{p,q}(t))$ with two parameters restricted by $0 < p < 2$, $1 < q < 2$, $H = 1 - 1/q + 1/(pq)$, $1/2 < H < \infty$, $E|\Delta_{p,q}(t)|^p = \infty$. Furthermore, $E|\Delta_{p,q}(t)|^r < \infty$ for any $r < p$, and in all cases they have Property I. Given (H,α) such that $0 < \alpha < 1$ and $1 < H < 1/\alpha$, one can find (p,q) such that $\alpha < p < 2$ and $1 < q < 2$ satisfying $H = 1 - 1/q + 1/(pq)$, and hence $E|\Delta_{p,q}(t)|^\alpha < \infty$. Here is the process with the claimed property. It is the process X_{12}, constructed by the LePage representation in the next section (see Theorem 4.1):

$$X_{12}(t) := \sum_{j=1}^{\infty} \varepsilon_j \Gamma_j^{-1/\alpha} \Delta_{p,q}^{(j)}(t),$$

where $(\Delta_{p,q}^{(j)})_{j=1}^{\infty}$ is a sequence of independent copies of $\Delta_{p,q}$ independent of $(\varepsilon_j)_{j=1}^{\infty}$ and $(\Gamma_j)_{j=1}^{\infty}$, and $(\varepsilon_j)_{j=1}^{\infty}$ and $(\Gamma_j)_{j=1}^{\infty}$ are as defined in the next section. It follows by the same argument as in [CNR] that X_{12} has either Property I or III, since $\Delta_{p,q}$ has Property I. Which of the two we are unable to decide at the moment.

The following problem was open. In C_1, does there exist an H-ss si SαS process other than an α-stable Lévy motion? However, during the

Workshop where this paper was presented, the authors were informed that Samorodnitsky and Taqqu recently solved the above problem negatively [ST2].

4. The LePage representation of SαS processes

We first explain the LePage representation, which is a series representation, of real-valued SαS processes.

Let $(\varepsilon_j)_{j=1}^{\infty}$ be a sequence of iid symmetric random variables with $E[|\varepsilon_1|^\alpha] = 1$, and let $(\Gamma_j)_{j=1}^{\infty}$ be a sequence of Poisson arrival times with unit rate, independent of (ε_j). Let $\mathbf{R}^{(T)}$ be the space of all finitely supported families $(a(t))_{t\in T}$ of real numbers. Suppose that there exists a positive measure μ on $\mathbf{R}^T$ equipped with the cylindrical σ-algebra such that for any $a \in \mathbf{R}^{(T)}$

$$E\exp\left\{i\sum_{t\in T} a(t)X(t)\right\} = \exp\left\{-\int\left|\sum_{t\in T} a(t)b(t)\right|^\alpha d\mu(b)\right\}. \tag{4.1}$$

This is the definition of α-stability of $(X(t))_{t\in T}$ adopted in [MP]. It is trivial that if $((X(t))_{t\in T}$ is α-stable in the above sense, it is SαS in the sense of our present paper. The converse needs more careful analysis.

Let $0 < \alpha < 2$. Marcus and Pisier [MP] showed that if μ is a finite measure and $(V_j)_{j=1}^{\infty}$ is a sequence of iid $\mathbf{R}^T$-valued random variables with probability distribution $\nu = \mu/\mu(\mathbf{R}^T)$ and independent of (ε_j) and (Γ_j), then

$$(X(t)) \stackrel{d}{=} \left(c(\alpha)\mu(\mathbf{R}^T)^{1/\alpha}\sum_{j=1}^{\infty}\varepsilon_j\Gamma_j^{-1/\alpha}V_j(t)\right), \tag{4.2}$$

where $\stackrel{d}{=}$ denotes equality of all finite-dimensional distributions of two processes and $c(\alpha) = [\int_0^\infty(\sin x/x^\alpha)dx]^{-1/\alpha}$. The series on the right-hand side of (4.2) converges a.s. for each t and we have $E|V_j(t)|^\alpha < \infty$. Furthermore, for any $a_1, \cdots, a_n \in \mathbf{R}$, $t_1, \cdots, t_n \in T$,

$$E\exp\left\{i\sum_{k=1}^{n} a_k\left[c(\alpha)\sum_{j=1}^{\infty}\varepsilon_j\Gamma_j^{-1/\alpha}V_j(t_k)\right]\right\} = \exp\left\{-E\left|\sum_{k=1}^{n} a_k V_1(t_k)\right|^\alpha\right\}. \tag{4.3}$$

The following proposition is a direct consequence of (4.3), and gives us a characterization of self-similar stable processes.

Theorem 4.1. *Let $X = (X(t))_{t\in T}$ be an $S\alpha S$ process represented by (4.2) and let $V_1 = (V_1(t))_{t\in T}$ be a process satisfying $E[|V_1(t)|^\alpha] < \infty$, for any $c > 0$,*

$$E\left|\sum_{k=1}^{n} a_k V_1(ct_k)\right|^\alpha = E\left|\sum_{k=1}^{n} a_k c^H V_1(t_k)\right|^\alpha,$$

and for every $h \in T$,

$$E\left|\sum_{k=1}^{n} a_k(V_1(t_k+h) - V_1(h))\right|^\alpha = E\left|\sum_{k=1}^{n} a_k V_1(t_k)\right|^\alpha.$$

Then X is H-ss si $S\alpha S$.

Now we want to know when an SαS process $(X(t))_{t\in T}$ satisfies (4.1) and (4.2). To answer this question, we follow [H]. We say that $(X(t))_{t\in T}$ satisfies condition (S) if there exists a countable set $T' \subset T$ such that for each $t \in T$ there is a sequence of $\{a_n, n \geq 1\} \subset \mathbf{R}^{(T')}$ for which $\sum_{t'\in T'} a_n(t')X(t')$ converges to $X(t)$ in probability. Then the following is known (cf.[H]). If $(X(t))_{t\in T}$ is an SαS process and satisfies condition (S), then there exists a set of functions $\{(f(t,s))_{s\in[0,1]}, t \in T\} \subset L^\alpha[0,1]$ such that

$$(X(t)) \stackrel{d}{=} \left(\int_0^1 f(t,s)dZ_\alpha(s)\right),$$

where $Z_\alpha(s), 0 \leq s \leq 1$, is an α-stable Lévy motion.

Furthermore, Rosinski [R] studies the series representation of infinitely divisible stochastic integral processes, and among other results he finds that

(4.4)

$$\left(\int_S f(t,s)dZ_\alpha(s)\right) \stackrel{d}{=} \left(C\sum_{j=1}^{\infty} \varepsilon_j \Gamma_j^{-1/\alpha} \varphi(\xi_j)^{-1/\alpha} f(t,\xi_j)\right), \quad S \subset \mathbf{R},$$

where (ξ_j) is a sequence of iid random variables having a probability measure m on S with density $\varphi(x)$, and C is a positive constant.

Combining all this, we conclude the following:

Proposition 4.1. *Let $0 < \alpha < 2$. If an $S\alpha S$ process $(X(t))_{t\in T}$ satisfies condition (S), then there exists an $\mathbf{R}^T$-valued random variable V such* *that*

$$(X(t)) \stackrel{d}{=} \left(c(\alpha) \sum_{j=1}^{\infty} \varepsilon_j \Gamma_j^{-1/\alpha} V_j(t) \right),$$

where (V_j) *are independent copies of* V.

We now turn to complex-valued processes. Although Marcus and Pisier [MP] study the LePage representation of both complex- and real-valued SαS processes, their Lemma 1.4 is restricted to the real-valued case. Because of this, their proof of Proposition 1.5 seems insufficient for the complex-valued case. We therefore start with proving the complex-valued version of Lemma 1.4 of [MP].

Lemma 4.1. *Let* $0 < \alpha < 2$. *Suppose* v *is a rotationally invariant complex-valued random variable so that* $v \stackrel{d}{\sim} e^{i\theta} v$ *for any* $\theta \in \mathbf{R}$. *Suppose also* $E[|\mathrm{Re}\ v|^\alpha] < \infty$ *and let* $(v_j)_{j=1}^{\infty}$ *be a sequence of independent copies of* v *independent of* (Γ_j). *If we put*

$$Y = \sum_{j=1}^{\infty} \Gamma_j^{-1/\alpha} v_j,$$

then Y *converges a.s. and is SαS(σ), where* $\sigma = c(\alpha)^{-\alpha} E[|\mathrm{Re}\, v|^\alpha]$.

Proof. First note that $E[|\mathrm{Re}\, v|^\alpha] < \infty$ implies $E[|\mathrm{Im}\, v|^\alpha] < \infty$, since v is rotationally invariant. Note also that $\mathrm{Re}\, v$ and $\mathrm{Im}\, v$ are symmetric real-valued random variables. It is known that if $(u_j)_{j=1}^{\infty}$ is a sequence of iid real-valued random variables with $E|v_1|^\alpha < \infty$, $0 < \alpha < 2$, then the series $\sum_{j=1}^{\infty} \Gamma_j^{-1/\alpha} u_j$ converges a.s. (See, e.g. [MP].)

Now we have

$$Y = \sum_{j=1}^{\infty} \Gamma_j^{-1/\alpha} v_j = \sum_{j=1}^{\infty} \Gamma_j^{-1/\alpha} \mathrm{Re}\, v_j + i \sum_{j=1}^{\infty} \Gamma_j^{-1/\alpha} \mathrm{Im}\, v_j, \tag{4.5}$$

and it follows from the above observations that both series in (4.5) converge a.s. in case $E[|\mathrm{Re}\, v|^\alpha] < \infty$.

Next we show that Y is SαS(σ). Note that if v is rotationally invariant, then for any $z \in \mathbf{C}$,

$$\mathrm{Re}\,(zv) \stackrel{d}{\sim} |z| \mathrm{Re}\, v. \tag{4.6}$$

Now we have

$$E\exp\{\mathrm{Re}\,(\bar{z}Y)\} = E\exp\left\{\sum_{j=1}^{\infty}\Gamma_j^{-1/\alpha}\mathrm{Re}\,(\bar{z}v_j)\right\}$$

$$= \exp\{-c(\alpha)^{-\alpha}E[|\mathrm{Re}\,(\bar{z}v)|^{\alpha}]\} \quad \text{(by Lemma 1.4 of [MP])}$$

$$= \exp\{-c(\alpha)^{-\alpha}|z|^{\alpha}E[|\mathrm{Re}\,v|^{\alpha}]\} \quad \text{(by (4.6))}$$

$$= \exp\{-\sigma|z|^{\alpha}\}.$$

Hence Y is SαS(σ). Q.E.D.

Now we consider the LePage representation of a complex-valued process X of the following form:

$$X(t) = \int_{-\infty}^{\infty} f(t,\lambda)d\tilde{M}_{\alpha}(\lambda),$$

where $\tilde{M}_{\alpha}$ has independent complex rotatinally invariant α stable increments and its spectral (control) measure is Lebesgue measure, so that

$$\text{(4.7)} \quad E\exp\left\{i\mathrm{Re}\left(\bar{z}\int f(\lambda)d\tilde{M}_{\alpha}(\lambda)\right)\right\} = \exp\left\{-|z|^{\alpha}\int |f(\lambda)|^{\alpha}d\lambda\right\}.$$

This $\tilde{M}_{\alpha}$ is called a complex SαS motion. Furthermore $f(t,\cdot)$ is a complex-valued measurable function and satisfies

$$\int_{-\infty}^{\infty}|f(t,\lambda)|^{\alpha}d\lambda < \infty.$$

Let ψ be an arbitrary probability measure equivalent to Lebesgue measure on $\mathbf{R}$ and let φ be its Radon-Nikodym derivative ($\psi(d\lambda) = \varphi(\lambda)d\lambda$). Let $\{\xi,\xi_j, j \geq 1\}$ be a sequence of iid random variables with distribution ψ, and let $\{g,g_j,j \geq 1\}$ be a sequence of iid rotationally invariant complex-valued random variables with $E[g] = 0$ and $E[|\mathrm{Re}\,g|^{\alpha}] = 1$. Also suppose that $(\Gamma_j),(\xi_j),(g_j)$ are mutually independent.

Theorem 4.2.

$$Y(t) = c(\alpha)\sum_{j=1}^{\infty} g_j\Gamma_j^{-1/\alpha}\varphi(\xi_j)^{-1/\alpha}f(t,\xi_j)$$

converges a.s. for each t and

$$(Y(t)) \stackrel{d}{=} (X(t)).$$

Proof. We apply Lemma 4.1 to $v := g\varphi(\xi)^{-1/\alpha} f(t,\xi)$. We have $E[|\mathrm{Re}\, v|^\alpha] < \infty$, because

(4.8)

$$\begin{aligned} &E[|\mathrm{Re}\,(g\varphi(\xi)^{-1/\alpha} f(t,\xi))|^\alpha] \\ &\quad = E_\xi[E_g[|\mathrm{Re}\,(g\varphi(\xi)^{1/\alpha} f(t,\xi))|^{-\alpha}]] \\ &\quad = E_\xi[\varphi(\xi)^{-1}|f(t,\xi)|^{-\alpha} E_g[|\mathrm{Re}\, g|^\alpha]] \qquad \text{(by (4.6))} \\ &\quad = E_\xi[\varphi(\xi)^{-1}|f(t,\xi)|^\alpha] \qquad (\text{since } E_g[|\mathrm{Re}\, g|^\alpha] = 1\ ) \\ &\quad = \int_{-\infty}^{\infty} \varphi(\lambda)^{-1}|f(t,\lambda)|^\alpha \varphi(\lambda) d\lambda \\ &\quad = \int_{-\infty}^{\infty} |f(t,\lambda)|^\alpha d\lambda < \infty. \end{aligned}$$

Next we have for $z_k \in \mathbf{C}, k = 1, 2, \cdots, n$,

(4.9)

$$\begin{aligned} &E \exp\left\{ i\mathrm{Re}\left(\sum_{k=1}^n \bar{z}_k X(t_k) \right) \right\} \\ &\quad = E \exp\left\{ i\mathrm{Re}\left(\int \sum_{k=1}^n \bar{z}_k f(t_k,\lambda) d\tilde{M}_\alpha(\lambda) \right) \right\} \\ &\quad = \exp\left\{ - \int \left| \sum_{k=1}^n \bar{z}_k f(t_k,\lambda) \right|^\alpha d\lambda \right\} \qquad \text{(by (4.7))}. \end{aligned}$$

On the other hand,

$$\begin{aligned} &E \exp\left\{ i\mathrm{Re}\left(\sum_{k=1}^n \bar{z}_k Y(t_k) \right) \right\} \\ &\quad = E \exp\left\{ ic(\alpha) \sum_{j=1}^\infty \Gamma_j^{-1/\alpha} \mathrm{Re}\left(g_j \varphi(\xi_j)^{-1/\alpha} \sum_{k=1}^n \bar{z}_k f(t_k,\xi_j) \right) \right\} \\ &\quad = \exp\left\{ -E\left[\left| \mathrm{Re}\left(g\varphi(\xi)^{-1/\alpha} \sum_{k=1}^n \bar{z}_k f(t_k,\xi) \right) \right|^\alpha \right] \right\}. \end{aligned}$$

By the same arguments as in (4.8),

$$\begin{aligned}
E&\left[\left|\operatorname{Re}\left(g\varphi(\xi)^{-1/\alpha}\sum_{k=1}^{n}\bar{z}_k f(t_k,\xi)\right)\right|^{\alpha}\right] \\
&= E_{\xi}\left[\varphi(\xi)^{-1}\left|\sum_{k=1}^{n}\bar{z}_k f(t_k,\xi)\right|^{\alpha} E_g[|\operatorname{Re} g|^{\alpha}]\right] \\
&= \int_{-\infty}^{\infty}\varphi(\lambda)^{-1}\left|\sum_{k=1}^{n}\bar{z}_k f(t_k,\lambda)\right|^{\alpha}\varphi(\lambda)d\lambda \\
&= \int_{-\infty}^{\infty}\left|\sum_{k=1}^{n}\bar{z}_k f(t_k,\lambda)\right|^{\alpha} d\lambda.
\end{aligned}$$

Hence,

$$\begin{aligned}
E\exp&\left\{i\operatorname{Re}\left(\sum_{k=1}^{n}\bar{z}_k Y(t_k)\right)\right\} \\
&= \exp\left\{-\int_{-\infty}^{\infty}\left|\sum_{k=1}^{n}\bar{z}_k f(t_k,\lambda)\right|^{\alpha} d\lambda\right\}.
\end{aligned} \tag{4.10}$$

The proof is thus completed by (4.9) and (4.10). Q.E.D.

We remark that Theorem 4.1 holds also for complex-valued processes.

5. Hölder continuity of sample paths of some ss si SαS processes

Nolan [N] gave a necessary and sufficient condition for the uniform Hölder continuity of sample paths of SαS processes when $0 < \alpha < 1$. However, we are not able to establish a corresponding result for the case of $1 \le \alpha < 2$. On the other hand, Takashima [Taka] studied the Hölder continuity of the linear fractional stable processes with continuous sample paths. In this section, we examine the Hölder continuity of sample paths of the harmonizable fractional stable processes X_6 as an application of the LePage representation in the previous section.

Theorem 5.1. *X_6 has a version W such that*

$$P\left\{\lim_{\delta\downarrow 0}\sup_{\substack{|s-t|\le\delta \\ 0\le s,t\le 1}}\frac{|W(s)-W(t)|}{|s-t|^{\gamma}} = 0\right\} = 1$$

for any $\gamma < H$, so the sample paths of W are uniformly γ-Hölder continuous.

Proof. Recall that

$$X_6(t) = \int_{-\infty}^{\infty} f(t,\lambda) d\tilde{M}_\alpha(\lambda),$$

where

$$f(t,\lambda) = \frac{e^{it\lambda} - 1}{i\lambda}\left(a\lambda_+^{1-H-1/\alpha} + b\lambda_-^{1-H-1/\alpha}\right).$$

Let

$$\varphi(\lambda) = \frac{a_\varepsilon}{|\lambda|^{1-\varepsilon} + |\lambda|^{1+\varepsilon}}, \tag{5.1}$$

where $0 < \varepsilon < \min\{H, 1-H\} \times 2\alpha/(2-\alpha)$ and a_ε is the normalization for $\int \varphi(\lambda) d\lambda = 1$. Define $Y = (Y(t))$ in Theorem 4.2 with the g_i being centered Gaussian random variables and ξ_i having distribution $\psi(d\lambda) = \varphi(\lambda)d\lambda$. By Theorem 4.2, it suffices to check the sample path property of Y.

Fix (Γ_j) and (ξ_j) in the representation of Y and regard Y as a conditionally Gaussian process. We have

$$\begin{aligned} &E_g[|Y(s) - Y(t)|^2] \\ &= c(\alpha)^2 E_g\left[\left|\sum_{j=1}^{\infty} \Gamma_j^{-1/\alpha} \varphi(\xi_j)^{-1/\alpha}(f(s,\xi_j) - f(t,\xi_j))g_j\right|^2\right] \\ &= c(\alpha)^2 E_g[|g|^2] \sum_{j=1}^{\infty} \Gamma_j^{-2/\alpha} \varphi(\xi_j)^{-2/\alpha} |f(s,\xi_j) - f(t,\xi_j)|^2. \end{aligned}$$

We are going to estimate expressions in terms of

$$d_{\Gamma,\xi}^2(s,t) = \sum_{j=1}^{\infty} \Gamma_j^{-2/\alpha} \varphi(\xi_j)^{-2/\alpha} |f(s,\xi_j) - f(t,\xi_j)|^2.$$

We now have

$$|f(s,\lambda) - f(t,\lambda)|^2 \le C\{(|s-t|\lambda)^2 \wedge 1\}|\lambda|^{-2(H+1/\alpha)}.$$

Hence for all small $\delta > 0$,

$$E_\xi \left[\sup_{\substack{|s-t|\le\delta \\ 0\le s,t\le 1}} \varphi(\xi)^{-2/\alpha} |f(s,\xi) - f(t,\xi)|^2 \right]$$
$$\le C \int_0^\infty \varphi(\lambda)^{-2/\alpha} \lambda^{-2(H+1/\alpha)} \{(\delta\lambda)^2 \wedge 1\} \varphi(\lambda) d\lambda$$
$$= \int_0^{1/\delta} + \int_{1/\delta}^\infty =: I_1 + I_2, \quad \text{(say)}.$$

Since $\varphi(\lambda) \sim c|\lambda|^{-1+\varepsilon}$ as $\lambda \to 0$ and $\varphi(\lambda) \sim c|\lambda|^{-1-\varepsilon}$ as $\lambda \to \infty$,

$$I_1 = \int_0^1 + \int_1^{1/\delta} \le C\delta^2 \int_0^1 \lambda^{\varepsilon(1-2/\alpha)-2H+1} d\lambda$$
$$+ C\delta^2 \int_1^{1/\delta} \lambda^{\varepsilon(2/\alpha-1)-2H+1} d\lambda.$$

Since $0 < \varepsilon < 2\alpha(1-H)/(2-\alpha)$, we have

$$I_1 \le C\delta^2 + C\delta^{\varepsilon(1-2/\alpha)+2H} \le C\delta^{2H+\varepsilon(1-2/\alpha)}.$$

Note that we have also chosen $\varepsilon > 0$ such that

$$h := 2H + \varepsilon(1 - \frac{2}{\alpha}) > 0.$$

As to I_2, we have

$$I_2 = C \int_{1/\delta}^\infty \varphi(\lambda)^{1-2/\alpha} \lambda^{-2(H+1/\alpha)} d\lambda$$
$$\le C \int_{1/\delta}^\infty \lambda^{-(1+\varepsilon)(1-2/\alpha)-2(H+1/\alpha)} d\lambda$$
$$= C \int_{1/\delta}^\infty \lambda^{\varepsilon(2/\alpha-1)-2H-1} d\lambda$$
$$\le C\delta^{\varepsilon(1-2/\alpha)+2H} = C\delta^h.$$

Thus

$$I_1 + I_2 \le C\delta^h.$$

Finally we have

$$E_\xi \left[\sup_{\substack{|s-t|\le\delta \\ 0\le s,t\le 1}} d^2_{\Gamma,\xi}(s,t) \right] \le C\delta^h \sum_{j=1}^\infty \Gamma_j^{-2/\alpha}$$

a.s. with respect to (Γ_j). Hence if we take $\delta = 2^{-n}$, then for all small $a > 0$,

$$E_\xi \left[\sum_{n=1}^{\infty} 2^{(1-a)hn} \sup_{\substack{|s-t| \le 2^{-n} \\ 0 \le s,t \le 1}} d^2_{\Gamma,\xi}(s,t) \right]$$

$$\le C \left(\sum_{n=1}^{\infty} 2^{-anh} \right) \left(\sum_{j=1}^{\infty} \Gamma_j^{-2/\alpha} \right) < \infty \quad \text{a.s. with respect to } (\Gamma_j).$$

Thus,

$$\sum_{n=1}^{\infty} 2^{(1-a)hn} \sup_{\substack{|s-t| \le 2^{-n} \\ 0 \le s,t \le 1}} d^2_{\Gamma,\xi}(s,t) < \infty \quad \text{a.s. with respect to } (\Gamma_j), (\xi_j).$$

Therefore, there exists a number n_0 (a.s. with respect to (Γ_j) and (ξ_j)) such that for any $n \ge n_0$

$$2^{(1-a)hn} \sup_{\substack{|s-t| \le 2^{-n} \\ 0 \le s,t \le 1}} d^2_{\Gamma,\xi}(s,t) < 1.$$

Hence, for any s and t with $|s-t| \le 2^{-n_0}$,

$$d^2_{\Gamma,\xi}(s,t) \le 2|s-t|^{(1-a)h},$$

and so

$$E_g[|Y(s) - Y(t)|^2] \le 2|s-t|^{2(1-a)H + \varepsilon(1-2/\alpha)(1-a)}.$$

Here we can take $a > 0$ and $\epsilon > 0$ arbitrarily small. Hence by a result assuring the Hölder continuity of Gaussian processes (see, e.g., [CL], Chapter 9), we arrive at the conclusion. Q.E.D.

Remark 5.1. If we apply Corollary 2 of [N] to our X_6 with $0 < \alpha < 1$, we also see that X_6 has a uniformly γ-Hölder continuous version with $\gamma < H$.

Remark 5.2. More delicate arguments involving slowly varing functions allow us to replace $\gamma < H$ in the statement of Theorem 5.1 by $\gamma = H$. This result will be published elsewhere in combination with others.

6. Local time and nowhere differentiability

For sample continuous stochastic processes, a natural further question addresses sample differentiability. In this section, we apply an argument in Berman [B] to prove that for $0 < H < 1/2$ an H-ss si SαS process is nowhere differentiable.

We start with a result on the local time of H-ss si SαS processes which is a direct consequence of an argument due to Berman [B]. For convenience we prove the results in this section only for real-valued processes.

Theorem 6.1. *Let $X = (X(t))_{t\in T}$ be an H-ss si process with $0 < H < 1$. Then X has L^2-local time a.s.*

Proof. Let $I = [a, b], -\infty < a < b < \infty$, and put

$$\mu_X(A) = \text{Leb}\{t \in I | X(t) \in A\}, \qquad A \in B(\mathbf{R}),$$

where Leb denotes Lebesgue measure. Let

$$h(\theta) = \int_{-\infty}^{\infty} e^{i\theta x} d\mu_X(x) = \int_I e^{i\theta X(t)} dt.$$

Since X is H-ss si SαS, we have

$$(6.1) \qquad \begin{aligned} E[|h(\theta)|^2] &= E\left[\int_{I\times I} e^{i\theta(X(t)-X(s))} dtds\right] \\ &= \int_{I\times I} e^{-c|t-s|^{\alpha H}|\theta|^\alpha} dtds, \end{aligned}$$

where c is a positive constant determined by

$$Ee^{i\theta X(1)} = e^{-c|\theta|^\alpha}.$$

Hence

$$E\left[\int_{-\infty}^{\infty} |h(\theta)|^2 d\theta\right] = \int_{I\times I} \int_{-\infty}^{\infty} e^{-c|u|^\alpha} du \frac{dtds}{|t-s|^H} < \infty,$$

if $0 < H < 1$. Therefore for a.a. $\omega \in \Omega$ (the underlying probability space) $h(\theta)$ is square integrable so that there exists an L^2-occupation density of the occupation measure $\mu_X(\cdot)$, which is the local time. Q.E.D.

Theorem 6.2. *Let $X = (X(t))_{t\in T}$ be an H-ss si SαS process with $0 < H < 1/2$, and let I be a finite interval. Then X satisfies*

$$\frac{(X_M - X_m)}{\text{Leb}(I)} \geq C|\log(X_M - X_m)|^\delta$$

for some positive constants C and δ, where

$$X_M = \sup_{t\in I} X(t) \quad \textit{and} \quad X_m = \inf_{t\in I} X(t).$$

Hence if $X(t)$ is continuous, it is nowhere differentiable, and if it is right continuous, it is nowhere differentiable from the right.

Proof. Fix $\omega \in \Omega$ such that $h(\theta)$ is square integrable. By the Fourier inversion formula,

$$\text{Leb}(I) = \mu_X((X_m, X_M)) = \int_{-\infty}^{\infty} \frac{e^{-i\theta X_M} - e^{-i\theta X_m}}{2\pi i\theta} h(\theta)d\theta.$$

Hence for any $\varepsilon > 0$,

$$\begin{aligned}
&\text{Leb}(I)^2 \\
&= \frac{1}{4\pi^2}\Big[\int_{-\infty}^{\infty} \left|\frac{e^{-i\theta(X_M - X_m)} - 1}{\theta}\right| \frac{1}{|\theta|^{1/2}(|\log|\theta|| + 1)^{(1+\varepsilon)/2}} \\
&\qquad \times |\theta|^{1/2}(|\log|\theta|| + 1)^{(1+\varepsilon)/2}|h(\theta)|d\theta\Big]^2 \\
&\le \int_{-\infty}^{\infty} \frac{|e^{-i\theta(X_M - X_m)} - 1|^2}{|\theta|^3(|\log|\theta|| + 1)^{1+\varepsilon}} d\theta \\
&\qquad \times \int_{-\infty}^{\infty} |\theta|(|\log|\theta|| + 1)^{1+\varepsilon}|h(\theta)|^2 d\theta \\
&=: I_1 \times I_2.
\end{aligned} \tag{6.2}$$

By (6.1),

$$\begin{aligned}
E[I_2] &= \int_{-\infty}^{\infty} |\theta|(|\log|\theta|| + 1)^{1+\varepsilon} E[|h(\theta)|^2]d\theta \\
&= \int_{I\times I}\int_{-\infty}^{\infty} |\theta|(|\log|\theta|| + 1)^{1+\varepsilon} e^{-c|t-s|^{\alpha H}|\theta|^\alpha} d\theta dt ds \\
&= \int_{I\times I}\int_{-\infty}^{\infty} |u|(|\log\frac{|t-s|^H}{|u|}| + 1)^{1+\varepsilon} e^{-c|u|^\alpha} du \frac{dtds}{|t-s|^{2H}} \\
&\le C\int_{I\times I} (|\log|t-s|| + 1)^{1+\varepsilon} \frac{dtds}{|t-s|^{2H}} < \infty,
\end{aligned}$$

if $H < 1/2$. Thus

$$I_2 < \infty \quad \text{for a.a. } \omega \in \Omega. \tag{6.3}$$

As to I_1,

$$I_1 = \int_{-\infty}^{\infty} \frac{|e^{-iu}-1|^2 (X_M - X_m)^2}{|u|^3(|\log|(X_M - X_m)/u|| + 1)^{1+\varepsilon}} du$$
$$\le C \frac{(X_M - X_m)^2}{|\log(X_M - X_m)|^{\varepsilon}}.$$

We thus have from (6.2)–(6.4) that for some positive constant C,

$$\frac{(X_M - X_m)}{\mathrm{Leb}(I)} \ge C|\log(X_M - X_m)|^{\varepsilon/2}.$$

Q.E.D.

Remark 6.1. Theorem 6.2 is restricted to H-ss si SαS processes with $0 < H < 1/2$. The same conclusion for the case $H \ge 1/2$ is supported by many examples in this paper, but we are not able to prove it in general setting. Rosinski [R1] characterized which SαS processes of the form $\int f(t,u)dZ_\alpha(u)$ possess absolutely continuous sample paths by conditions on the kernel f. By checking these conditions (cf. Theorem 6.1 in [R1]), we see that X_5 with $1/\alpha < H < 1$ and X_6 with $0 < H < 1$ do not have absolute continuous sample paths.

ACKNOWLEDGEMENT

The authors would like to thank Wim Vervaat for his valuable comments on the earlier versions of this paper.

REFERENCES

[B] S.M. Berman, *Harmonic analysis of local times and sample functions of Gaussian processes*, Trans. Amer. Math. Soc. **143** (1969), 269–281.

[CM] S. Cambanis and M. Maejima, *Two classes of self-similar stable processes with stationary increments*, Stoch. Proc. Appl. **32** (1989), 305–329.

[CNR] S. Cambanis, J. Nolan and J. Rosinski, *On the oscillation of infinitely divisible processes*, Stoch. Proc. Appl. (1990), to appear.

[CL] H. Cramér and M.R. Leadbetter, "Stationary and Related Stochastic Processes," Wiley, 1967.

[H] C.D. Hardin,Jr., *On the spectral representation of symmetric stable processes*, J. Multivar. Anal. **12** (1982), 385–401.

[KM] Y. Kasahara and M. Maejima, *Weighted sums of i.i.d. random variables attracted to integrals of stable processes*, Probab. Th. Rel. Fields **78** (1988), 75–96.

[KMV] Y. Kasahara, M. Maejima and W. Vervaat, *Log-fractional stable processes*, Stoch. Proc. Appl. **30** (1988), 329–339.

[KS] H. Kesten and F. Spitzer, *A limit theorem related to a new class of self similar processes*, Z. Wahrsch. verw. Geb. **50** (1979), 5–25.

[Mae1] M. Maejima, *On a class of self-similar processes*, Z. Wahrsch. verw. Geb. **62** (1983), 235-245.

[Mae2] M. Maejima, *A remark on self-similar processes with stationary increments*, Canadian J. Statist. **14** (1986), 81–82.

[MV] B.B. Mandelbrot and J.W. Van Ness, *Fractional Brownian motions, fractional noises and applications*, SIAM Rev. **10** (1968), 422–437.

[Mar] M.B. Marcus, *Hölder conditions for Gaussian processes with stationary increments*, Trans. Amer. Math. Soc. **134** (1968), 29–52.

[MP] M.B. Marcus and G. Pisier, *Characterization of almost surely continuous p-stable random Fourier series and strongly stationary processes*, Acta Math. **152** (1984), 245–301.

[N] J.P. Nolan, *Continuity of symmetric stable processes*, J. Multivar. Anal. **29** (1989), 84–93.

[R1] J. Rosinski, *On stochastic integral representation of stable processes with sample paths in Banach spaces*, J. Multivar. Anal. **20** (1986), 277–302.

[R2] J. Rosinski, *On path properties of certain infinitely divisible processes*, Stoch. Proc. Appl. **33** (1989), 73–87.

[ST1] G. Samorodnitsky and M.S. Taqqu, *The various linear fractional Lévy motions*, Probability, Statistics, and Mathematics (T.W. Anderson, K.B. Athreya and D.L. Iglehart, eds.), Academic Press (1989), 261-270.

[ST2] G. Samorodnitsky and M.S. Taqqu, $1/\alpha$*-self-similar* α*-stable processes with stationary incremnets*, preprint (1990).

[S] Y. Sato, *Joint distributions of some self-similar stable processes*, preprint (1989).

[Taka] K. Takashima, *Sample path properties of ergodic self-similar processes*, Osaka J. Math. **26** (1989), 159–189.

[Take] S. Takenaka, *Integral-geometric construction of self-similar stable processes*, preprint (1989).

[TW] M.S. Taqqu and R. Wolpert, *Infinite variance self-similar processes subordinate to a Poisson measure*, Z. Wahrsch. verw. Geb. **62** (1983), 53–72.

[V] W. Vervaat, *Sample path properties of self-similar processes with stationary increments*, Ann. Probab. **13** (1985), 1–27.

Institute of Mathematics, Yoshida College, Kyoto University, Kyoto 606, Japan
Department of Mathematics, Keio University, Hiyoshi, Yokohama 223, Japan

A Stochastic Integral Representation for the Bootstrap of the Sample Mean

JOHN KINATEDER[1]

Introduction.

Assume that F is the distribution function of a symmetric distribution in the domain of attraction of an α-stable distribution. Let $X_1, X_2, \ldots$ be a sequence of independent random variables distributed according to F. As Feller [3] shows, if $0 < \alpha < 2$, then there exist constants $a_n > 0$ such that $n(1 - F(a_n y)) \to y^{-\alpha}$ as $n \to \infty$, and for such a_n, $S_n = a_n^{-1}(X_1 + \cdots + X_n)$ converges in distribution to an α-stable random variable. If $\alpha = 2$, then choose a_n such that $a_n^{-1}(X_1 + \cdots + X_n)$ converges in distribution to the standard normal.

Let F_n be the empirical distribution function of $X_1, \ldots, X_n$:

$$F_n(x) = \frac{1}{n} \sum_{k=1}^{n} I(X_k \leq x).$$

For each observation of the data $X_1, \ldots, X_n$, we consider the distribution of

$$S_n^* = a_n^{-1} \sum_{j=1}^{n} (X_j^* - \bar{X}_n)$$

where $X_1^*, \ldots, X_n^*$ are independent and distributed according to F_n. This is equivalent to resampling from the data, $X_1, \ldots, X_n$, with replacement, and applying the same statistic to the resampled data as we would to the original data. This is called a bootstrapped version of the statistic.

It was shown by Athreya [1] that the bootstrapped version of S_n does not converge weakly to a constant distribution along all sample sequences. He showed that it the conditional distribution of S_n^* given $X_1, \ldots, X_n$ converges in distribution (with respect to the weak topology on the space of bounded measures) to a random distribution. He gave a representation for this random limit distribution in terms of random Poisson measures.

Later, Knight [4] gave a different representation for this limit law. He showed that

$$\mathcal{L}(S_n^* \,|\, X_1, \ldots, X_n) \xrightarrow{d} \mathcal{L}(\sum_{k=1}^{\infty} \epsilon_k \Gamma_k^{-1/\alpha}(M_k^* - 1) \,|\, \epsilon_k \Gamma_k^{-1/\alpha}, k \geq 1)$$

[1]Research partially supported by ONR Grant N000014-85-K-0150

where $\epsilon_1, \epsilon_2, \ldots$ are independent and identically distributed, $P(\epsilon_1 = 1) = \frac{1}{2} = P(\epsilon_1 = -1)$, $\Gamma = \langle \Gamma_1, \Gamma_2, \ldots \rangle$ are the arrival times of a Poisson process with unit rate, $\Gamma_k = \sum_{j=1}^{k} \xi_i$ where $P(\xi_i \geq x) = e^{-x}$ for all i, and $M_1^*, M_2^*, \ldots$ are independent Poisson mean 1 random variables. $\{\epsilon_j\}, \{\Gamma_j\}, \{M_{nj}^*\}$ are mutually independent.

1. Results. The following theorem gives a new representation for the bootstrap and its limit law. A quick look at Proposition 1 shows that this representation extends to the finite variance case.

Theorem 1(a).

$$\mathcal{L}(S_n^* \mid \widetilde{X}) = \mathcal{L}\left(\int_{-\infty}^{\infty} t\, W^n \circ H^n(dt) \mid \widetilde{X}\right)$$

where
(i) $\widetilde{X} = (X_{(1)}, \ldots, X_{(n)})$ *are the absolutely ordered observations*

$$|X_{(1)}| \geq \cdots \geq |X_{(n)}|.$$

(ii) W^n *is the scaled partial sum process associated with* $(X_1, \ldots, X_n)$:

$$W^n(t) = a_n^{-1} \sum_{k=1}^{\lfloor nt \rfloor} X_k, \qquad t \in [0,1].$$

(iii) $(M_{n1}^*, \ldots, M_{nn}^*)$ *is a multinomial vector independent of* W^n, *with parameters* $(\frac{1}{n}, \ldots, \frac{1}{n})$, *and* H^n *is the empirical distribution function of* $(M_{n1}^* - 1, \ldots, M_{nn}^* - 1)$.

It should be pointed out here that in reference to the bootstrap, since resampling with replacement has no dependence on the order of the data,

$$\mathcal{L}(S_n^* \mid \widetilde{X}_n) = \mathcal{L}(S_n^* \mid X_1, \ldots, X_n).$$

But this alternative conditioning is not valid for the stochastic integral representation;

$$\mathcal{L}(\int r\, W^n \circ H^n(dr) \mid \widetilde{X}_n) \neq \mathcal{L}(\int r\, W^n \circ H^n(dr) \mid X_1, \ldots, X_n).$$

Theorem 1(b).

$$\mathcal{L}(\sum_{k=1}^{\infty} \epsilon_k \Gamma_k^{-1/\alpha}(M_k^* - 1) \mid \epsilon_k \Gamma_k^{-1/\alpha}, k \geq 1)$$
$$= \mathcal{L}\left(\int_{-\infty}^{\infty} t\, W \circ H(dt) \mid \epsilon_k \Gamma_k^{-1/\alpha}, k \geq 1\right)$$

where
(i) W is the homogeneous independent increments symmetric α–stable process (LePage [5]):

$$W(t) = \sum_{k=1}^{\infty} \epsilon_k \Gamma_k^{-1/\alpha} I(T_k \leq t), \qquad t \in [0,1].$$

(ii) H is the distribution function of a centered Poisson (mean 1) random variable: $H(x) = P(M_1^* - 1 \leq x)$.
(iii) $T_1, T_2, \ldots$ are independent uniformly distributed on $(0,1)$.

2. Proof of Results.

Proof of Theorem 1(a): Let $(M_{n1}, \ldots, M_{nn}^*)$ and H^n be as in the hypothesis. Define

$$\widetilde{S}_n^* = a_n^{-1} \sum_{k=1}^{n} X_k (M_{nk}^* - 1).$$

Notice that conditionally on the data $X_1, \ldots, X_n$, $\widetilde{S}_n^*$ has the same distribution as S_n^*.

To write $\widetilde{S}_n^*$ as a stochastic integral, for real t let

$$A^n(t) = a_n^{-1} \sum_{k=1}^{n} X_k I(M_{nk}^* - 1 \leq t).$$

Then

$$\begin{aligned} \widetilde{S}_n^* = a_n^{-1} \sum_{k=1}^{n} X_k (M_{nk}^* - 1) &= \sum_{r=-1}^{n-1} r \left(a_n^{-1} \sum_{k=1}^{n} X_k I(M_{nk}^* - 1 = r) \right) \\ &= \sum_{r=-1}^{n-1} r(A^n(r) - A^n(r-)) \\ &= \int r\, A^n(dr). \end{aligned}$$

If we show that the A^n process has the same joint distribution with $\widetilde{X}$ as the $W^n \circ H^n$ process, then the proof will be complete.

Lemma 1.1. *For each n, $\langle \widetilde{X} \,;\, A^n(t), t \in R \rangle \stackrel{d}{=} \langle \widetilde{X} \,;\, W^n \circ H^n(t), t \in R \rangle$.*

Proof: Since both H^n and A^n are right continuous and constant except for possible jumps at $r = -1, 0, 1, \ldots, n-1$, it suffices to show that

$$\langle \widetilde{X}, W^n \circ H^n(-1), \ldots, W^n \circ H^n(n-1) \rangle \stackrel{d}{=} \langle \widetilde{X}, A^n(-1), \ldots, A^n(n-1) \rangle.$$

It is enough to show that the vector of differences in adjacent components in the above vectors have the same distribution. Condition on $(M_{n1}^*, \ldots, M_{nn}^*)$ and use the exchangeability of $(X_1, \ldots, X_n)$ to show that

$$\left\langle \tilde{X}, \sum_{j=1}^{nH^n(-1)} X_j, \sum_{j=nH^n(-1)+1}^{nH^n(0)} X_j, \ldots, \sum_{j=nH^n(n-2)+1}^{n} X_j \right\rangle$$

$$\stackrel{d}{=} \left\langle \tilde{X}, \sum_{j=1}^{n} X_j I(M_{nj}^* = 0), \ldots, \sum_{j=1}^{n} X_j I(M_{nj}^* = n) \right\rangle.$$

This completes the proof of the lemma. Q.E.D.

To prove part (b) of the theorem, we will need the following result provided by LePage [5].

Let ϵ, Γ be as above.

Theorem L. *Suppose $Y_1, Y_2, \ldots$ is a sequence of independent and identically distributed random variables with $E|Y_1|^\alpha < \infty$. Then the following three results hold:*

(L.1) The sequence of partial sums

$$\sum_{j=1}^{n} \epsilon_j \Gamma_j^{-1/\alpha} Y_j$$

converges almost surely to a symmetric α-stable random variable.

(L.2) Its limit, which we denote by the left hand side below, has the following distribution:

$$\sum_{j=1}^{\infty} \epsilon_j \Gamma_j^{-1/\alpha} Y_j \stackrel{d}{=} (E|Y_1|^\alpha)^{1/\alpha} \sum_{j=1}^{\infty} \epsilon_j \Gamma_j^{-1/\alpha}.$$

(L.3) If $A_1, \ldots, A_n$ are disjoint measurable subsets of R then the following random variables are mutually independent:

$$\sum_{j=1}^{\infty} \epsilon_j \Gamma_j^{-1/\alpha} I(Y_j \in A_1), \ldots, \sum_{j=1}^{\infty} \epsilon_j \Gamma_j^{-1/\alpha} I(Y_j \in A_n).$$

Lemma 1.2. *With probability 1,*

$$\sum_{k=1}^{\infty} \epsilon_k \Gamma_k^{-1/\alpha} (M_k^* - 1) = \sum_{r=-1}^{\infty} r \sum_{k=1}^{\infty} \epsilon_k \Gamma_k^{-1/\alpha} I(M_k^* - 1 = r). \tag{2.2}$$

Proof: Let $Y = \sum_{k=1}^{\infty} \epsilon_k \Gamma_k^{-1/\alpha}$. Since Y is symmetric α–stable, it has characterisic function $e^{-\sigma^\alpha|\theta|^\alpha}$ for some $\sigma > 0$. Consider the partial sums

$$\sum_{r=-1}^{K} r \sum_{k=1}^{\infty} \epsilon_k \Gamma_k^{-1/\alpha} I((M_k^* - 1) = r). \tag{2.3}$$

By (L.1) and (L.2)

$$\sum_{k=1}^{\infty} \epsilon_k \Gamma_k^{-1/\alpha} M_k^* I(M_k^* = r) = r \sum_{k=1}^{\infty} \epsilon_k \Gamma_k^{-1/\alpha} I(M_k^* = r)$$
$$\stackrel{d}{=} r(P(M_1^* = r))^{1/\alpha} Y.$$

By (L.3), as r varies, the inner sums in (2.2) are independent. Therefore, the characteristic function of the left hand side in (2.3) is

$$\prod_{r=-1}^{K} e^{-\sigma^\alpha |r(P(M_k^*=r))^{1/\alpha}\theta|^\alpha} = e^{-\sigma^\alpha|\theta|^\alpha \sum_{r=-1}^{K} |r|^\alpha P(M_1^*=r)}.$$

As $K \to \infty$ this tends to $e^{-\sigma^\alpha|\theta|^\alpha E|M_1^*|^\alpha}$. This is the characteristic function of $\left(E|M_1^*|^\alpha\right)^{1/\alpha} Y$. Thus as $K \to \infty$,

$$\sum_{r=-1}^{K} r \sum_{k=1}^{\infty} \epsilon_k \Gamma_k^{-1/\alpha} I(M_k^* = r) \xrightarrow{d} \left(E|M_k^*|^\alpha\right)^{1/\alpha} Y. \tag{2.4}$$

Since the left hand side is the partial sum of independent random variables (see (L.3)), the sum must also converge almost surely (see Breiman [2], Prop. 8.36).

The double sum in (2.4) equals $\sum_{k=1}^{\infty} \epsilon_k \Gamma_k^{-1/\alpha} M_k^* I(M_k^* \le K)$ because for each r, the inner sum converges almost surely by (L.1). But, using (L.1) it can be shown that for each K,

$$\sum_{k=1}^{\infty} \epsilon_k \Gamma_k^{-1/\alpha} M_k^* = \sum_{k=1}^{\infty} \epsilon_k \Gamma_k^{-1/\alpha} M_k^* I(M_k^* \le K) + \sum_{k=1}^{\infty} \epsilon_k M_k^* I(M_k^* > K) \text{ a.s.,}$$

and since the last term on the right has the same distribution as

$$(E|M_1^*|^\alpha I(M_k^* > K))^{1/\alpha} Y,$$

which converges in probability to zero as $K \to \infty$,

$$\sum_{r=-1}^{K} r \sum_{k=1}^{\infty} \epsilon_k \Gamma_k^{-1/\alpha} I(M_k^* = r) \to_p \sum_{k=1}^{\infty} \epsilon_k \Gamma_k^{-1/\alpha} M_k^*. \tag{2.5}$$

Coupled with the a.s. convergence of the left hand side of (2.5) already established, the lemma is proved. Q.E.D.

Proof of Theorem 1(b): By definition of W,

$$\int_{-\infty}^{\infty} t\,W \circ H(dt) = \sum_{r=-1}^{\infty} r \sum_{k=1}^{\infty} \epsilon_k \Gamma_k^{-1/\alpha} I(M_k^* - 1 = r).$$

Hence, by Lemma 1.2, it is enough to show that

$$\left\langle \sum_{k=1}^{\infty} \epsilon_k \Gamma_k^{-1/\alpha} I(M_k^* - 1 = r), r \geq -1\,;\, \epsilon_k \Gamma_k,\, k \geq 1 \right\rangle$$

$$\stackrel{d}{=} \left\langle \sum_{k=1}^{\infty} \epsilon_k \Gamma_k^{-1/\alpha} I(T_k \in (H(r-1), H(r)]), r \geq -1\,;\, \epsilon_k \Gamma_k,\, k \geq 1 \right\rangle.$$

For each k, $\langle I(M_k^* - 1 = r), r \geq -1 \rangle$ has the same distribution as $\langle I(T_k \in (H(r-1), H(r)]), r \geq -1 \rangle$. Also, for different k, these processes are independent of eachother. Since both processes are independent of ϵ_k, Γ_k, $k \geq 1$, Theorem 1(B) follows. Q.E.D.

3. Extension from finite variance case. The following proposition shows that the above representation for the limit law carries over naturally to the finite variance case.

Proposition 1. *If W is a Brownian motion, then $\int t\,W \circ H(dt)$ has the standard normal distribution.*

Proof: To see this, look at the characteristic functions again. The stochastic integral can be written as $\sum_{r=-1}^{\infty} r[W(H(r)) - W(H(r-))]$. By independence of the increments of Brownian motion, its characteristic function is

$$\lim_{n \to \infty} \prod_{-1}^{n} e^{-\frac{r^2 t^2 [H(r) - H(r-)]}{2}} = \lim_{n \to \infty} e^{-\frac{t^2}{2} \sum_{r=-1}^{n} r^2 [H(r) - H(r-)]}$$

$$= e^{-\frac{t^2}{2} \int r^2\, dH(r)}$$

$$= e^{-\frac{t^2}{2}}.$$

This is the characteristic function of the standard normal distribution, hence the assertion is proved. Q.E.D.

Since the paths of Brownian motion are continuous with probability one, conditioning on the ordered jumps provides us with no additional information. That is,

$$\mathcal{L}\left(\int t\,W \circ H(dt) \mid \textit{ ordered jumps of } W\right) = \mathcal{L}\left(\int t\,W \circ H(dt)\right).$$

By applying Bickel and Freedman's results, using the proper scaling, $a_n = n^{1/2}\sigma$, we see that in the symmetric case the result concerning a distribution in the domain of attraction of an infinite variance stable random variable can be viewed as an extension of what was already known in the finite variance case.

4. Remark. Since $\widetilde{X}^n$ is a functional of the partial sum process W^n, and S_n^* is a functional of the bootstrapped partial sum process $W^n \circ H^n$, it is clear that the behavior of S_n^* jointly with $\widetilde{X}^n$ is dependent on the behavior of $W^n \circ H^n$ jointly with W^n. In a forthcoming paper, the author has established an invariance principle for the sequence of joint processes $(W^n \circ H^n, W^n)$ which helps explain this behavior for large n.

Acknowledgements: The author would like to thank Professor Raoul LePage for the suggestion of the problem and all of the helpful discussions.

References

[1] Athreya, K.B. (1987) Bootstrap of the mean in the infinite variance case. Ann. Statist. 15 724-731.
[2] Breiman, L. (1968) *Probability.* Addison-Wesley, California.
[3] Feller, W. (1971). *An Introduction to Probability Theory and Applications.* John Wiley, New York.
[4] Knight, K. (1989). On the bootstrap of the sample mean in the infinite variance case. Ann. Statist. 17 1168-1175.
[5] LePage, R. (1980). Multidimensional infinitely divisible variables and processes. Part I: Stable Case. Tech. Rep. no.292, Stanford, 1980.

John Kinateder
Department of Statistics and Probability
Michigan State University
East Lansing, Michigan

Multiple stable integrals appearing in weak limits

JERZY SZULGA

Abstract

Limit laws of certain finite multilinear random forms coincide with distributions of stable multiple integrals.

1 Preliminaries

By a random measure on a Borel space T we understand a countably additive vector measure with values in $L_0 = L_0(\Omega, \mathcal{A}, \mathsf{P})$. A random measure F is called simple if it is of the form

$$F = \sum_{i=1}^{n} Y_i \, \delta_{\tau_i} \tag{1.1}$$

where $\tau_1, \ldots, \tau_n$ are T-valued random variables and $Y_1, \ldots, Y_n$ are $\mathbb{R}^d$-valued random variables.

When $Y_i \equiv 1$, F is the empirical measure based on a sequence of random observations (τ_n).

We define the integral with respect to a simple random measure F as a random functional

$$F(h) = \int h \, dF = \sum h(\tau_i) \, Y_i, \quad h : T \to \mathbb{R}.$$

which is continuous, i.e., $h_k \to h$ in measure implies $F(h_k) \xrightarrow{\mathsf{P}} F(h)$.

Consider now a simple $\mathbf{R}^d$-valued random measure

$$\underline{F} = \sum_{i=1}^{n} \delta_{\tau_i} \underline{Y_i},$$

where $\underline{Y_i} = (Y_1^1, \ldots, Y_i^d)$ is an $\mathbf{R}^d$-valued random variable. Then a d-fold integral admits a usual definition :

$$\mathsf{F}(h) = \sum_{i_1,\ldots,i_d \leq n} h(\tau_{i_1}, \ldots, \tau_{i_d}) \cdot Y_{i_1}^1 \cdots Y_{i_d}^d.$$

Throughout the paper, (Y_i) and (τ_i) are independent sequences of i.i.d. random variables. Moreover, Y_i's are symmetric. Also, without loss of generality, we can assume that $T = [0, 1]$ and τ_i's are uniformly distributed on T.

We shall indentify the measure of a set A and the integral of the function $1\!\!1_A$, e.g., $\mathsf{F}(A) = \mathsf{F}(1\!\!1_A)$.

From the point of view of empirical measures, a simple process of type (1.1), with (τ_i) and (Y_i) being independent sequences of i.i.d. random variables, can be recognized as a "resampled" empirical measure (the term was introduced in [1]). Resampling by random variables with a finite variance is a rather known motive (cf., e.g. [2]). Limiting distributions, invariance principles, etc., are of major interest. There are close connections of the discussed subject to U-statistics and von Mises' functionals (cf., e.g., [2,5]). The latter are certain multiple integrals, where integrators are some "nice" processes.

If variances are infinite, not much is known about the relevant topics. In case of symmetric α-stable random variables Y_i's the matter was investigated in [1] but only for $\alpha > 1$. The latter restriction was necessary in presence of martingale methods.

Our goal is to show that distributions of multiple stochastic integrals become natural limits of laws of suitably normalized multidimensional empirical measures resampled by using random variables from the domain of normal attraction of a symmetric stable distribution. We shall extend some results from [1].

In this paper all discussed functions will be tacitly assumed to be Borel measurable. We choose the functional

$$||X||_0 = \mathsf{E}\, 1 \wedge ||X||$$

to metrize the convergence in probability in the space of $\mathbb{R}^d$-valued random variables, where $||\cdot||$ denotes the Euclidean norm in $\mathbb{R}^d$. Note that the L_1-norm of $1 \wedge ||X||$ can be replaced by any L_p-norm.

To metrize weak convergence of distributions we shall use a minimal metric

$$L(\mu,\nu) = L(X,Y) = \inf\{||X-Y||_0 : \mathcal{L}(X) = \mu,\, \mathcal{L}(Y) = \nu\}$$

(cf. e.g. [15]). We shall write $X \underset{C}{\asymp} Y$ if $C^{-1}Y \leq X \leq CY$.

2 Multiple integrals

The definition of a multiple stochastic integral used in this paper is adopted from [8]. It is essentially independent of a specific method of a construction of a multiple integral.

Let $\underline{X} = (X_1,\ldots,X_d)$ be an $\mathbf{R}^d$-valued random measure generated by a pure jump Lévy process with values in $\mathbf{R}^d$. We say that a multiple stochastic integral $\mathbf{X}(h) = X_1 \cdots X_d h = \int \cdots \int h\, dX_1 \ldots dX_d$ exists on a certain class $\mathcal{H}$ of function $h : [0,1]^d \to \mathbf{R}$ if the following conditions hold:

(i) $\mathcal{H}$ is a vector space containing all bounded functions with bounded supports and the mapping $\mathbf{X} : \mathcal{H} \to L_0$ is linear;
(ii) For any collection of non-overlaping intervals $(a_j, b_j]$, $j = 1, \ldots, d$,

$$\mathbf{X}((a_1, b_1] \times \cdots \times (a_d, b_d]) = \prod_{j=1}^{d} X_j((a_j, b_j]) \qquad \text{a.s.};$$

(iii) If $h = \lim_n h_n$ a.e. and $|h| \le g \in \mathcal{H}$ then $h_n, h \in \mathcal{H}$ and $\mathbf{X}(h_n) \xrightarrow{\mathsf{P}} \mathbf{X}(h)$.

The maximum space $\mathcal{H}$ satisfying the above three conditions is called the space of $\mathbf{X}$-integrable functions and denoted by $\mathcal{H}_{\mathbf{X}}$.

Let $\mathcal{H}_0$ be the algebra spanned by indicators $\mathbb{I}_A$ (more precisely, by functions $h_A(t_1, \ldots, t_d) = \mathbb{I}_A(t_j)$ indexed by Borel sets $A \subset [0, \infty)$)

Lemma 2.1 *Let $\underline{F}_n$ be a sequence of simple random measures whose finite dimensional distributions converge weakly to finite dimensional distributions of a random measure $\underline{X}$. Then*

$$(\mathbf{F}_n(h) : h \in \mathcal{H}_0) \xrightarrow{\mathcal{D}} (\mathbf{X}(h) : h \in \mathcal{H}_0). \tag{2.1}$$

Proof. Obvious. ∎

Lemma 2.2 *Let $\underline{X} = (X_n(i_1, \ldots, i_d))$ be a sequence of d-dimensional random tetrahedral matrices and $\underline{\varepsilon} = (\varepsilon_n)$ be a Rademacher sequence. Suppose that $\underline{X}$ and $\underline{e}$ are independent. Define*

$$\begin{aligned} \Theta_n &= \sum_{i_1, \ldots, i_d} X_n(i_1, \ldots, i_d) \varepsilon_{i_1} \cdots \varepsilon_{i_d}, \\ S^2(\Theta_n) &= \mathsf{E}[\,|\Theta_n|^2 \,|\, \underline{X}\,] = \sum_{i_1, \ldots, i_d} |X_n(i_1, \ldots, i_d)|^2 \end{aligned}$$

Then

(a) $\Theta_n \xrightarrow{\mathsf{P}} 0 \iff S^2(\Theta_n) \xrightarrow{\mathsf{P}} 0$;
(b) $(\Theta(n))$ is tight $\iff$ $(S^2(\Theta_n))$ is tight.

Proof. (b) is a straightforward consequence of (a). Since

$$\mathsf{E}\,1 \wedge |\Theta| \le \left(\mathsf{E}\,1 \wedge S^2(\Theta)\right)^{1/2}, \tag{2.2}$$

the implication "$\Longleftarrow$" of (a) is obvious.

Suppose $\Theta(n) \xrightarrow{\mathsf{P}} 0$. Then $\Theta(n') \to 0$ a.s. along a subsequence (n'). From Fubini's Theorem and the fact that Rademacher forms converge or diverge in L_0 and in L_2 at the same time (cf. [9]), we infer that $S^2(\Theta)(n') \to 0$ a.s. This proves the implication. ■

Theorem 2.3 *Let $\underline{F}_n$ be a sequence of simple random measures whose finite dimensional distributions converge weakly to finite dimensional distributions of a random measure $\underline{X}$. If $(\mathsf{F}_n(h))$ is tight then $\mathbf{X}(h)$ exists.*

Suppose that $h \notin \mathcal{H}_{\mathbf{X}}$. We can choose a sequence of functions from $\mathcal{H}_0$ such that $h_k \to h$ a.e., $|h_k| \nearrow |h|$, and $|\mathbf{X}(h_k)| \xrightarrow{\mathsf{P}} \infty$ (cf. Theorem 4.1 in [8]). Therefore $|\mathbf{X}(h_k)| \wedge t \xrightarrow{\mathsf{P}} t$ for every $t > 0$. Hence

$$\sup_k \mathsf{E}\; |\mathbf{X}(h_k)| \wedge t = t, \qquad t > 0.$$

By Lemma 2.1

$$\lim_n \mathsf{E}\; |\mathsf{F}_n(h_k)| \wedge t = E\; |\mathbf{X}(h_k)| \wedge t, \qquad t > 0.$$

Combining the above identities, we infer that

$$\sup_n \sup_k \mathsf{E}\; |\mathbf{X}(h_k)| \wedge t = t, \qquad t > 0.$$

Now, using an inequality similar to (2.2), we see that

$$\sup_n \sup_k \mathsf{E}\; S^2(\mathsf{F}_n(h_k)) \wedge t^2 = t^2.$$

By monotone convergence,

$$\sup_n \mathsf{E}\; S^2(\mathsf{F}_n(h)) \wedge t^2 = t^2, \quad t > 0,$$

and thus

$$\lim_{n'} \mathsf{E}(t^2 - S^2(\mathsf{F}_{n'}(h))^+ = 0, \qquad t > 0$$

along a subsequence (n'). Since $a \mapsto (t^2 - a)^+$ is a decreasing function, then putting $K = t^2/2$, we obtain

$$\limsup_n K\, P(S^2(\mathsf{F}_{n'}(h)) \leq K) = 0, \quad t > 0$$

which shows that the sequence $(S^2(\mathsf{F}_{n'}(h)))$ is not tight. By Lemma 2.2, the sequence $(\mathsf{F}_n(h))$ cannot be tight which completes the proof. ■

Theorem 2.4 *Let $\underline{F}_n$ be a sequence of simple random measures whose finite dimensional distributions converge weakly to finite dimensional distributions of a random measure $\underline{\mathbf{X}}$. Let $\mathbf{X}(h)$ exist and assume that*

$$\forall\, \epsilon > 0 \;\; \exists\, g \in \mathcal{H}_0 \quad \sup_n \|\mathsf{F}_n(h-g)\|_0 < \epsilon. \tag{2.3}$$

Then

$$(\mathsf{F}_n(h) : h \in \mathcal{H}_{\mathbf{X}}) \xrightarrow{\mathcal{D}} (\mathbf{X}(h) : h \in \mathcal{H}_{\mathbf{X}}).$$

Let $\mathbf{X}(h)$ exist. For an arbitrary but fixed ϵ we can find a simple function g from $\mathcal{H}_0$ such that $\|\mathbf{X}(h) - \mathbf{X}(g)\|_0 < \epsilon$ and $\sup_n \|F_n(h-g)\|_0 \leq \epsilon$.

The proof follows classically from the triangle inequality for the metric L

$$\begin{aligned} L(\mathsf{F}_n(h), \mathbf{X}(h)) &\leq L(\mathsf{F}_n(h), \mathsf{F}_n(g)) + L(\mathsf{F}_n(g), \mathbf{X}(g)) + L(\mathbf{X}(g), \mathbf{X}(h)) \\ &\leq L(\mathsf{F}_n(g), \mathbf{X}(g)) + 2\epsilon \end{aligned}$$

Notice that the weak convergence of finite dimensional distributions follows by linearity of processes $\mathbf{F}_n$ and $\mathbf{X}$. ■

For more information concerning specific multiple stochastic integrals we refer to [3,6] (Wiener and more general L_2-processes), [10,12,13] (stable and strictly stable processes), [8] (symmetric Poisson and Lévy processes), and [7] (assymetric Lévy processes).

3 Resampling by stable random variables.

Let Y be a symmetric random variable belonging to the domain of normal attraction of a (symmetric) α-stable random variable. In other words, the distributions of the normalized sums

$$\frac{1}{n^{1/\alpha}} \sum_{k=1}^{n} Y_k$$

converge to an α-stable symmetric distribution. We write $Y \in DNA_\alpha(c)$ if the stable random variable has the scale parameter c.

Clearly, if $Y \in DNA_\alpha(1)$ then $cY \in DNA_\alpha(c)$. If Z is a random variable independent of Y such that $\mathsf{E}|Z|^\alpha < \infty$ then $Z \cdot Y \in DNA_\alpha(\|Z\|_\alpha)$. One may use, e.g., the following characterization of $DNA_\alpha(c)$ ([4]): a symmetric random variable $Y \in DNA_\alpha(c)$ if and only if

$$\lim_{x \to \infty} x^\alpha \mathsf{P}(|Y| > x) = c.$$

Let (Y_i) be independent copies of Y and (τ, τ_i) be a sequence of i.i.d. random variables, independent of (Y_i). Define

$$F_n(h) = \frac{1}{n^{1/\alpha}} \sum_{k=1}^{n} h(\tau_k) Y_k.$$

Consider an α-stable integral $X(h) = \int h\, dX$ with the characteristic function $\mathsf{E}\exp\{iX(h)\} = \exp\{-\mathsf{E}|h(\tau)|^\alpha\}$, where $\mathsf{E}|h(\tau)|^\alpha < \infty$. Clearly,

$$F_n \xrightarrow{\mathcal{D}} X.$$

That is, stability of the limit of single values $F_n(h)$ of the process is obvious, so is appropriateness of the scale parameter $\|h(\tau)\|_\alpha$. The convergence of finite dimensional distributions follows from linearity of processes $\mathbf{F}_n$ and X.

In particular, the *resampled empirical measures*

$$F_n = \frac{1}{n^{1/\alpha}} \sum_{i=1}^{n} Y_i \, \delta_{\tau_i} \tag{3.1}$$

converge weakly to the distribution of a stable random measure X.

Recall that we may consider τ_i's which are i.i.d. random variables uniformly distributed on $[0,1]$. Also, we can assume that $X([0,1])$ is a standard symmetric α-stable random variable.

For a fixed function h such that $\int |h|^\alpha < \infty$ the sequence $(F_n(h))$ is bounded in probability (tightness) and thus it is bounded in L_q for any $q < \alpha$. The latter statement is a straighforward consequence of hypercontractivity of random variables in DNA_α (cf. [9,11])

Denote the d-dimensional stable integral by $\mathbf{X}(h) = \int \cdots \int h\, d\mathbf{X}$. Let (τ_i), (Y_i) be as before. $F_n(h)$ acquires its d-dimensional counterpart

$$\mathsf{F}_n(h) = \frac{1}{n^{d/\alpha}} \sum_{1 \le i_1, \ldots, i_d \le n} h(\tau_{i_1}, \ldots, \tau_{i_d}) \cdot Y_{i_1} \cdots Y_{i_d}.$$

In order to see that the resampled empirical d-dimensional measures converge to the distribution of a multiple stochastic integral we have to prove that the empirical measure $\mathsf{F}_n(h)$ can be uniformly approximated in probability by the empirical measures $\mathsf{F}_n(g)$, where $g \in \mathcal{H}_0$. Unfortunately, at this moment we are not able to show that this feature is guaranteed merely by existence of the multiple stable integral. Our immediate task is to extend the result from [1], where a quite narrower space L_r, $r > \alpha$, of integrands h was used, and the stability index was restricted to values greater than 1. We instead use the "weakest practical" restriction (at the current time), requiring h to belong to a suitable Orlicz space. Denote, for $\alpha \in (0,2)$,

$$\Phi(x) = x(1 + \ln_+^{d-1} x), \quad x > 0; \qquad \Phi_\alpha(x) = \Phi(|x|^\alpha).$$

Theorem 3.1 *Assume that* $\int |h|^\alpha (1 + \ln_+^{d-1} |h|) < \infty$. *Then the resampled empirical measures* $\mathsf{F}_n(h)$ *converge weakly to the distribution of the multiple stable integral* $\mathbf{X}(h) = \int \cdots \int h\, dX^d$.

The assumption guarantees existence of the stable integral (cf. [12]). We must only derive a uniform estimate of $\mathbf{F}_n(h)$ in the L_0-quasinorm. We switch to the equivalent L_q-norm for some $q < \alpha$. The equivalence, which just has been observed in case $n = 1$, carries over to multidimensional forms and integrals (cf [9]).

Observation 1. We can assume that the random variable Y in the definition of the resampled empirical measure has the symmetric α-stable distribution. In fact, since the tail of a symmetric random variable from DNA_α has the same order as the tail of a symmetric α-stable random variable then, for some choice of a constant $C > 0$,

$$\mathsf{E}|J_n(h)|^q \text{ (with } Y_i\text{'s from DNA}_\alpha) \quad \underset{C}{\asymp} \quad \mathsf{E}|J_n(h)|^q \text{ (with } \alpha\text{-stable } Y_i\text{'s)}$$

in which Fubini's theorem is combined with one of classical comparison results (cf., e.g., [9], Theorem 2.13).

Then we can write

$$J_n(h) = \int \cdots \int H_n \, d\mathbf{X},$$

where

$$H_n(\omega, t_1, \ldots, t_d) = \sum_{i_1, \ldots, i_d} h(\tau_{i_1}, \ldots, \tau_{i_d}) \mathbb{I}_{A_{i_1}}(t_1) \cdots \mathbb{I}_{A_{i_d}}(t_d)$$

and $A_i, i = 1, \ldots, n$, are disjoint with $|A_i| = 1/n$.

Observation 2. For symmetric α-stable processes, let us write $\mathbf{X} = \mathbf{X}^{(\alpha)}$ to disclose the stability index. Notice that $\int h \, d\mathbf{X}^{(\alpha)}$ exists if and only if $\int h^2 \, d\mathbf{X}^{(\alpha/2)}$ exists.

To see this, recall that $Y^{(\alpha)} \stackrel{\mathcal{D}}{=} \gamma\sqrt{Z^{(\alpha/2)}}$, where Z is a positive $\alpha/2$-stable random variable and γ is a symmetric Gaussian random variable (cf. [14]). Then the multilinear forms $\sum a_{i_1,\ldots,i_d} Y_{i_1}^{(\alpha)} \cdots Y_{i_d}^{(\alpha)}$ and $\sum a_{i_1,\ldots,i_d}^2 Z_{i_1}^{(\alpha/2)} \cdots Z_{i_d}^{(\alpha/2)}$ diverge or converge in probability (or in L_q) simultaneously. In the latter random multilinear form, positive $\alpha/2$-stable random variables can be replaced by symmetric $\alpha/2$-stable random variables. Indeed, linear forms generate the same space $\ell_{\alpha/2}$ and then decoupling principle can be applied (see, e.g. [8], Corollary 6.4) to extend the equivalence to multilinear forms. A fortiori, the equivalence can be expressed by using two-side estimates for L_q-norms of both multilinear forms involving an absolute constant C which continue to hold for their limits, i.e. for integrals

$$\|\int h\, d\mathbf{X}^{(\alpha)}\|_q \underset{C}{\asymp} \|\,|\int h^2\, d\mathbf{X}^{(\alpha/2)}|^{1/2}\,\|_q = \|\int h^2\, d\mathbf{X}^{(\alpha/2)}\|_{q/2}^{1/2}.$$

Now, randomizing h and taking the expectation, we obtain

$$\|\int H_n\, d\mathbf{X}^{(\alpha)}\|_q \underset{C}{\asymp} \|\int H_n^2\, d\mathbf{X}^{(\alpha/2)}\|_{q/2}^{1/2}.$$

Therefore, in the remainder of the proof we can fix any $\alpha \leq 1$ or $\alpha > 1$.

Observation 3. We choose $\alpha > 1$. For a deterministic function h, using convexity and sub-multiplicativity of Φ_α, we obtain the estimate

$$\begin{aligned}\|\int h d\mathbf{X}\|_q \leq C\|h\|_{\Phi_\alpha} \leq\ & C\left(\frac{1}{K} + \frac{1}{K}\int \Phi_\alpha(Kh)\right)) \\ \leq\ & C\left(\frac{1}{K} + \frac{\Phi_\alpha(K)}{K}\int \Phi_\alpha(h))\right)\end{aligned}$$

for any number $K > 0$. We are still free to pick any $q < \alpha$ and we do choose $q < 1$. Now, replacing h by the random function H_n, taking the expectation and applying Fubini's Theorem, we arrive at the requested uniform estimation

$$\sup_n \|J_n(h)\|_q \leq C(\frac{1}{K} + \frac{\Phi_\alpha(K)}{K}\int \Phi_\alpha(h)),$$

This completes the proof. ■

The properties used in the proof concerned the continuity of certain embedding mappings. Notice how important it was to have inequalities in order to randomize the integrands without loosing continuity. We conjecture a strengthened version of Theorem 3.1, with the weaker assumption that $\mathbf{X}(h)$ exists.

Acknowledgement. I appreciate discussions with Rafal Sztencel concerning Orlicz norms.

REFERENCES

[1] Dehling, D., Denker, M., Woyczyński, W. A. (1988). Resampling U-statistics using p-stable laws. Preprint (to appear in J. Mult. Anal.)

[2] Dehling, H., Denker, M., Philipp, W. (1984) Invariance principles for sums of Banach space valued random elements and empirical processes. Z. Wahr. verw. Geb. **62** 509-552.

[3] Engel, D. D. (1982). The multiple stochastic integral. Mem. Amer. Math. Soc. **38** # 265.

[4] Feller W. (1971). *An Introduction to Probability Theory and its Applications.* Vol.2, 2nd ed., Wiley, New York.

[5] Filippova, A.A. (1962). Mises' theorem on asymtotic behavior of functionals of empirical distribution functions and its statistical applications. Theory Prob. Appl. **7** 24-57.

[6] Itô, K. (1956). Spectral type of shift transformations of differential processes with stationary increments, Trans. Amer. Math. Soc. **81** 253-263.

[7] Kallenberg, O. (1989). Some uses of point processes in miltiple stochastic integration. Tech. Report **#269**, UNC, Chapel Hill.

[8] Kallenberg, O., Szulga (1989). Multiple stochastic integrals with respect to Poisson and Levy processes, to appear in Probab. Th. Rel. Fields.

[9] Krakowiak, W., Szulga, J. (1988). Hypercontraction principle and random multilinear forms. Probab. Th. rel. Fields **77** 325-342.

[10] Krakowiak, W., Szulga, J. (1988). A multiple stochastic integral with respect to a strictly p-stable random measure. Ann. Probab. **16** 764-777.

[11] Kwapień, S., Szulga, J. (1989) Hypercontraction methods in moment

inequalities for series of independent random variables in normed spaces. to appear in Ann. Probab.

[12] Rosiński, J., Woyczyński, W. A. (1986). On Itô stochastic integration with respect to p-stable motion: Inner clock, integrability of sample paths, double and multiple integrals. Ann. Probab. **14** 271-286.

[13] Surgailis, D. (1985). On the multiple stable integral. Z. Wahrschein. verw. Gebiete **70** 621-632.

[14] Zolotarev, V.M. (1986). *One dimensional stable distributions.* Transl. of Math. Monographs, Amer. Math. Soc., Providence, Rhode Island.

[15] Zolotarev, V.M., (1983). Probability metrics. Theory Probab. Appl. 28 278-302.

Jerzy Szulga
Mathematics ACA
Auburn University
Auburn, AL 36849

Characterizations of ergodic stationary stable processes via the dynamical functional.

KRZYSZTOF PODGÓRSKI AND ALEKSANDER WERON*

Dedicated to K.Urbanik on his 60th birthday

Abstract

For a stationary process $\mathbb{X}$ we introduce the dynamical functional Φ by the formula $\Phi(Y,t)=E\exp(i(Y\circ S_t-Y))$, where Y belongs to the closure of the span $\{X_t: t\in\mathbb{R}\}$ in the topology of convergence in measure and S_t is the shift transformation. It is proved that a stochastically continuous process $\mathbb{X}$ is ergodic if and only if for each Y we have

$$\lim_{T\to\infty} \frac{1}{T}\int_0^T \Phi(Y,t)\,dt = |Ee^{iY}|^2.$$

This characterization is applied to symmetric stable processes to reprove and unify two independent equivalent conditions for ergodicity presented in [2] and [7].

Preliminaries

Let $\mathbb{X}$ be a real stochastic process $(X_t)_{t\in\mathbb{R}}$ with a continuous time parameter. Let B_∞ denote the product σ-field

* Supported in part by the Grant CPBP 01.02.

on $\mathbb{R}^{\mathbb{R}}$ generated by the class of all Borel measurable cylinders and $P_{\mathbb{X}}$ be the distribution of $\mathbb{X}$ on $\mathcal{B}_{\infty}$. A stochastic process $(x_t)_{t\in\mathbb{R}}$ on the probability space $(\mathbb{R}^{\mathbb{R}},\mathcal{B}_{\infty},P_{\mathbb{X}})$ such that for $t\in\mathbb{R}$ and $x\in\mathbb{R}^{\mathbb{R}}$ $x_t(x)=x(t)$ is called the canonical representation of $\mathbb{X}$.The complete metric space of equivalence classes of all measurable real functions on $(\mathbb{R}^{\mathbb{R}},\mathcal{B}_{\infty},P_{\mathbb{X}})$ with the topology of convergence in measure we shall denote by $\mathcal{L}(\mathbb{X})$. For each $t\in\mathbb{R}$ we denote by S_t the shift transformation on $\mathbb{R}^{\mathbb{R}}$ such that for $x\in\mathbb{R}^{\mathbb{R}}$ and $t\in\mathbb{R}$ $(S_t x)(s)=x(s+t)$. In this paper we will only consider stationary processes, i.e. such that all finite dimensional distributions of $\mathbb{X}$ are invariant with respect to the shift S_t for each $t\in\mathbb{R}$. Thus $(S_t)_{t\in\mathbb{R}}$ is a group of measure-preserving transformations on $(\mathbb{R}^{\mathbb{R}},\mathcal{B}_{\infty},P_{\mathbb{X}})$. This group induces a group $(T_t)_{t\in\mathbb{R}}$ of operators on $\mathcal{L}(\mathbb{X})$ such that for each $t\in\mathbb{R}$ and $Y\in\mathcal{L}(\mathbb{X})$ $T_tY=Y\circ S_t$.

Let $\mathcal{L}_0(\mathbb{X})$ denote the closure of $\text{lin}\{x_t:t\in\mathbb{R}\}$ with respect to the topology of convergence in measure. It is easy to check that $\mathcal{L}_0(\mathbb{X})\subseteq\mathcal{L}(\mathbb{X})$ and for all $t\in\mathbb{R}$ $T_t(\mathcal{L}_0(\mathbb{X}))=\mathcal{L}_0(\mathbb{X})$.

Let $\mathcal{I}_{\mathbb{X}}=\{A\in\mathcal{B}_{\infty}:\text{for each } t\in\mathbb{R}\ P_{\mathbb{X}}(A\Delta S_tA)=0\}$. Then $\mathcal{I}_{\mathbb{X}}$ is a σ-field and elements of $\mathcal{I}_{\mathbb{X}}$ are called invariant sets.

Definition.(see [4]) A stationary stochastic process $\mathbb{X}$ is called ergodic if and only if, for each $A\in\mathcal{I}_{\mathbb{X}}$ $P_{\mathbb{X}}(A)=0$ or 1.

For $Y\in\mathbb{L}_1(\mathbb{R}^{\mathbb{R}},\mathcal{B}_{\infty},P_{\mathbb{X}})$ let $A_TY=\frac{1}{T}\int_0^T T_tY\,dt$. It follows from the Birkhoff ergodic theorem that a measurable stochastic process is ergodic if and only if for each $Y \in \mathbb{L}_1(\mathbb{R}^{\mathbb{R}}, \mathcal{B}_\infty, P_{\mathbb{X}})$ $\lim_{T\to\infty} \mathbf{A}_t Y = EY$ a.e.

We assume throughout the paper that we deal with measurable processes only. We shall use the following characterization of ergodicity for measurable stochastic processes.

Lemma 1. *Let $\mathbb{X}$ be a measurable process. The following conditions are equivalent*

(i) $\mathbb{X}$ is ergodic;

(ii) for each $Y \in \mathcal{L}(\mathbb{X})$

$$\lim_{T\to\infty} E \quad [(A_T e^{iY}) e^{-iY}] = |Ee^{iY}|^2;$$

(iii) for each $Y \in lin\{x_t : t \in \mathbb{R}\}$

$$\lim_{T\to\infty} E\ [(A_T e^{iY}) e^{-iY}] = |Ee^{iY}|^2.$$

Proof. Suppose first that $\mathbb{X}$ is ergodic. Clearly, for each $Y \in \mathcal{L}(\mathbb{X})$ $|(A_T e^{iY}) e^{-iY}| \le 1$. By the Lebesgue convergence theorem and the Birkhoff theorem it follows that

$$\lim_{T\to\infty} E\ [(A_T e^{iY}) e^{-iY}] = E\ [(\lim_{T\to\infty} A_T e^{iY}) e^{-iY}] =$$

$$= E\ [(Ee^{iY}) e^{-iY}] = |Ee^{iY}|^2.$$

The implication (ii)⇒(iii) is obvious.

Finally ,let us suppose that the condition (iii) is satisfied. It is sufficient to show that for each $A \in \mathcal{I}_{\mathbb{X}}$ $P^2_{\mathbb{X}}(A) = P_{\mathbb{X}}(A)$. By the Birkhoff theorem

$$\lim_{T\to\infty} E\,[(A_T e^{iY})e^{-iY}]=E\,[E(e^{iY}|\mathcal{I}_{\mathbf{X}})e^{-iY}].$$

By the assumption we have that for each $Y\in\mathrm{lin}\{x_t:t\in\mathbb{R}\}$

$$E\,[E(e^{iY}|\mathcal{I}_{\mathbf{X}})e^{-iY}]=|Ee^{iY}|^2.$$

So

$$E|E(e^{iY}|\mathcal{I}_{\mathbf{X}})|^2=|Ee^{iY}|^2.$$

By the convergence of Fourier series we have that for each measurable rectangles A of the form $\bigcap_{i=1}^{n} x_{t_i}^{-1}(a_i,b_i)$ we have

$$E|E(1_A|\mathcal{I}_{\mathbf{X}})|^2= P^2(\mathbf{X}\in A).$$

And consequently by the definition of $\mathcal{B}_\infty$ the above equality holds for any $A\in\mathcal{B}_\infty$. For $A\in\mathcal{I}_{\mathbf{X}}$ we have

$$|E(E(1_A|\mathcal{I}_{\mathbf{X}})|^2= E\,1_A^2 = P(\mathbf{X}\in A).$$

Q.E.D.

Definition.(see[5] and [9]) A process $\mathbf{X}$ is called symmetric α-stable for $\alpha\in(0,2]$ (abbrevated SαS) if for arbitrary $n\in\mathbb{N}, a_1,...,a_n\in\mathbb{R}, t_1,...,t_n\in\mathbb{R}$ each random variable $\sum_{i=1}^{n} a_i X_{t_i}$ has a symmetric α-stable distribution.

If $\mathbf{X}$ is a stationary SαS process, then there exists on $\mathrm{lin}\{X_t:t\in\mathbb{R}\}$ a norm in the case when $\alpha\in[1,2]$ and a quasi-norm when $\alpha\in(0,1)$ denoted by $\|\cdot\|_\alpha$ such that for each $Y\in\mathrm{lin}\{X_t:t\in\mathbb{R}\}$ $Ee^{itY}=\exp(-|t|^\alpha\|Y\|_\alpha^\alpha)$. The convergence in $\|\cdot\|_\alpha$ is the same as the convergence in measure. In this situation the space $\mathcal{L}_0(\mathbf{X})$ is denoted by $\mathcal{L}_\alpha(\mathbf{X})$. For each $Y\in\mathcal{L}_\alpha(\mathbf{X})$ we have $Ee^{itY}=\exp(-|t|^\alpha\|Y\|_\alpha^\alpha)$. Moreover for each $Y\in\mathcal{L}_\alpha(\mathbf{X})$ and $t\in\mathbb{R}$ $\|T_tY\|_\alpha=\|Y\|_\alpha$. So $(T_t)_{t\in\mathbb{R}}$ is a group of isometries on

$\mathcal{L}_\alpha(\mathbf{X})$. For details see [8] and [9].

When $\alpha=2$ we deal with a stationary Gaussian process with mean zero. For each $Y\in\mathcal{L}_2(\mathbf{X})$ we have $\|Y\|_2^2=\mathrm{Var}_{\mathbf{X}}Y$. It is well known that if $Y\in\mathcal{L}_2(\mathbf{X})$, then there exists a finite symmetric measure μ_Y absolutely continuous with respect to the spectral measure μ such that $\mathrm{Cov}(T_tY,Y)=\int e^{it\lambda}\,d\mu_Y(\lambda)$. Conversely, if μ_0 is a finite symmetric measure absolutely continuous with respect to μ then there exists $Y\in\mathcal{L}_2(\mathbf{X})$ such that $\mathrm{Cov}_{\mathbf{X}}(T_tY,Y)=\int e^{it\lambda}\,d\mu_0(\lambda)$.

Dynamical functional

Let $\mathbf{X}$ be a stationary stochastically continuous process. Let us notice that for each $Y\in\mathcal{L}_0(\mathbf{X})$ $E\exp(i(T_tY-Y))$ exists and is finite. The map $\Phi:\mathcal{L}(\mathbf{X})\times\mathbb{R}\to\mathbb{C}$ given by the formula $\Phi(Y,t)=E\exp(i(T_tY-Y))$ will be called the dynamical functional for a stochastic process $\mathbf{X}$, since it describes the dynamical behaviour of the process through the increment T_tY-T_0Y. It is easy to check that Φ is continuous with respect to the product topology on $\mathcal{L}_0(\mathbf{X})\times\mathbb{R}$ and for each $Y\in\mathcal{L}_0(\mathbf{X})$ $\Phi(Y,\cdot)$ is a positive-definite function.

Proposition. *Let $\mathbf{X}$ be a stationary and stochastically continuous process. Then $\mathbf{X}$ is ergodic if and only if for each $Y\in\mathcal{L}_0(\mathbf{X})$*

$$\lim_{T\to\infty}\frac{1}{T}\int_0^T\Phi(Y,t)\,dt=|Ee^{iY}|^2.$$

Proof. Let $\mathbf{X}$ be ergodic. Then, by Lemma 1, for an arbitrary

$Y \in \mathcal{L}(\mathbf{X})$ $\lim_{T\to\infty} E \frac{1}{T}\int_0^T \exp(i(T_t Y - Y))\, dt = |Ee^{iY}|^2$. By the Fubini Theorem we have $\lim_{T\to\infty} \frac{1}{T}\int_0^T E\exp(i(T_t Y-Y))\, dt = |Ee^{iY}|^2$, and the first part of proposition follows from the fact that $\mathcal{L}_0(\mathbf{X}) \subseteq \mathcal{L}(\mathbf{X})$.

Next, let us suppose that for each $Y \in \mathcal{L}_0(\mathbf{X})$

$$\lim_{T\to\infty} \frac{1}{T}\int_0^T \Phi(Y,t)\, dt = |Ee^{iY}|^2.$$

Since $\mathrm{lin}\{x_t : t\in\mathbb{R}\} \subseteq \mathcal{L}_0(\mathbf{X})$ we obtain that for each $Y \in \mathrm{lin}\{x_t : t\in\mathbb{R}\}$

$$\lim_{T\to\infty} \frac{1}{T}\int_0^T E\exp(i(T_t Y-Y))dt = |Ee^{iY}|^2,$$

which is equivalent to the ergodicity of $\mathbf{X}$ by Lemma 1.

Q.E.D.

Example. In the case of a Gaussian process i.e. $\alpha=2$ we have $\|T_t Y - Y\|_2^2 = \mathrm{Var}_{\mathbf{X}}(T_t Y - Y) = 2\|Y\|_2^2 - \mathrm{Cov}_{\mathbf{X}}(T_t Y, Y) = \exp(\int [1-e^{it\lambda}]\, d\mu_Y(\lambda))$.

Thus, $\Phi(Y,t) = e^{2\int [e^{it\lambda}-1]\, d\mu_Y(\lambda)} = e^{-2\|Y\|_2^2}\, e^{2\hat{\mu}_Y(t)}$.

It is well known from the Grenander-Fomin-Maruyama results that $\mathbf{X}$ is ergodic if and only if its spectral measure μ has no atoms (see [3] and [4]).Now we shall present a simple proof of this fact by using the criterion for ergodicity formulated above.

For each $Y \in \mathcal{L}_2(\mathbf{X})$, $|Ee^{iY}|^2 = e^{-2\|Y\|_2^2}$. This implies that $\lim_{T\to\infty} \frac{1}{T}\int_0^T \Phi(Y,t)\, dt = |E^{iY}|^2$ if and only if $\lim_{T\to\infty} \frac{1}{T}\int_0^T e^{2\hat{\mu}_Y(t)} dt = 1$ or equivalently $e^{2\mu_Y}(\{0\}) = 1$. Hence, by the Proposition, $\mathbf{X}$ is ergodic if and only if for each $Y \in \mathcal{L}_2(\mathbf{X})$ and $k \in \mathbb{N}$ $\mu_Y^{(k)}(\{0\}) = 0$.

Let us suppose that μ has no atoms. Since μ_Y is

absolutely continuous with respect to μ , for each $k\in\mathbb{N}$ $\mu_Y^{(k)}$ has no atoms. Consequently, $\mathbb{X}$ is ergodic.

Note that μ_Y is symmetric for each $Y\in\mathcal{L}_2(\mathbb{X})$. So if μ_Y has an atom, then $\mu_Y^{(2)}$ has an atom at zero. This implies that if for each $k\in\mathbb{N}$ $\mu_Y^{(k)}(\{0\})=0$ then μ_Y has no atoms.

So, if we suppose that $\mathbb{X}$ is ergodic, then for each $Y\in\mathcal{L}_2(\mathbb{X})$ μ_f has no atoms. Equivalently, for all symmetric $h\in\mathbb{L}_2(\mathbb{R},\mathcal{B}_{\mathbb{R}},\mu)$ a measure $\eta_h(\cdot)=\int_{(\cdot)} h\, d\mu$ has no atoms. Since μ is symmetric it implies that μ also has no atoms.

Ergodic symmetric stable processes

Now, using the proposition, we obtain the characterization of ergodic SαS processes which clarifies the relation between the characterization given in [7] and the Cambanis et al. result [2].

We will need the following technical lemma.

Lemma 2. *Let ϕ be a real valued measurable and bounded function on $\mathbb{R}$. The following conditions are equivalent:*

(i) for each positive number c $\lim_{T\to\infty} \frac{1}{T}\int_0^T e^{c\phi(t)}dt = 1$;

(ii) for any $\varepsilon>0$ $\lim_{T\to\infty} \frac{1}{T} \left|\{t\in[0,T]:|\phi(t)|>\varepsilon\}\right| = 0$;

(iii) there exist natural numbers $n,k,n\neq k$ and a real number b such that

$$\lim_{T\to\infty} \frac{1}{T}\int_0^T (\phi(t)+b)^n\, dt = b^n$$

and

$$\lim_{T\to\infty} \frac{1}{T}\int_0^T (\phi(t)+b)^k \, dt = b^k \;;$$

(iv) $$\lim_{T\to\infty} \frac{1}{T}\int_0^T |\phi(t)| \, dt = 0 \quad .$$

Proof. First we prove the implications (i)⇒(ii) and (iii)⇒(ii). For each T>0 let λ_T be a probabilistic measure on $\mathbb{R}$ such that for $A\in\mathcal{B}_{\mathbb{R}}$ $\lambda_T(A)=\frac{|A\cap[0,T]|}{T}$. Since ϕ is bounded so for each T>0 $\lambda_T\circ\phi^{-1}([-M,M]^c)=0$, where $M=\sup_{t\in\mathbb{R}}|\phi(t)|$. Thus the family of probabilistic measures $\{\lambda_T\circ\phi^{-1}\}_{T>0}$ is tight. This implies (see [1]) that for any sequence of positive numbers $(T_n)_{n\in\mathbb{N}}$ there exists a subsequence $(T'_k)_{k\in\mathbb{N}}$ and a measure ω such that $(\lambda_{T'_k}\circ\phi^{-1})_{k\in\mathbb{N}}$ converges weakly to ω. Thus for any continuous and bounded function f on $\mathbb{R}$

$$\lim_{k\to\infty}\int f\, d\lambda_{T'_k}\circ\phi^{-1} = \lim_{k\to\infty}\frac{1}{T'_k}\int_0^{T'_k} f(\phi(t))\, dt = \int f d\omega.$$

Since for each T>0 $\lambda_T\phi^{-1}([-M,M]^c)=0$ the function f can be, in general, unbounded.

Let us suppose that (iii) is satisfied. Thus there exist $n,k\in\mathbb{N}$ $n>k$ $\int(u+b)^k d\omega(u) = b^k$ and $\int(u+b)^n d\omega(u) = b^n$. This implies that $\left(\int(u+b)^k d\omega(u)\right)^{n/k} = \int\left((u+b)^k\right)^{n/k} d\omega(u)$. But for a strictly convex function we have an equality in the Jensen inequality only in the case when the function is constant a.e. Thus $(u+b)^k=b^k$ ω-a.e. which implies that $\omega=\delta_{\{0\}}$ and ω does not depend on the sequence $(T'_k)_{k\in\mathbb{N}}$. So $\lambda_T\circ\phi^{-1}$ tends weakly to $\delta_{\{0\}}$ when T tends to infinity. Consequently

$$\lim_{T\to\infty} \lambda_T \circ \phi^{-1}(\{t\in\mathbb{R}: |t|>\varepsilon\}) = 0$$

which is equivalent to (ii).

Now, if the condition (i) is satisfied, we replace the functions $(u+a)^k$ and $(u+a)^n$ by the functions e^{cu} and $e^{c'u}$ and then repeat the above proof.

Next we prove the implications (ii)⇒(i), (ii)⇒(iii) and (ii)⇒(iv). For a nondecreasing continuous at zero function f and a measurable bounded function η we have the following inequalities

$$f(-\varepsilon)\frac{1}{T}\left|\{t\in\mathbb{R}:\eta(t)>-\varepsilon\}\cap[0,T]\right|+f(-M)\frac{1}{T}\left|\{t\in\mathbb{R}:\eta(t)\le-\varepsilon\}\cap[0,T]\right|\le$$

$$\le \frac{1}{T}\int_0^T f(\eta(t))\,dt \le$$

$$\le f(\varepsilon)\frac{1}{T}\left|\{t\in\mathbb{R}:\eta(t)<\varepsilon\}\cap[0,T]\right|+f(M)\frac{1}{T}\left|\{t\in\mathbb{R}:\eta(t)\ge\varepsilon\}\cap[0,T]\right|.$$

If $\lim_{T\to\infty}\frac{1}{T}\left|\{t\in\mathbb{R}:|\eta(t)|\ge\varepsilon\}\cap[0,T]\right| = 0$, then

$$f(-\varepsilon)\le \liminf_{T\to\infty} \frac{1}{T}\int_0^T f(\eta(t))\,dt \le \limsup_{T\to\infty}\frac{1}{T}\int_0^T f(\eta(t))\,dt \le f(\varepsilon)$$

and since f is continuous at zero, $\lim_{T\to\infty} \frac{1}{T}\int_0^T f(\eta(t))\,dt=f(0)$.

Applying this fact to apropriate functions we obtain desired conditions.

The implication (iv)⇒(ii) is obvious by the Tchebysev inequality.

Q.E.D.

Theorem. *Let $\mathbf{X}$ be a stationary measurable SαS process. Then the following conditions are equivalent:*

(i) $\mathbb{X}$ *is ergodic,*

(ii) *for each* $Y\in\mathcal{L}_\alpha(\mathbb{X})$

$$\lim_{T\to\infty} \frac{1}{T}\int_0^T \exp(2\|Y\|_\alpha^\alpha - \|T_t Y - Y\|_\alpha^\alpha)\, dt = 1;$$

(iii) *for each* $Y\in\mathcal{L}_\alpha(\mathbb{X})$ *and* $\varepsilon>0$

$$\lim_{T\to\infty} \frac{1}{T} \left|\{t\in[0,T] : |\|T_t Y-Y\|_\alpha^\alpha - 2\|Y\|_\alpha^\alpha| > \varepsilon\}\right| = 0;$$

(iv) there exist natural numbers $n,k,n\neq k$ *such that for each* $Y\in\mathcal{L}_\alpha(\mathbb{X})$

$$\lim_{T\to\infty} \frac{1}{T}\int_0^T \|T_t Y-Y\|_\alpha^{\alpha n}\, dt = 2^n\|Y\|_\alpha^{\alpha n}$$

and

$$\lim_{T\to\infty} \frac{1}{T}\int_0^T \|T_t Y-Y\|_\alpha^{\alpha k}\, dt = 2^k\|Y\|_\alpha^{\alpha k};$$

(v) *for each* $Y\in\mathcal{L}_\alpha(\mathbb{X})$

$$\lim_{T\to\infty} \frac{1}{T}\int_0^T |2\|Y\|_\alpha^\alpha - \|T_t Y-Y\|_\alpha^\alpha|\, dt = 0 .$$

Proof. First we prove the equivalence (i)⇔(ii). In the case of a SαS process $Ee^{iY}=e^{-\|Y\|_\alpha^\alpha}$ for each $Y\in\mathcal{L}_\alpha(\mathbb{X})$. By the proposition, $\mathbb{X}$ is ergodic if and only if for each $Y\in\mathcal{L}_\alpha(\mathbb{X})$ $\lim_{T\to\infty}\frac{1}{T}\int_0^T \Phi(Y,t)\, dt= = e^{-2\|Y\|_\alpha^\alpha}$ or equivalently

$$\lim_{T\to\infty} \frac{1}{T}\int_0^T \exp(2\|Y\|_\alpha^\alpha)\, \Phi(Y,t)\, dt = 1.$$

The conclusion follows from the form of the dynamical functional for SαS case given by the equality

$$\Phi(Y,t)=\exp(-\|T_t Y-Y\|_\alpha^\alpha).$$

Since for any positive number c $\exp(2\|cY\|_\alpha^\alpha - \|T_t(cY)-cY\|_\alpha^\alpha) = \exp(c\ (2\|Y\|_\alpha^\alpha - \|T_t Y-Y\|_\alpha^\alpha))$, thus the equivalence of the next conditions follows at once from Lemma 2.

Q.E.D.

Remark. The equivalence between (i) and (iv) for $n=1$ and $k=2$ reduces to the Cambanis-Hardin-Weron theorem [2], and the equivalence between (i) and (ii) is given by Podgórski [7].

REFERENCES

[1] P.Billingsley, *Convergence of Probability Measures*, (Wiley,New York,1968).

[2] S.Cambanis,C.D.Hardin Jr.and A.Weron,*Ergodic properties of stationary stable processes*.Stochastic Process.Appl. 24(1987)1-18

[3] H.Dym and H.P.McKean, *Gaussian Processes, Function Theory and the Inverse Spectral Problem*.(Academic Press, New York,1976).

[4] S.V.Fomin,I.P.Cornfeld and Ya.G.Sinai, *Ergodic theory*. (Springer-Verlag,Berlin,New York,1982)

[5] C.D.Hardin Jr., *On the spectral representation of symmetric stable processes*.J.Multivariate Anal.12(1982) 384-401

[6] R.LePage,*Multidimensional infinite divisible variables and processes,Part I:Stable case*.Lecture Notes in Math. 1391,(Springer-Verlag,Berlin,1989),148-163.

[7] K.Podgórski,*A new characterization of ergodic symmetric stable processes*.(to appear)

[8] G.Samorodnitsky and M.Taqqu, *Stable processes*. (in preparation).

[9] A.Weron,*Stable processes and measures;A survey*.Lecture Notes in Math.1080,(Springer-Verlag,Berlin,New York, 1984),306-364

[10] A.Weron,*Harmonizable stable processes on groups:Spectral, ergodic and interpolation properties*.Z.Wahr.verw.Gebiete, 68(1985),473-491.

K.Podgórski and A.Weron
Institute of Mathematics
Technical University of Wroclaw
50-370 WROCLAW
POLAND

Progress in Probability

Progress in Probability includes all aspects of probability theory and stochastic processes, as well as their connections with and applications to other areas such as mathematical statistics and statistical physics. Each volume presents an in-depth look at a specific subject, concentrating on recent research developments. Some volumes are research monographs, while others will consist of collections of papers on a particular topic.
We encourage preparation of manuscripts in LaTex or AMS TeX for delivery in camera-ready copy, which leads to rapid publication, or in electronic form for interfacing with laser printers or typesetters.
Proposals should be sent directly to the series editors or to Birkhäuser Boston, 675 Massachusetts Avenue, Cambridge, MA 02139.